OSCILLATION-BASED TEST IN MIXED-SIGNAL CIRCUITS

FRONTIERS IN ELECTRONIC TESTING

Consulting Editor
Vishwani D. Agrawal

OSCILLATION-BASED TEST IN MIXED-SIGNAL CIRCUITS

by

Gloria Huertas Sánchez
IMSE-CNM CSIC-Universidad de Sevilla, Spain

Diego Vázquez García de la Vega
IMSE-CNM CSIC-Universidad de Sevilla, Spain

Adoración Rueda Rueda
IMSE-CNM CSIC-Universidad de Sevilla, Spain

José Luis Huertas Díaz
IMSE-CNM CSIC-Universidad de Sevilla, Spain

 Springer

A C.I.P. Catalogue record for this book is available from the Library of Congress.

ISBN-10 1-4020-5314-2 (HB)
ISBN-13 978-1-4020-5314-6 (HB)
ISBN-10 1-4020-5315-0 (e-book)
ISBN-13 978-1-4020-5315-3 (e-book)

Published by Springer,
P.O. Box 17, 3300 AA Dordrecht, The Netherlands.

www.springer.com

Printed on acid-free paper

A Pablete,

que fue testigo del final de este libro "desde dentro".

To Pablete,

who has been witness to the finishing of this book "from within".

Contents

Preface

Driven by the need of reducing the defective circuits to a minimum, present-day fabrication technologies require design techniques been complemented by effective test procedures. In the case of digital ICs, there are many procedures to cope with test problems in an effective manner. However, analog integrated circuits or the analog part of a mixed-signal integrated circuit bring enormous difficulties when dealing with the problem of how to test them adequately.

Analog circuits are difficult to test because there is a wide variety of analog building blocks, their specifications are very broad, and there is a strong dependency of circuit parameters on component variations. For mixed-signal ICs, where analog circuits must coexist with digital components, testing difficulties increase substantially because the access to both analog and digital blocks is severely restricted. The consequences are a reduced fault coverage, a higher test application time and a longer test development time.

In mixed-signal IC's, the most difficult components to test are the analog cores, since analog test is based on checking functional specifications, what can be conflicting when test time has to be kept small, the number of available pins is reduced and full access to input/output core terminals can not be granted. Furthermore, functional test techniques greatly differ depending on the involved analog components and/or their application field, turning almost impossible to define a general (functional) test methodology applicable to any analog block.

Experience forged from the test of digital circuits encourages researchers to try structural or fault-driven test methods for analog components and explore Built-In Self-Test (BIST) alternatives as well. This has to be done in a manner that increases accessibility to provide core isolation and test resources access, but it might have a high cost in terms of area overhead, power wasting, performance degradation and/or noise and parasitic penalties. But neither moving from functional to structural testing nor incorporating BIST are trivial issues in what analog circuitry is concerned, and are still far from a wide acceptance by the designer community. This acceptance will depend on several factors like compatibility with functional test approaches, test efficiency, test confidentiality and additional design effort.

Among the emerging structural test solutions, the so-called Oscillation-Based Test (OBT) technique is very appealing. It is conceptually simple, does not demand strong circuit modifications during testing and can handle BIST (called in this case OBIST) without the penalty of dedicated, additional on-chip signal generation hardware. In broad terms, when OBT is employed no external test stimuli are required, some few simple measurements are used, and can be combined with a multiplexing scheme to probe internal nodes, thus complying with some of the factors above.

The purpose of this book is to provide the reader with a deep understanding of OBT and OBIST. The basic concepts underlying OBT/OBIST are presented, as well as the principles for applying this test methodology to complex integrated circuits. Detailed examples and practical implementation details are provided throughout the book in order to help the interested engineer to evaluate whether this technique may or may not be used for a particular appliaction. Our aim is to provide the reader with an overview of the lights and shadows this test technique offers nowadays.

Chapter 1 focuses attention on a mixed-signal structural testing methodology called Oscillation-Based Test (OBT). The state-of-the-art is reviewed, given an overview of the past, the present and the future expectations of this test method. The goal of this Chapter is to define the basics of a new improved OBT concept and overcome some of its main limitations.

Chapter 2 describes a simple, practical and intuitive mathematical approach to model the oscillators required in the OBT strategy: the Describing-Function (DF) technique. The aim of this Chapter is to provide an acceptable theoretical OBT solution which allows us to accurately predict the oscillation parameters.

Chapter 3 discusses a systematic way to apply the OBT approach to discrete-time filters. A particular discrete-time filter structure (the Fleischer and

Laker (FL) biquad) is studied in detail. The objective of this Chapter is to extrapolate the obtained conclusions in order to establish general guidelines for employing OBT to generic discrete-time filter structures.

Chapter 4 discusses a systematic way of applying the OBT approach to discrete-time Sigma-Delta ($\Sigma\Delta$) modulators. The goal of this Chapter is to establish conclusions defining a general OBT procedure for generic discrete-time $\Sigma\Delta$ modulators.

Chapter 5 reviews the OBT implementation in some practical discrete-time filter examples. A generic biquadratic filter is studied using both, symbolic expressions and specific numerical data. The aim of this Chapter is to extract conclusions on the establishment of the test parameters, the validation of the oscillator model, the fault coverage, the test quality, etc.

Chapter 6 presents some practical considerations for the application of the Oscillation-Based Built-In-Self-Test (OBIST) to a Dual-Tone Multi-Frequency (DTMF) embedded macrocell. The objective of this Chapter is to describe an example of the integration of the OBT-OBIST technique into the frame of analog-core-based design of complex mixed-signal ICs.

Chapter 7 reports experimental results extracted by two circuit demonstrators in which the OBT/OBIST approach has been implemented. The aim of this Chapter is to experimentally validate the OBT/OBIST methodology in mixed-signal ICs.

J.L. HUERTAS DIAZ
Instituto de Microelectrónica de Sevilla

Chapter 1

Oscillation-Based Test Methodology
Basic concepts and state of art

IN MIXED-SIGNAL ICs, where analog blocks coexist with digital components and where there is a restricted access to both analog and digital parts, an efficient testing of the entire mixed-signal circuit is currently considered as a challenging task, specially as the complexity of the analog portion of the chip increases. Consequently, testing is a determining factor in the final product cost, and may compromise the economical feasibility of future System-on-Chip markets. These last years, analog testing methods have been object of study for many researchers [1]-[98], [130], in such a way that different types of test techniques and strategies have been proposed and diverse Design-for-Testability (DfT) procedures (in conjunction with these mentioned test techniques) have been employed.

This Chapter focus its attention on an analog testing methodology called Oscillation-Based Test (OBT). In particular, the first part of the Chapter reviews the beginnings of the OBT concept, highlighting its main advantages and shortcomings. Then, a second part is devoted to define a more general and practical OBT approach which can be successfully applied to a large number of systems and then, accepted as a good test solution by the test community.

1.1 LINKING OSCILLATION WITH TESTING

1.1.1 Point of origin: Early OBT

The idea of using built-in self-oscillations to determine faulty behaviours in systems is a relatively new test strategy dating from 1995 [1]-[3], when Karim Arabi and Bozena Kaminska established the basic definitions of a test methodology they called "Oscillation-Based-Test" or simply OBT. Since then, the OBT concept has been interpreted in different manners in such a way that several Oscillation-Based-Test Methologies can be found in the literature [1]-[3], [5], [8]-[11], [13]-[15], [21], [25]-[41], [49]-[53], [96]-[98].

Fig. 1.1 displays a graphical description of the OBT method presented in [1] and Table 1.1 reviews its basic principles, summing up the main steps to apply this technique. Basically, this approach can be applied to analog and mixed-signal circuits and is based on splitting any (complex) System Under Test (SUT) in simpler functional building blocks which are separately tested (Step_1). During the test mode, each of these Blocks Under Test (BUTs) is converted in an oscillator (Step_2). When there is no fault in the BUT, the oscillator inherently produces a test output signal whose oscillation frequency is related to the fault-free structure of the specific BUT, in a way that its value (within a tolerance margin) may be considered as the test parameter (Step_3). So, a fault in a component of such a block could be detected by measuring the oscillation frequency and by checking out whether it deviates from its expected nominal value (Step_4). That means, discrepancies between the measured oscillation frequency of a BUT and its previously estimated nominal value indicate potential faults.

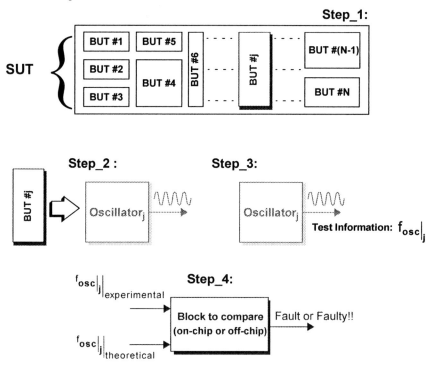

Figure 1.1: OBT Approach

Step_1	Rearranging the SUT into BUTs easier to test.
Step_2	Transforming each BUT (or a combination of these BUTs) into an oscillator producing an oscillatory output signal whose frequency value, f_{osc}, is related to the BUT internal structure.
Step_3	Measuring the oscillation frequency.
Step_4	Detecting a fault when the measured oscillation frequency deviates from the nominal frequency.

Table 1.1: Steps to apply the Arabi and Kaminska OBT approach [1]

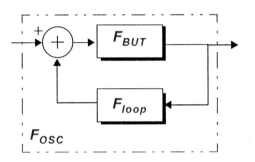

Figure 1.2: Converting a BUT in an oscillator

The idea behind Arabi and Kaminska's OBT technique involves the operation of converting the structure of every BUT into an oscillator by modifying its internal circuitry. Arabi and Kaminska propose several mechanisms to do that. Among them, one of the most efficient methods[1] employs a feedback loop including a transfer function F_{loop} and an adder (Fig. 1.2). The resulting closed-loop system is an oscillator, F_{OSC}, [100], [104]. However, this scheme only can be successfully applied to convert second-order active filters into oscillators by making the quality factor, Q, infinite. That is -from a mathematical point of view- shifting the system poles to the imaginary axis [100].

On the other hand, the oscillation test strategy presented by Arabi and Kaminska, introduces, likewise, a test structure at system level as a general solution for the application of the OBT concept as a DfT technique [1], [7]. In the test mode, analog multiplexers sort out the oscillatory output of the BUTs and their involved frequencies are evaluated and interpreted either externally, using a suitable test equipment, or internally using, for example, a frequency to number converter.

---

[1] Another solution proposed by Arabi and Kaminska was, for example, to use heuristic circuit techniques to build up an oscillator from the original BUT.

Fig. 1.3 displays a version of the OBT structure proposed by Arabi and Kaminska [1], [7]. In this proposal, the *Control Test Mode Signals* configure the system in its test mode, but before starting the test procedure, the test circuitry is checked out by activating the *Control Self-Test Mode Signals*. Other *Control Signals* are introduced, if necessary, to regulate the *Evaluation and Interpretation Unit*. Such a unit either consists of an external block or is embedded into the on-chip test circuitry. In any case, the *Control Logic Block* is fed by the corresponding output of this unit.

Basically, the *Evaluation and Interpretation Unit* converts the oscillation frequency of each test output signal into a representative number. Then, the *Control Logic Block* compares such a representative number to a previously calculated nominal number. A fault occurs when the representative number deviates from the given, nominal number. Obviously, a very delicate issue here is the accurate establishment of the frequency nominal number and its tolerance.

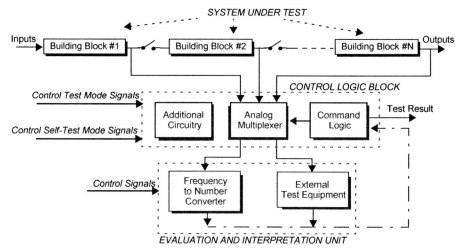

Figure 1.3: Basic OBIST structure

The OBT methodology is very appealing to ease the testing problems mainly due to four reasons:

1.- Test Signal Generation: the test signals are internally generated. It avoids specific hardware from a tester or dedicated resources on chip, eluding the problem of generating the test vectors. Moreover, this property is more important as the effectiveness of many testing methods is severely based on the choice of appropriate test vectors. Mainly, as the complexity of the SUT increases, the problem of producing suitable test vectors (guaranteeing a high fault coverage) even becomes more crucial. The time employed in the process of selecting a satisfactory set of excitation signals has to be added to the whole test time. Therefore, this ***vector-less*** test method represents an improvement not only because eliminates the problem associated with the complexity of the test vector generation but also because the time savings implicit in this test generation process.

2.- Test Evaluation: a simple measurement is used, the oscillation frequency. In this case, the test reference value is a frequency which can be easily transformed to a number, minimizing the accuracy degradation, making easier the test signature evaluation, and removing, likewise, the error sources in the testing process related to the determination of voltage references and signatures. Consequently, due to the digital nature of the oscillation frequency, it can be easily interfaced to boundary scan devices and additionally, can be combined with a multiplexing scheme to probe internal nodes.

3.- Test Time: the test time is reduced due to the fact that only a limited number of oscillation frequencies has to be evaluated for each BUT (at least as might seem at a first glance, if the start-up time is not taken into account).

4.- Extension to BIST: the OBT approach can be easily extended to BIST since this method does not require specific external test stimuli generators and application procedures and, as any digital output test strategy, it can be simply integrated with test methods dedicated to the digital part of the chip.

But in spite of the above advantages, there are several practical drawbacks limiting the OBT applicability. These shortcomings refer, mostly, to the practical modifications needed for obtaining feasible OBT results (well-established and sustained oscillations, more than a test oscillation parameter, high fault coverage, reduced test time, etc...). Thus, all these issues will be studied in this Chapter. But, apart from them, other two basic problems can be pointed out when dealing with OBT. One is related to the application of this technique to complex circuits, where a unified decomposition scheme is still lacking. In fact, it has been applied to small building blocks (biquadratic filters, ADC blocks) [1]-[3], [5], [8]-[11], [13]-[15], [21], [49]-[53], [96]-[98]; just only recently, an example has been discussed on its application to a higher complexity level [25]-[28], [30]-[31], [37]. The other difficulty refers to both the access to different internal nodes and the conversion of the significant test information into a single-digit signal. For the sake of convenience, these two weaknesses will be studied in Chapter 6.

1.1.2 Evolution of the OBT concept

With the aim of setting this thesis within a framework, we summarize all the main references dealing with the OBT strategy published in the last years [1]-[3], [5], [8]-[11], [13]-[15], [21], [25]-[41], [49]-[53], [96]-[98]. Our goal herein is to outline the different versions of the OBT concept which forms the backdrop of this book.

Let us thus distinguish three stages in the evolution of the OBT concept:

First stage: the prior-art of the OBT concept

From 1995 to 1997 the OBT concept emerges [1]-[3], [5], [8]-[10], [49]-[51]. Basically, in this stage the SUT is converted in a linear sinusoidal oscillator and, therefore, only a single oscillation frequency (whose nominal value is determined by linear mathematical analysis) is measured. Moreover, the only documented results are obtained by simulation or discrete circuit realizations. In this stage, researchers overlook some important practical considerations, oscillatory behavior is made strongly dependent on the extra elements added for test purposes, and the only used test parameter is the oscillation frequency.

Second stage: the state-off- art of the OBT concept

In subsequent years, some authors [11], [13]-[15], [52]-[53] continue with the same philosophy and the same line of reasoning than before. However, a new trend emerges as well. Another authors wonder whether only one test parameter (the oscillation frequency) is or not enough to both identify the fault locations and to obtain a high fault coverage [25]-[31], [96]-[97]. Therefore, a new OBT concept arises where the SUT is transformed in a non-linear oscillator which guarantees robust oscillations providing the control over other test parameters (such as the amplitude of the oscillations). Such a new OBT concept successfully solves most of the problems set out in the first stage of the OBT concept.

Different OBT approaches for digital and analog circuits were proposed during this second stage [11], [13]-[15], [25]-[31], [52]-[53], [96]-[97]. Some of the proposals affirmed that this technique could be considered as a general test solution, applicable to most of the basic blocks composing typical complex analog and mixed-signal circuits (amplifier, operational amplifier, comparator, schmitt trigger, filter, voltage reference, oscillator, phase-lock loop, analog-to-digital converters, etc...) or to a combination of these blocks. However, a more detailed study of the existing OBT strategies -examining all their difficulties and shortcomings- guides us to point out, at that time, there was not a general approach of the OBT methodology to be employed for a majority of the analog and mixed-signal systems. This last assertion was based mainly on three reasons:

1.- First of all because, if the related literature is reviewed, we find out that only some kinds of circuits were considered (several types of filters, operational amplifiers, and even analog-to-digital converters). Moreover, for every case, the employed algorithm and the required mechanism to force oscillations were different, and strongly determined by the explicit features of the specific Circuit Under Test (CUT).

2.- A second reason is that there were not enough works regarding the problems of the OBT technique when applied to an embedded complex system. Only some references [36]-[37] can be considered as a first attempt to overcome the mentioned practical system-level problems. In these works, a complex SUT was considered and the difficulties related to the implementation of the OBT strategy were studied.

3.- A third reason, and possibly the most important, is that there was not enough experimental support to postulate that the OBT technique guaranteed a high fault coverage for the studied examples. Furthermore, it was unclear that it could be applied in any situation, with the exception of those works reported elsewhere [33]. Only some practical discrete realizations were described, trying to confirm the robustness of a specific OBT strategy [7].

Third stage: the general OBT concept

All the above-mentioned issues led us to work out an OBT generalization, aiming to deal with different mixed-signal circuits. This relies on measuring frequency but may also use other oscillation parameters for more accurate and relaxed testing. Oscillator robustness is addressed during the test design of the SUT, and non-linear analysis techniques are employed to get an accurate model of the oscillator behavior [25]-[41]. This extended concept is what will be further developed in this book. Table 1.2 outlines the features of the new OBT concept.

Year	1999, 2000, 2001, 2002
Authors	G. Huertas, et al.
Practical Examples SUTs	SC Filters Mixed-Signal Macrocells Oversampling A/D Converters
OBT Mechanism	Transforming the SUT in an oscillator with amplitude control by limitation
Area Overhead	Additional Circuitry to convert the SUT into an oscillator Extra Circuitry to implement the test process and the test evaluation and its interpretation (Analog Multiplexer, Frequency-to-number Converter, Control Logic, Swopamps)
Required Test Measurements	Oscillation Frequency and Other Oscillation Parameters
Determining Nominal Test Parameters	By non-linear mathematical analysis and more accurate non-linear simulation
Test Results	By simulation and integrated prototypes, including experimental data
Practical Considerations	Sustained oscillations High Fault Coverage Accuracy of the theoretical oscillation parameters

Table 1.2: The new OBT concept

1.1.3 Critical analysis of the OBT concept

One of the first incentives to begin this wok was to resolve the drawbacks arose by the first versions of the OBT method. Our goal was to develop some improvements to convert the OBT method in a more efficient test strategy. Let us focus on how the OBT basics have to be modified in order to increase its efficiency in analog testing. Obviously, the original points are the same premises as the first versions of OBT. The same objective is pursued: to transform a mixed-signal SUT (or a part of it) in an oscillator and then to use the achieved oscillations as the test information to detect faults.

We agree with the first versions of OBT in three main points:

1.- The necessity of splitting the whole SUT into meaningful simpler blocks under test. It is essential in the case of having a complex high-order macrocell. Handling a high-order oscillator could be quite complex [100] and thus, the analytical design of the resulting oscillator may be quite difficult. Of course, the overall SUT may oscillate but then, relating the oscillatory behaviour to specific faults would be a very difficult task and, even sometimes, an unmanageable problem, making test interpretation very complex as well.

The step of dividing the SUT into BUTs will be dealt with in next chapters. Different solutions at system level will be proposed and the reasons that made us opt for them will be explained.

2.- The consideration of, at least, two operation modes for the SUT: *a normal mode* where the system performs in its regular way, and *a test mode* where all the BUTs and the remaining elements of the whole SUT (such as switches, buffers, control circuitry....) may be tested simultaneously and/or sequentially.

3.- The requirement that the mechanism to carry out the OBT methodology must involve no significant changes in the SUT structure. That is, changes in the SUT have to be reduced to a minimum in order to prevent significant degradation in the normal operation mode.

But, observing the main features of the prior-art of the OBT concept, we have to reconsider the following issues:

1.- According to the Arabi and Kaminska's approach, any system can be reconfigured as an oscillator by only adding to its structure a linear feedback mechanism that compensates poles. It is, from a mathematical viewpoint, placing at least one pair of complex poles on the imaginary axis (for continuous-time circuits). Apparently, it is enough to adjust the linear feedback elements to establish and sustain oscillations. However, in this Chapter, we will demonstrate the need for imposing some additional restrictions on: the feedback type, the number of feedback elements and the location of the extra components. The required conditions to obtain robust and well-characterized oscillations from the proposed closed-loop system (see Fig. 1.2) will be studied.

2.- On the other hand, it will be shown (herein and in other Chapters), that for most cases only the frequency deviation does not enable the full detection of all catastrophic and/or all parametric faults and does not ensure a high fault coverage. Other measurements may and must be performed.

3.- It will be also described in next chapters, how to relate the frequency (and/or other oscillation parameters) to the elements and/or the specifications of the BUT and how to obtain an accurate nominal value. It is very important to define a good mathematical oscillator model. Moreover, in some particular cases, it will be neccesary to complement the mathematical analysis with some additional simulations or some previous experimental results[2] that help to predict the oscillation parameters with accuracy.

4.- At system level, it will be studied in Chapter 6, a systematic way to apply OBT to a complex macrocell. It will be shown that it is feasible to divide a complex system into simpler building blocks, even if its structure relies on a core whose components are firmly embedded. We will examine what it really means (using switches and/or analog multiplexers) in terms of impact over the system nominal performance and area overhead.

5.- Finally, issues about the test application time will be also analysed in Chapter 6. In the previous versions of OBT, authors did not consider any start-up condition. Moreover, the proposed system-level manner to analyze the different achieved frequencies was to sequencially multiplex the results, and it was a very time consuming strategy. On the other hand, no approach about the time required for the test evaluation and interpretation has been made up to date.

In short, we have questioned in this book the basis of OBT. The goal is to give another outlook on the OBT approach. We will keep the idea of modifying the SUT structure to generate an output (oscillatory) test signal, reflecting the SUT performance. However, we will try to do so by solving all the drawbacks explained above.

1.2 THE OBT OSCILLATOR

Let us consider, first of all, some weak points in the previous OBT concept. That is, those points dealing with the means in what the SUT is transformed in an oscillator. In this Section, RC oscillators will be studied. The goal is to refute some ideas behind the OBT concept and to point out some details which must be taken into account when the OBT strategy is

---

[2.] In the validation or prototyping phase.

applied. In particular, some features of the so-called Wien-bridge oscillator will be discussed, demonstrating that it is not enough to build up a linear sinusoidal oscillator (as was presented in some references [1]-[3]) to achieve an "acceptable" OBT solution. On the contrary, any non-linear mechanism for stabilizing the amplitude of the oscillations is mandatory not only to fulfil the oscillation conditions, but also to maintain steady oscillations and then, to establish another valuable test oscillation parameters apart from the oscillation frequency.

1.2.1 Direct approach: classical linear oscillator

In general, a linear sinusoidal oscillator can be defined as any structure with a pair of imaginary-axis poles (in what the s-domain is being considered). But the poles of any RC-active network can be effortlessly placed on the imaginary axis by, for example, adjusting a coefficient or a gain of the network component.

Let us consider, for the sake of simplicity, the very popular linear KRC oscillator (so-known as the Wien-bridge oscillator) depicted in Fig. 1.4. Many authors who have dealt with the OBT concept have employed modified versions of this structure [1]-[3], [96]-[97] to test either operational amplifiers or RC filters. Even such devices have been considered as benchmark circuits in many references. However, let us point out in this book some important issues that must be carefully analysed.

Let us formulate two preliminary questions:

Oscillator Linear Model:

Q1. Does it allow testing both Operational Amplifiers and/or RC Networks?

Q2. Does it allow predicting the values of the oscillation parameters?

In Fig. 1.4-(a) the operational amplifier (whose gain is supposed infinite so far) in conjunction with the feedback resistors R_a and R_b plays the role of the K amplifier (Fig. 1.4-(b)). The feedback to the positive terminal is by means of the RC network. On the other hand, Fig. 1.4-(c) shows the Wien-bridge oscillator equivalent circuit.

In order to determine the pole placement, the loop-gain (which is called here, $G(K)$) must be calculated first (Fig. 1.4-(b)). Breaking the loop in the point **(1)** (Fig. 1.4-(a) and inspecting the resulting circuit, the obtained loop-gain will be a band-pass function of s, given by

$$G(K) = \frac{K\dfrac{s}{R_2 C_1}}{s^2 + s\left(\dfrac{1}{R_1 C_1} + \dfrac{1}{R_2 C_2} + \dfrac{1}{R_2 C_1}\right) + \dfrac{1}{R_1 R_2 C_1 C_2}} \tag{1.1}$$

with its magnitude peak at $\omega_0 = \dfrac{1}{\sqrt{R_1 R_2 C_1 C_2}}$.

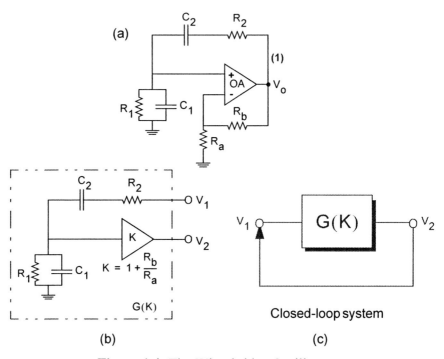

Figure 1.4: The Wien-bridge Oscillator

However, the network has poor selectivity (low quality factor). This can be seen by calculating the 3-dB bandwidth of the loop gain

$$BW = \left(\sqrt{\frac{R_1 C_1}{R_2 C_2}} + \sqrt{\frac{R_2 C_2}{R_1 C_1}} + \sqrt{\frac{R_1 C_2}{R_2 C_1}}\right)\omega_0 \tag{1.2}$$

The R's and C's are normally fixed equal, and thus, $BW = 3\omega_0$. Therefore, the BW is more than twice the center frequency.

The poles of the oscillator are the roots of the characteristic equation of the closed-loop system in Fig. 1.4-(c).

$$1 - G(K) = 0$$

$$s^2 + s\left(\frac{1}{R_1C_1} + \frac{1}{R_2C_2} + \frac{1}{R_2C_1}(1 - K)\right) + \frac{1}{R_1R_2C_1C_2} = 0 \tag{1.3}$$

It can also be written as

$$s^2 + b(K)s + \omega_0^2 = 0 \tag{1.4}$$

where

$$b(K) = \frac{1}{R_1C_1} + \frac{1}{R_2C_2} + \frac{1}{R_2C_1}(1 - K)$$

$$\omega_0^2 = \frac{1}{R_1R_2C_1C_2} \tag{1.5}$$

And the two pole solutions will come given by

$$s_{1,2} = \frac{-b}{2} \pm \frac{\sqrt{b^2 - 4\omega_0^2}}{2} = Re \mp jImg$$

$$Re = \frac{-b}{2} \qquad Img = \sqrt{4\omega_0^2 - b^2} \tag{1.6}$$

To place the poles on the imaginary axis the coefficient of the s-term must be zero in (1.4). This is controlled by setting the gain K to the critical value, K_c

$$K_c = 1 + \frac{R_2}{R_1} + \frac{C_1}{C_2} \tag{1.7}$$

Notice that (1.7) corresponds to the oscillation condition.

We have, then, that when $K = K_c$ the poles are located at $\pm j\omega_0$, and the system will oscillate with the frequency ω_0. Notice that this frequency, ω_0, is explicitly defined only by the values of the RC elements and does not depend on any element of the involved amplifier. Therefore, at least as might seem at a first glance, the answer to **Q1** is:

A1. An oscillator linear model, considering an ideal opamp, does not relate the test parameter (the oscillation frequency) with the operational amplifier structure.

Nevertheless, if a more realistic model of the operational amplifier is used, the oscillation frequency becomes a function of the opamp characteristic as will be discussed later (see also [49]). However, evaluating such a dependency shows that the sensitivity of the test parameter with the RC elements is always very high. An undesirable fact if we are interested in testing only the operational amplifier.

On the other hand, from this linear analysis and considering an ideal opamp, the oscillation magnitude is unknown (although indirectly determined by the nonlinear properties of the amplifier. That is, limited by the saturation levels of the amplifier). Therefore,

---

A2. An oscillator linear model does not predict the value of the oscillation amplitude.

On the other hand, let us formulate a third question:

Oscillator Linear Model:

Q3. Does it really allow testing RC Filters?

We have considered the operational amplifier as ideal. Then, the Wien-bridge Oscillator in Fig. 1.4, under the oscillation condition (1.7), will oscillate with an oscillation frequency given by $\omega_0 = 1/\sqrt{R_1 R_2 C_1 C_2}$. Observe that all the RC elements exert the same influence on this parameter (frequency). If we calculate the deviation of the frequency for a $\pm X\%$ deviation in any RC element, we have always a value given by the following expression:

$$\frac{\Delta\omega_0}{\omega_0} = \left(\frac{10}{\sqrt{100 \pm X}} - 1\right) \cdot 100\% \tag{1.8}$$

It means that regardless of the nominal value of the oscillation frequency, a deviation window of $\pm 10\%$ in any RC element causes a deviation window of $[-4.6, +5.4]\%$ in the oscillation frequency (observe Fig. 1.5-(a)).

On the other hand, Fig. 1.5-(b) shows that, depending on the specific nominal value of ω_0, the value in hertz of the such a frequency deviation is higher or smaller. Notice, for example, from Fig. 1.5-(b), that as the nominal value of ω_0 increases, the value in hertz to be discriminated, increases as well. However, for small values of the oscillation frequency, the value in hertz of its deviation, under a deviation in any RC element, is also small. The detection of this deviation is determined by the precision of the involved tester. It leads us to assert that:

A3. Depending on the value range of the RC elements, the single oscillation frequency, ω_0, could not be sufficient to detect a deviation of the such elements.

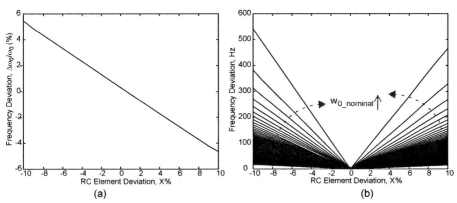

Figure 1.5: Frequency Deviations depending on the RC elements

Let us formulate a fourth question:

Amplifier Model:
Q4. Does it change the oscillation condition and the estimated oscillation frequency?

The goal now is to reveal how the oscillation frequency and the critical value of the gain K_c are significantly altered if, for example, a one-pole model is used for the operational amplifier. The answer of this question highlights the consequent problem: how accurate must be the amplifier model in order to obtain a good estimation of the oscillation mode, a key point when the OBT concept is being applied (a similar problem was also formulated in [49] and [125]).

To solve this question let us assume again that the R's and the C's are equal. In this case, the closed-loop characteristic equation when the amplifier is ideal, (1.3), is transformed into

$$s^2 + \frac{s}{RC}(3 - K) + \frac{1}{R^2 C^2} = 0 \qquad (1.9)$$

Now, the gain of the non-inverting amplifier, when a one-pole roll-off model[3] is used for the operational amplifier would be

$$K = \frac{GB}{s + \dfrac{GB}{K_0}} \tag{1.10}$$

where $K_0 = 1 + \dfrac{R_b}{R_a}$.

Thus, we obtain a modified characteristic equation

$$s^3 + s^2\left(\frac{GB}{K_0} + \frac{3}{RC}\right) + s\left[\frac{GB}{RC}\left(\frac{3}{K_0} - 1\right) + \frac{1}{R^2C^2}\right] + \frac{GB}{K_0}\frac{1}{R^2C^2} = 0 \tag{1.11}$$

$$s^3 + a_2 s^2 + a_1 s + a_0 = 0$$

Then, to produce imaginary-axis poles, a_i coefficients must fulfil[4] $a_1 a_2 = a_0$. Therefore

$$(GB \cdot RC)^2(3 - K_0) + 3K_0(GB \cdot RC)(3 - K_0) + 3K_0^2 = 0 \tag{1.12}$$

Let $K_0 = 3 + \Delta K$ and simplify[5]

$$(GB \cdot RC)^2\Delta K + 3K_0(GB \cdot RC)\Delta K - 3(3 + \Delta K)^2 = 0 \tag{1.13}$$

When $GB \cdot RC \gg 9$ and $\Delta K \ll 3$, the last equation can be approximated by $(GB \cdot RC)^2\Delta K - 27 = 0$, resulting in:

$$\Delta K = \frac{27}{(GB \cdot RC)^2} = 27\left(\frac{\omega_0}{GB}\right)^2 \tag{1.14}$$

Notice, from this last equation, that the critical value of the gain to sustain oscillations must be set higher than for the ideal case. And, therefore, the

---

[3.]   The amplifier one-pole model is $a_v(s) = (-a_v\omega_a)/(\omega_a + s)$, where ω_a represents the open-loop band-width, a_v is the dc-gain, and GB is the gain-bandwidth product that is related to ω_a, a_v in such a way that $GB = a_v\omega_a$.

[4.]   In this case, the third-order polynomial could be also written as $(s + p_1) \cdot (s^2 + \omega_{om}^2) = 0$, where p_1 will be the real remaining pole due to the amplifier. By equating terms in both expressions, this condition is achieved. Moreover, if we check the location of p_1, it is always placed on the left half of the s-plane.

[5.] $K_0 = K_C + \Delta K = 3 + \Delta K$.

higher the oscillation frequency, the higher the required gain for oscillations. Consequently, unless additional circuitry is used, the oscillator will drop out of oscillation as the frequency has a higher value. Therefore:

A4. An accurate operational amplifier model is needed to achieve a good approximation of the gain critical value and a good estimation of the oscillation frequency.

On the other hand, to obtain the modified oscillation frequency, we have that $\omega_{0m} = \sqrt{a_0/a_2} = a_1$. That is (assuming that K_0 is adjusted to the critical value)

$$\omega_{0m} = \sqrt{\frac{\dfrac{GB}{K_0}\omega_0^2}{\dfrac{GB}{K_0} + 3\omega_0}} = \omega_0 \sqrt{\frac{1}{1 + 3K_0\dfrac{\omega_0}{GB}}} \approx \omega_0\left(1 - \frac{9}{2}\frac{\omega_0}{GB}\right) \tag{1.15}$$

And then

$$\frac{\Delta\omega_0}{\omega_0} \approx -\frac{9}{2}\frac{\omega_0}{GB} \tag{1.16}$$

Therefore, the higher the ω_0/GB ratio, the lower becomes the modified oscillation frequency, ω_{om} compared to its ideal value, ω_o and the higher becomes the error of considering an ideal operational amplifier.

Notice, moreover, that a new oscillation condition can be deduced from (1.15), it is that $GB > \frac{9}{2}\omega_0$ to guarantee a correct estimation of the oscillation frequency.

If

$$GB = \alpha \cdot \omega_o \tag{1.17}$$

is considered (of course, fulfilling $\alpha > \frac{9}{2}$), then the error of supposing the ideal frequency value, ω_o, instead of the modified frequency value, ω_{om} will be (according to (1.15))

$$|E_{\omega_o}| = \frac{9}{2\alpha} \cdot 100 \tag{1.18}$$

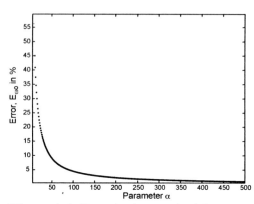

Figure 1.6: Error in the nominal frequency

This relationship is graphically shown in Fig. 1.6. As can be observed, if $\alpha > 500$, then the error made in estimating the oscillation frequency is less than 1%. In this range of α, it can be considered than the oscillation frequency is independent of the amplifier gain-bandwidth product, GB. But, it is no advisable when the OBT strategy is applied to an operational amplifier. In this case, it is necessary that the characteristics of the amplifier appear explicitly in the oscillation frequency expression[6]. Nevertheless, in this last case, a very accurate operational amplifier model is required in order to obtain a good estimation of the nominal frequency, a critical point because this nominal value must be supposed as the reference value for evaluating the test output.

Let us, then, formulate a fifth question:

Amplifier Model:
Q5. Does a more complex amplifier model allow testing operational amplifiers?

In practice, there exists a clear dependence between the modified oscillation frequency and the important characteristics of the operational amplifier under test. But, it is only reflected when a non-ideal operational amplifier model is considered. Specifically, those features implicitly contained in the GB expression (such as the amplifier pole placement, ω_a and the amplifier dc-gain, a_v). However, from (1.15) it can be observed that the modified oscillation frequency depends on both the ideal oscillation frequency (exclusively related to the RC-network whose circuitry is included only for reconfiguring the amplifier), and the amplifier GB (whose value deviation must be detected

---

6. However, if the OBT strategy is being applied, for example, to a filter stage (given by the RC-network) the condition $\alpha > 500$ is a good choice for the amplifier characteristics in order to detect faults only focused on the RC-filter components (that is, resistors, capacitors, connections and amplifiers but at high level) ignoring the actual implementation of the amplifier.

for testing the amplifier). But deviations in each one of these parameters do not cause the same impact in the new nominal frequency value.

For example, for a $\pm 10\%$ deviation range of the ideal frequency, ω_o (which depends on the RC- network components) and varying the α parameter (defined in (1.17)) from 5 to 500, the changes in the actual frequency are displayed in Fig. 1.7. It should be clear from that when α

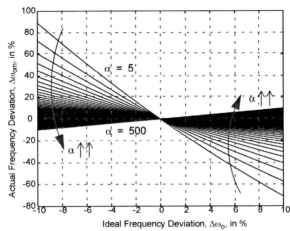

Figure 1.7: Frequency Deviations in relation to α

exceeds the limit of 500, a deviation of the ideal frequency translates into a similar deviation of the actual frequency, regardless of the value of α.

On the other hand, for any given ideal oscillation frequency, if parameter α is deviated around a $\pm 10\%$ range for a wide area of nominal values (sweeping these nominal values from 5.1, to verify the oscillation condition $\alpha > \dfrac{9}{2}$, until 459, in order not to exceed the upper limit of 500). Then, it can be seen from Fig. 1.8 that as α is regarded smaller, a deviation of its value translates into a higher deviation of the actual frequency and, of course, it occurs independently the value of the ideal frequency value.

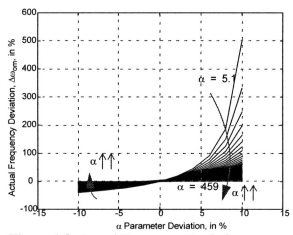

Figure 1.8: Actual Frequency Deviation versus α

Finally, the objective is to find a trade-off between these two possibilities; if α is defined small, the actual oscillation frequency strongly depends on such a value (it means that the actual oscillation frequency is severely related to the operational amplifier characteristics) but, in this case, the actual oscillation frequency also depends severely on the ideal oscillation frequency (that is, on the RC-network components). On the other hand, if α is defined very high, then the relation between the actual oscillation frequency and the operational amplifier characteristics disappears. But this fact must be avoided if we need to obtain some information about the operational amplifier to be capable of testing it. In this case, we only can say that a visible deviation in the oscillation frequency can be owing to either the value of α is under 500 (see again Fig. 1.7) or the extra RC elements contain any fault. The conclusion is that:

A5. Depending on both the amplifier model and the specific value of the RC elements an operational amplifier can or cannot be tested using a classical sinusoidal oscillator.

Let us finally formulate a last question:

Oscillation Linear Model:
Q6. Does it allow achieving more than a test parameter?

Observe, considering again the closed-loop system of Fig. 1.4 and a simplified amplifier model, that if $K < K_c$, the poles are always in the left half-plane, and thus, once initiated, oscillations cannot be sustained. The sine-wave amplitude decays exponentially. Whereas, if $K > K_c$, the poles are in the right half-plane, and thus oscillations grow exponentially in amplitude until going beyond the linear dynamic range of the amplifier (observe Fig. 1.9).

An illustrative case is when the R's and C's are identical. In this particular case, the output takes the form

$$V_o(t) \approx V_m e^{\frac{1}{2}\Delta K \omega_0 t} \sin(\omega_0 t + \phi) \tag{1.19}$$

where ΔK is the deviation of the gain K from the critical value, that for this case is $K_c = 3$. Due to term ΔK, the pole placement is not exactly at $\pm j\omega_0$, but approximately at $\frac{1}{2}\Delta K \omega_0 \pm j\omega_0$. As seen in Fig. 1.9, when $\Delta K = 0$, the output is a sine wave of amplitude V_m, when $\Delta K > 0$, the output waveform is an overdamped sinusoid and, finally, when $\Delta K < 0$, the output response is a underdamped sinusoid. The time constant associated with the growth or decay of the two last cases is

$$\tau = \frac{2}{\Delta K \omega_0} = \frac{T}{\pi \Delta K} \tag{1.20}$$

where T is the period of the sine waveform. For example, if K decreases by 0.01% compared to the critical value, then $\frac{\tau}{T} = \frac{100}{\pi \times 3 \times 0.01} \approx 1000$.

This means that after 1000 periods, the amplitude of oscillations decreases to 37% of its initial value. But this fact is not very significant if OBT approach only requires the frequency as test parameter [1]. However, more recent studies show that the oscillation amplitude is also necessary to achieve high fault coverage and/or increase the observability of the fault locations [25]-[41], [96]-[97]. Thus, in general, robust oscillations are required for test purposes.

Because the network is not driven, (1.19) expresses the natural response of the network. Any disturbance, such as the application of the dc sources to activate the amplifier, excites this response. But, in this point, a significant problem comes out:

Is it really feasible to design this kind of oscillators (classical linear oscillators)?

Is it possible to maintain this response with $\Delta K = 0$ and V_m constant?

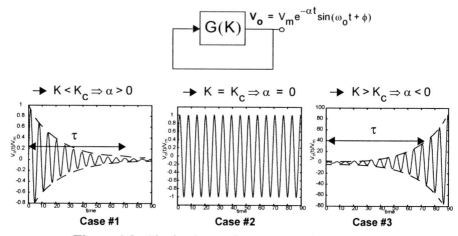

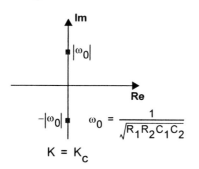

Figure 1.9: Obtained waveforms. Three possible cases

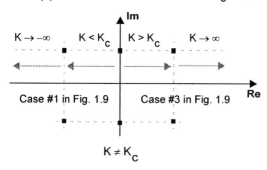

From Fig. 1.10, the answer to this question can be inferred[7]. It is clear that if the value of the gain K turns a little off the critical value, the system evolves to a non-oscillatory response without finding, in theory, any mechanism that forces it to return to the oscillatory state.

In order to keep always $\Delta K = 0$, the amplifier gain must be held precisely at the critical value given by (1.7). But this is impossible to achieve in practice (at least from a theoretical viewpoint, using a completely linear anal-

Figure 1.10: Poles Behavior

---

7. Fig. 1.10 displays the behavior of the involved pair of complex poles responsible for oscillations in the closed-loop system of Fig. 1.4.

ysis) and thus, additional circuitry must be used to perpetuate oscillations.

Thus, the network oscillates with ω_0 only if $K = K_c$. But it involves that in the physical implementation, the resistor and capacitor ratios must be accurately the required ones. If these ratios vary slightly, the system will not have any way to compensate that effect and, depending on the initial condition, the system will be unstable or stable and the response will be overdamped or underdamped (see Fig. 1.9).

From a practical viewpoint, this linear analysis should be complemented with some additional consideration regarding the oscillation maintenance. This is traditionally done assuming a nonlinear amplifier model which takes into account the inherent non-linearities of the operational amplifier when it is working in the saturation region [125]. Indeed, if a non-linear amplifier is considered, the previous study is not valid and another results can be extracted. Only one important conclusion can be derived from all the previous classical linear analysis:

A6. Some form of nonlinearity (inherent or intentionally introduced to the structure) has to be employed in order to guarantee the stability of the output signal amplitude.

That is, the nonlinearity allows adjusting the non-oscillatory behavior (Fig. 1.11) in such a way that if K turns off the critical value K_c, the own system is able to return the poles to the imaginary axis through the nonlinear mechanism.

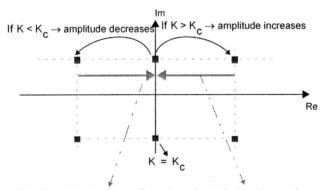

Figure 1.11: Magnitude stabilization phenomenon

In fact, if we use a completely linear analysis of the operational amplifier, this oscillation magnitude stabilization phenomenon shown in Fig. 1.11 is disguised, and, likewise, the oscillation magnitude value cannot be calculated [125]. Otherwise, only a nonlinear analysis allows the entire study of the oscillation magnitude stabilization phenomenon and the calculation of the oscillation magnitude value as well. But there is an intrinsic difficulty in achieving the required sustained oscillators by only considering the non-linear behavior of the operational amplifier and it is, that the accurate mathematical analysis of a non-linear system is very complex and most of the times impracticable. Even designers when are designing amplifiers do not pay attention to the non-linear region of the circuit because no reliable information can be obtained when that region is examined and, exclusively, they focus attention on the linear region where the system must operate. In fact, there are written evidences of this problem: for a similar oscillator structure if one uses two different amplifiers (either OTA-based or CFOA-based), the circuit have not the same behavior (see [125]). For one of them, the amplifier inherent nonlinearity can stabilize the magnitude of the oscillations whereas, for the other one, the amplifier inherent nonlinearity forestall the stabilization phenomenon.

Therefore,

Additional non-linear techniques are required not only to provide control of the oscillation amplitude (allowing to employ the amplitude as a test parameter) but also to predict accurate oscillation parameters.

That is, a non-linear amplitude stabilization mechanism is needed. Two schemes are traditionally employed to sustain oscillations at constant amplitude. One scheme introduces a controllable nonlinearity in the amplifier's gain characteristic. The other scheme provides for an automatic adjustment of the gain characteristic. Next, these schemes will be briefly reviewed.

1.2.2 Second approach: oscillator using non-linear methods

From the above Section a main issue has been concluded: the necessity of a non-linear oscillator with well-characterized response and whose model provides accurate information about some other appealing oscillation parameters such as the oscillation amplitude. Let us introduce the problem of how to achieve robust oscillators. As was already discussed, the only *safe* way to assure sustained oscillations would be a mechanism with a non-linear part (placed on purpose in the system) whose functionality can be well-controlled and analysed theoretically.

If the existing bibliography is reviewed, two methods can be first reported to obtain robust oscillations. These are either oscillators with non-linear amplifier gain or oscillators with automatic gain control [100]. In both cases, the stabilization of the oscillation signal is not attained in the saturation range of the amplifier because other factors determine this phenomenon. As mentioned before, by introducing an extra nonlinear mechanism, we have two parameters to play with. Two test parameters would be available to distinguish if a fault is in the amplifier or in the added circuitry, for example. Nevertheless, these nonlinear mechanisms require to handle complex nonlinear analysis techniques to establish with precision the steady-state output [100]. And the more complex the amplifier model is, the more complicated these techniques are. This is an important point, which makes these techniques not appealing for OBT. Obviously, researchers look for test methods which do not demand a lot of design effort. Therefore, for OBT, a third nonlinear robust oscillator is then proposed in this thesis, whose oscillations can be described by a linearized model.

-Automatic gain control oscillator:

A better control of poles is attained through automatic gain control where the output amplitude is continuously supervised and compared to a fixed nominal level [100]. If for any reason a change in amplitude takes place, the amplifier gain is compensated until it is returned to its expected nominal value. Under equilibrium conditions, operation is in the linear range of the amplifier, and therefore, distortion is very low.

Then, automatic gain control is extensively exploited in the design of oscillators in order to obtain constant oscillation amplitude with low distortion. In this case, the amplifier gain is governed by the oscillation amplitude and the non-oscillatory behavior becomes as illustrates Fig. 1.12.

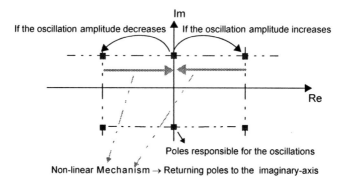

Figure 1.12: Amplitude Stabilization Phenomenon

This gain (see Fig. 1.13) is forced to be high enough to start the oscillations either through the natural system evolution or through a start-up technique. As the oscillation amplitude grows up, the gain is automatically reduced to the necessary value to sustain oscillations. Fig. 1.13 shows the required amplifier characteristic to achieve it.

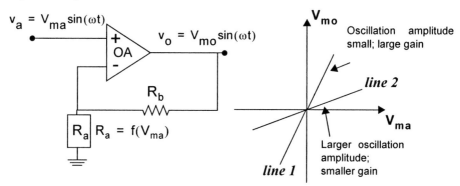

Figure 1.13: Amplifier with automatic gain adjustment

If the amplifier gain is $1 + R_b/R_a$, it can be modified by changing either R_a or R_b. In Fig. 1.13, R_a is made a function of the input signal amplitude. The practical dependence is such that the larger V_{ma}, the larger R_a, and thus, the smaller the gain. Therefore, the gain characteristic displayed in Fig. 1.13 is achieved. When the peak values of the input signal are small (low-amplitude input signals) the amplifier operation goes along *line 1* (high gain), whereas when the peak values of the input signal are large (high-amplitude input signals) the amplifier operation goes along *line 2* (low-gain). As a result, the amplifier gain is adjusted according to the oscillation amplitude behavior. Besides, as V_{ma} is gradually increased, amplifier operation moves from *line 1* to *line 2* in a continuous manner.

When inserting this amplifier in the equal-R, equal-C Wien-bridge network, R_a and R_b are selected in such a way that the gain is 3 (oscillation condition) for the pursued oscillation amplitude. If for any reason the oscillation amplitude declines slightly from its nominal value, then it also reduces the value of R_a. Consequently, the gain is increased and the reduction of the oscillation amplitude is, thus, counteracted. On the contrary, if the oscillation amplitude increases, then R_a increases too, reducing the gain and counteracting again the effect of the original change. In fact, the technique to control the oscillation amplitude is accomplished since R_a changes automatically to produce the neccesary critical gain to force oscillations[8]. As far as the steady-state operation is along a straight line (see Fig. 1.13), there is no significant distortion in the output.

-Oscillator with non-linear amplifier gain (see [100] for details):

A second non-linear method would be to employ an amplifier with a settled but nonlinear gain characteristic instead of the linear gain K considered above (see Fig. 1.14). For small amplitudes of the output signal, the gain, now called m_1, is higher than the critical value, to guarantee the starting and the growing of oscillations. Whereas, for larger levels of the output signal, the incremental gain ($\Delta v_o / \Delta v_a = m_2$) is smaller. As a result, the oscillation amplitude is finally stabilized. Therefore, this procedure forces the system evolution to a steady oscillation when it starts from a non-oscillatory state.

---

[8] Such a value may be altered with the operation frequency (due to imperfect tracking of capacitors or due to other reasons).

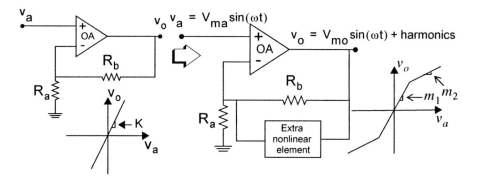

Figure 1.14: Non-linear amplifier gain

A practical circuit for an oscillator based on an amplifier with non-linear gain characteristic is shown in Fig. 1.15 (see [100]). The back-to-back zeners are used to achieve a break-point in the gain characteristic.

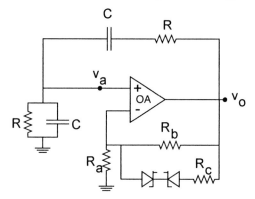

Figure 1.15: Oscillator with nonlinear gain

The amplifier gain $m_1 = 1 + R_b/R_a$ is chosen slightly greater than 3 to guarantee that the output builds up exponentially (and no additional mechanism is required to start up the oscillations) until the amplifier is driven into the m_2 range of operation (near the peak swing of the output). As a result, both v_a and v_o grow into nonsinusoidal, periodic signals. A quantitative analysis can be only made assuming low output distortion and, therefore, considering v_a perfectly sinusoidal. As (1.1) shows, the RC-feedback network at the noninverting terminal of the amplifier is a band-pass network that performs a reasonable filtering action. As was debated in the previous section, provided that the amplifier can be regarded ideal, the oscillation frequency coincides with the peak response (that is, $\omega_o = 1/RC$). Thus, the selective filtering action of the RC-network results in a signal at v_a (see Fig. 1.15) which is somewhat more nearly sinusoidal than the signal at v_o. Therefore, to a first-order approximation, the signal at v_a may be considered sinusoidal. When this sinusoidal signal passes through the amplifier, its top portion is distorted symmetrically about the peak value (indicating the presence of odd harmonics only), according to the gain characteristic displayed in Fig. 1.14 (which can be interpreted graphically as is shown in Fig. 1.16)[9].

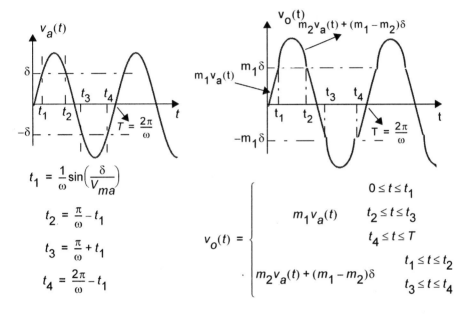

Figure 1.16: Non-linear gain amplifier effects

Assuming $v_a = V_{ma}\sin(\omega t)$, and then finding the fundamental component of the resulting output, v_o. we can calculate v_o, which is then expanded in its Fourier series to achieve the fundamental component, $v_{of} = V_{mo}\sin(\omega t)$. If it is worked out, the following expression for the gain of the fundamental component is achieved

---

9. In practice, because v_a is not in fact sinusoidal, a slight skewing is also present in the top portion of the output waveform.

$$\frac{V_{mo}}{V_{ma}} = \begin{cases} m_1 & \text{if } V_{ma} < \delta \\ & \text{and} \\ m_2 + \frac{2}{\pi}(m_1 - m_2)\left[\sin\left(\frac{\delta}{V_{ma}}\right)^{-1} + \frac{\delta}{V_{ma}}\sqrt{1 - \left(\frac{\delta}{V_{ma}}\right)^2}\right] & \text{if } V_{ma} > \delta \end{cases} \tag{1.21}$$

If (1.21) is examined and drawn for two start-up values of gain m_1 considering m_2 as a parameter, Fig. 1.17 is obtained. Obviously, as was remarked above, the gain m_1 has to be chosen somewhat higher than 3 in order to guarantee the start of oscillations. As is shown in Fig. 1.17, as long as the peak value of v_a, V_{ma}, is less than the break-point voltage in the amplifier characteristic, δ, the gain at the fundamental frequency is $m_1 > 3$ and, therefore, the system operates in such a way that keeps oscillations growing. But this effect is counteracted when V_{ma} exceeds its limit value, δ, and the amplifier goes in the m_2 region. Then, provided that the selected value of m_2 is under 3, the amplitude of oscillations tends to diminish until the effective gain is reduced to 3.

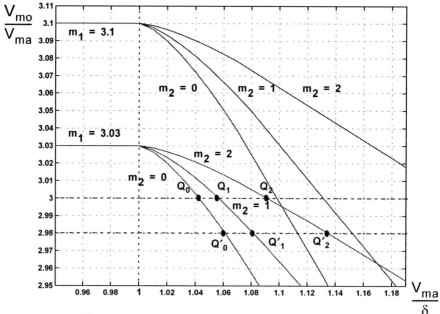

Figure 1.17: The gain at the fundamental frequency

Notice from Fig. 1.17, that depending on the exact value of gain m_2, for each value of m_1, equilibrium establishes at a different point. But, certainly, the higher m_2, the smaller the distortion in the output (see Fig. 1.16). On the other hand, if K_c decreases from 3 to 2.98 (due to, for example, imperfect capacitor matching), the operation stabilizes at points $Q'_{0, 1, 2}$ rather than at $Q_{0, 1, 2}$. Consequently, in this case, the output amplitude will be larger than before. Clearly, this result may be minimized if m_2 is reduced. Finally, as the start-up gain m_1 is selected larger than 3, the output must spend a likewise longer time in the m_2 region, until reaching the equilibrium state where the gain is 3 for the fundamental component. In this case, the oscillation amplitude is higher, and the output exhibits more distortion.

From the above analysis we may conclude that the selection of m_1 and m_2 could be conflictive because, as the amplifier-gain characteristic is more abrupt, the output signal distortion becomes higher. However, in the test framework, the distortion can be wisely used as test parameter. In practice, for test purposes, low-distorsion is not strictly needed. Furthermore, frequency sensitivity and time to build-up the oscillation are much more important for testing.

With this simplified non-linear study, only the amplitude of the oscillations is examined since the operational amplifier is considered ideal. In this case, the oscillation frequency is only regulated by the RC-network components and does not depend on the non-linear amplifier characteristic. This just an approximation, which can be improved by using a more complex amplifier model, althought in this latter case all analytical results must be corrected, leading to expressions of a higher complexity [125].

1.2.3 Proposed approach: amplitude controlled by limitation

As was discussed, with the above-mentioned non-linear oscillation methodologies, at least a new oscillation parameter is gained for testing purposes due to the fact that, in this case, the oscillation amplitude can be accurately estimated. However, observing the results of the two previous sections, we can assert that neither of the two described non-linear methods can be considered a feasible OBT scheme. A significant reason can be emphasized: a complex non-linear mathematical theory would have to be handled.

In short, a non-linear method is needed which allows not only to place correctly the poles of the SUT and to implement a mechanism to stabilize the oscillations, but also to handle easily the oscillation expressions. As was said, oscillations can be forced by either of the nonlinearities existing in any block of the oscillator architecture. Therefore, all the nonlinearities must be under control in such a way that we can disregard those we are not interested in. Moreover, the proposed strategy must not be limited by the precision of the additional circuitry.

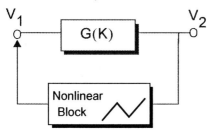

Figure 1.18: Non-Linear Oscillator

A simpler method to proceed would be to built an oscillator with amplitude control by limitation. Let us modify the closed-loop system of Fig. 1.4-(c), obtaining the new closed-loop system shown in Fig. 1.18. This scheme is the base of such a kind of oscillators.

The characteristic equation would be now

$$1 + N(A)G(K) = 0 \qquad (1.22)$$

being $N(A)$ an equivalent linearized transfer function of the nonlinear element with respect to the first harmonic amplitude, A, of its output (providing that one employs the so-called describing-function approach[10]) and $G(K)$ the loop gain of the Wien-bridge oscillator of Fig. 1.4-(a).

Because of this nonlinear mechanism, if the oscillation conditions are fulfilled [11] the system oscillates with a frequency, ω_0 and an amplitude, A_0, satisfying the following expressions

$$\omega_0 = \frac{1}{\sqrt{R_1 R_2 C_1 C_2}} \qquad N(A_0) = \frac{1 + \dfrac{R_2}{R_1} + \dfrac{C_1}{C_2}}{K_0} \qquad (1.23)$$

---

[10] The reader is encouraged to go to Chapter 2 for an in-depth treatment of the describing-function approach.

[11] Chapter 2 is again referred for the way to determine the oscillation conditions.

being $K_0 = 1 + \dfrac{R_b}{R_a}$.

Two test parameters are now available to test the RC elements. Again, supposing an ideal model for the amplifier, the oscillation parameters do not reflect any influence from it. And, if we assume that the R's and the C's are equal, then

$$\omega_0 = \frac{1}{RC} \qquad N(A_0) = \frac{3}{K_0} \qquad (1.24)$$

Moreover, if again we suppose $K_0 = 3$, then we are in the same situation than in the case of Fig. 1.4-(c) but with the proviso that now we have an (almost) perfect control of the oscillation amplitude and the stabilization mechanism.

Observe that unlike the frequency, the amplitude gain, $N(A_0)$ does not depend on all the RC elements in the same way (see (1.23)). If we calculate the deviation of the equivalent linearized gain, $N(A_0)$ for a $\pm X\%$ deviation in the RC elements (R_1, R_2, C_1, C_2), we obtain Table 1.3.

$\dfrac{\Delta[N(A_0)]\big|_{R_2}}{N(A_0)}$	$\dfrac{\Delta[N(A_0)]\big|_{C_1}}{N(A_0)}$	$\dfrac{\Delta[N(A_0)]\big|_{R_1}}{N(A_0)}$	$\dfrac{\Delta[N(A_0)]\big|_{C_2}}{N(A_0)}$
$\dfrac{\pm\frac{R_2}{R_1}X}{1+\frac{R_2}{R_1}+\frac{C_1}{C_2}}\%$	$\dfrac{\pm\frac{C_1}{C_2}X}{1+\frac{R_2}{R_1}+\frac{C_1}{C_2}}\%$	$\dfrac{\frac{R_2}{R_1}\left(\frac{\mp X}{100\pm X}\right)}{1+\frac{R_2}{R_1}+\frac{C_1}{C_2}}\%$	$\dfrac{\frac{C_1}{C_2}\left(\frac{\mp X}{100\pm X}\right)}{1+\frac{R_2}{R_1}+\frac{C_1}{C_2}}\%$

Table 1.3: Deviation of the equivalent linearized gain for a deviation of a RC element

It means that depending on the nominal value of the oscillation gain (that is, depending on the specific values of R_1, R_2, C_1, C_2), a deviation window of $\pm10\%$ in one of the RC elements causes a deviation window in the oscillation gain larger or smaller. Let us study, as an illustrative example, the case of a deviation in the element R_2. It can be seen from Fig. 1.19 that the value of the gain deviation is strongly determined by the value of the remaining elements

(R_1, C_1, C_2). And something similar results when deviations in the other elements are considered. A main fact can then be emphasized: $N(A_o)$ (and in fact, the oscillation amplitude) exhibits a great versatility to detect faults under certain conditions whereas the frequency only allows to cover faults depending on the required accuracy.

Clearly, the dependencies of the oscillation amplitude with respect to the RC elements have the peculiar feature of being determined by all those elements whereas the oscillation frequency only depends on a specific element at the same time. Therefore, we can take advantage of the oscillation amplitude for testing.

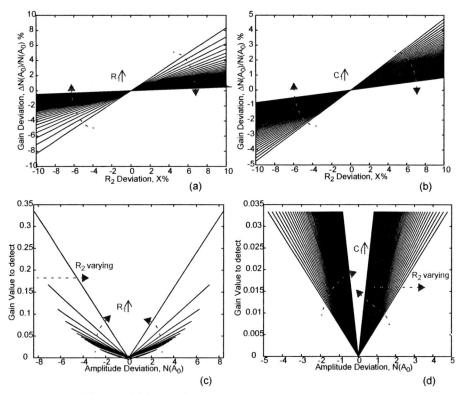

Figure 1.19: Equivalent Linearized Gain Deviation

Assuming, for example, a very specific case $R_1 = R_2 = R = 10k\Omega$ and $C_1 = C_2 = C = 100e^{-9}F$, we would have the situation shown in Fig. 1.20-(a). Therefore, the oscillation frequency measurement would provide more information than the oscillation amplitude measurement under a deviation of any RC element (for example, a deviation in R_1). However, simply moving the value of one of the other RC elements (for example, R_2 under and over $10k\Omega$) this situation changes (observe Fig. 1.20-(b)). If $R_2 > 20k\Omega$ the amplitude measurement begins to provide more flexibility than the frequency measurement.

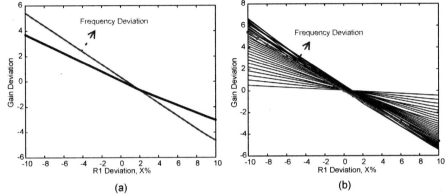

(a)   (b)

Figure 1.20: Amplitude Deviation Considerations

The above-discussed results were obtained by using a completely ideal model of the amplifier. However, if a more realistic model of the amplifier is used (one-pole model, for example), the characteristic equation is transformed to (considering $R_1 = R_2 = R$ and $C_1 = C_2 = C$)

$$s^2 + \frac{s}{RC}\left(3 - N(A)\frac{GB}{s + \frac{GB}{K_0}}\right) + \frac{1}{R^2 C^2} = 0 \qquad (1.25)$$

The oscillation parameters are now

$$\omega_{0m} = \sqrt{\frac{\frac{GB}{K_0}\omega_0^2}{\frac{GB}{K_0} + 3\omega_0}} = \omega_0 \sqrt{\frac{1}{1 + 3K_0\frac{\omega_0}{GB}}}$$

$$N(A_{0m}) = N(A_0)\left(1 + \frac{K_0^2}{\left(\frac{GB}{\omega_0}\right)^2 + 3K_0\frac{GB}{\omega_0}}\right)$$
(1.26)

And, assuming that K_0 is adjusted again to 3

$$\omega_{0m} \approx \omega_0\left(1 - \frac{9}{2}\frac{\omega_0}{GB}\right)$$

$$N(A_{0m}) = N(A_0)\left(1 + \frac{9}{\left(\frac{GB}{\omega_0}\right)^2 + 9\frac{GB}{\omega_0}}\right)$$
(1.27)

Now, we have two parameters depending on both the amplifier character-istic and the RC elements. However, by directly observing (1.27) the dependence on the amplitude with the amplifier GB is, in fact, smaller than its dependence on the frequency. But, both parameters, frequency and ampli-tude can be simultaneously used to also identify faults in the amplifier. The steps to test the amplifier are the following:

1.- Measuring the frequency and the amplitude of oscillations in the test output. Two data would be, thus, obtained: $data_1 = \omega_{0m}\big|_{measured}$ and

$data_2 = N(A_{0m})\big|_{measured} / N(A_0)\big|_{expected}$.

2.- Solving the involved two equations ((1.27)) with two unknown factors

$$data_1 = x\left(1 - \frac{9}{2}y\right)$$

$$data_2 = \left(1 + \frac{9y^2}{1 + 9y}\right)$$
(1.28)

being $x = \omega_0\big|_{measured}$ and $y = \frac{\omega_0}{GB}\bigg|_{measured}$.

We would have a way to distinguish if the fault is in the amplifier or in the added circuitry (related to ω_0). That is, if both parameters deviate in the same proportion from their nominal values, then we can suspect that the fault is in the extra circuitry. On the contrary, if only the factor y deviates we have sufficient evidence to affirm that the fault is in the amplifier.

This result is valid provided that we can consider $N(A_0)$ does not deviate practically under slight variations of the RC elements. That is, $N(A_0)$ and the RC elements are practically uncorrelated whereas the relationship between ω_0 and the RC elements is stronger. This is the case we are considering, when $R_1 = R_2 = R$ and $C_1 = C_2 = C$, and $R_2 \ll R_1$ or $C_1 \ll C_2$.

Something is clear: the non-linear oscillator shown in Fig. 1.18 provides more flexibility than its linear version (Fig. 1.4) because other oscillation parameters are straightforwardly achieved.

Hence, our proposal for the OBT oscillator is the scheme shown in Fig. 1.18 can guarantee three important points:

1.- Maximum insensitivity of the oscillation parameters with the extra circuitry added for test purposes. As will be seen, this issue has been ignored or at least overlooked in many references [1]-[3].

2.- Accuracy in estimating the reference values of the oscillation parameters. As will be seen this issue is related to both the model chosen to describe the test oscillator and the selected analysis method to study it.

3.- High fault coverage. This issue is related to the number of the involved oscillation parameters and as will be seen it is closely linked to the features of the selected oscillator designed for test purposes.

On the other hand, we can observe that there is not a good and simple mechanism to test exclusively the amplifier. Even supposing more than one test parameter, the dependence on such test parameters respect to the amplifier features is disguised with the dependence respect to other elements of the oscillator. Therefore, from our viewpoint, the first level of applicability of OBT are blocks with a functionality more complex than an amplifier [129]. Moreover, the OBT technique must not only rely on the possible existing inherent nonlinearities in the structure. The OBT technique must involve an additional nonlinear mechanism.

1.3 THE OBT CONCEPT REVISITED: PROPOSAL FOR ROBUST OBT

In this section, the principles of a new OBT concept will be lay down. Some of the different aspects pointed above such as the characteristics of the OBT oscillator and the general circuit modifications (type of feedback, added or removed components, number and kind of extra components, etc), the start-up problem, the measurements that must be carried out, and the fault coverage, will be at length considered in this Chapter whereas, for the sake of convenience, other issues such as the system partitioning, the application cost, the required test support at system/subsystem level, the compatibility with functional approaches, etc..., will be studied in other chapters.

1.3.1 The oscillator

As was discussed above, self-starting and self-sustained oscillations are required. It forces us to carefully thinking on a general feedback mechanism valid in any case. That is, on how to make any system oscillating independently of its transfer function and (if possible) using a common feedback element. That mechanism cannot be linear in practical circuits and must also take into account those non-linearities inherent to the operation (fault-free and faulty) of active components. A type of nonlinear feedback element has to be selected capable of generating robust oscillations. This problem has been extensively considered in [25]-[41], where a general and practical solution for building up the OBT oscillator was proposed (Fig. 1.21). When this is the case self-maintained oscillations can be guaranteed although conditions for starting up oscillations need a separate consideration. The so-called start-up problem will be discussed in coming sections.

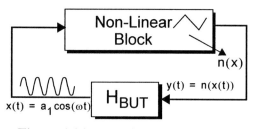

Figure 1.21: Generic OBT oscillator

In general, the block diagram of Fig. 1.21 is portraying a harmonic oscillator with amplitude control by limitation, provided that H_{BUT} is representing a linear block and $n(x)$ the nonlinear element responsible for the amplitude control. Some non-linear devices used to obtain these controllable characteristics will be studied in Chapter 2, where the techniques and tools used to analyze this kind of oscillator structures will be reviewed.

Many different types of non-linearities can be considered. They can be inherent in the feedback loop as well as deliberately inserted within. For the sake of convenience, only inherent non-linearities will be studied in this thesis. Examples of such non-linearities are, among others, saturation, dead space, hysteresis and relay (Fig. 1.22).

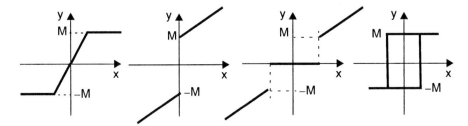

Figure 1.22: Different kinds of non-linearities

However, let us put the emphasis on finding a universal nonlinear feedback element, which can be shared by any circuit no matter which would be its transfer function. In this thesis, we are interested in an oscillator which can be standardized.

1.3.2 General circuit modifications

Fig. 1.21 is applicable provided that the BUT meets certain requirements and thus, the closed-loop system satisfies the oscillation conditions. For instance, for most second-order band-pass filters it can be possible to find out a feedback loop that, during the test mode, makes the quality factor of the filter very high or equivalently, moves a pair of poles to the imaginary axis (at least from a theoretical viewpoint). However, not all BUT can fulfil the oscillation conditions by only connecting a nonlinear feedback loop to the BUT function. So, it can also be required to modify the circuit structure by either adding or removing some passive components (see Fig. 1.23). Clearly, in this strategy it has to be distinguished two operational modes: a *normal mode,* when the system is connected to its regular input, all the additional components needed for testing are removed, and that circuitry took off for test purposes enabled again, and a *test mode,* where the feedback loop is closed around the BUT and the regular input is disconnected. Obviously, adding extra elements to the BUT must be made carefully in order to prevent problems in the operational mode (taking care of not affecting critical nodes and/or critical components). Moreover, the option of removing parts of the BUT circuitry must be avoided in order to guarantee that the maximum number of BUT components are covered in the test mode.

In broad terms, the proposal is to achieve oscillators formed by the block under test (or a part of it) plus or minus a set of components followed by a feedback loop also formed by additional components. Fig. 1.23 shows a general block diagram corresponding to a system modified to apply the OBT concept. The only modification affecting the signal path is a switching mechanism to separate operational and test modes. During the test mode, a feedback loop and some extra elements (either within the loop or within the BUT) are added to produce self-sustained and well- established oscillations.The purpose is to establish the basics of doing the extra circuit more or less fixed and more or less independent on the particular BUT and extensible to many kinds of BUTs.

Designers must view implementing the system and added circuitry as a global design problem. Besides achieving the system's functional specifications, they must strive to build a robust yet precise oscillator that exists around the system when the feedback loop is closed.

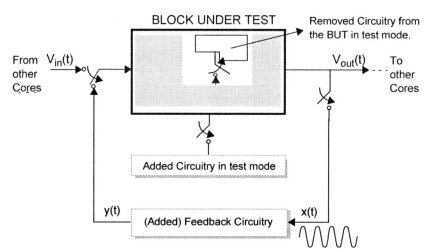

Figure 1.23: Converting any BUT into an oscillator

1.3.3 Start-up problem

The core of the OBT approach is the adaptation of the system circuitry to transform it in an oscillator and then, to measure the oscillation parameters. Therefore, it should be obvious that the OBT viability can be seriously compromised if a fast and safe start-up is not ensured.

In fact, two main problems can be clearly emphasized when an OBT scheme is examined:

-First, it may occur that under certain conditions (initial state, offset values...) oscillations may either start very slowly or even not to start at all [102]. In this sense, we have to provide a mechanism that guarantees *safe* start-up.

-Second, the oscillation parameter measurements have to be performed once the steady-state has been reached. Therefore, a short test time is a requirement of vital importance. Consequently, OBT requires also a start-up strategy which may reduce the transient time as much as possible.

In summary, in most cases we have to provide a *fast* and *safe* start-up strategy which ensures oscillations. We will come back to this problem in different parts of this thesis.

1.3.4 Requiring more test information

The frequency-only measurements were postulated as sufficient in [1], but our experience shows that, in many cases, they lead to relatively poor fault coverage rates [25]-[41]. From a rigorous defect-driven viewpoint, it has been demonstrated elsewhere [25]-[41] and discussed in previous sections that measuring the oscillation frequency may not be enough to achieve a reasonable fault coverage.

In fact, there is also a clear evidence of this matter in Arabi and Kaminska references [2], where authors set out that a single oscillation frequency is not enough to cover all target faults. They even propose a solution: implementing a programmable feedback element capable to generate different oscillation frequencies because with only a test mode configuration one can not guarantee a high fault coverage. But, obviously, the use of more than one oscillation mode has a big impact in the test area overhead (needs of employing extra control circuitry) and in the test time cost.

One of the main objectives of the new OBT concept is to increase the fault coverage with only one test configuration for each BUT. As was explained, the idea is to measure also another oscillation parameters as amplitude, signal distortion, DC-level, etc... Many positive results were shown in [25]-[41] where the fault coverage given by the oscillation frequency measurements is dramatically increased by measuring the oscillation amplitude. In fact, although amplitude measurements would require additional dedicated effort, it has been shown that the accuracy needed in the measurements can be relaxed if both parameters, frequency and amplitude, are considered. Furthermore, measuring more than one oscillator parameters increases the capability of identifying fault locations [25]-[41], [96]-[97].

However, it is not straightforward to extend the OBT strategy to incorporate other measurements. Fortunately, the own nature of the nonlinear oscillator scheme in our proposal (Fig. 1.21) eliminates this problem.

1.3.5 Characterizing the test oscillator

It is very important to model in detail the resulting oscillator. One of the difficulties of the OBT approach relies on that the oscillation parameters have to be accurately predicted either by analytical calculations or by simulations. Otherwise the test interpretation (basically, a comparison between the fault-free oscillation waveforms and those of a faulty circuit) could not be successfully performed. So, a key point of the new OBT concept is to define a robust non-linear model sufficiently accurate for predicting the nominal values of the test output signals. An efficient OBT concept must pay special attention to express the oscillation parameters as a function of the BUT elements and/or its performance.

There is not a general method to deal with any non-linear system. Non-linear differential equations can not be studied by a general methodology. Accurate solutions can be only given in certain kind of (relatively) simple non-linear differential equations. But, most practical non-linear differential equations can only be solved with a rough estimate.

In this way, if we can not apply a general method, we have to take individually each non-linear equation and then, try to develop a method of analysis for that particular equation.

One way to study a non-linear system where the degree of non-linearity is small, is to use techniques of equivalent linearization and solve the resulting linear problem. The describing-function method is one of these equivalent linearization methods. In many practical cases, the main interest is the stability of the system, and analytical solutions of the non-linear differential equations are not needed (settling down stability criteria is normally much more feasible than obtaining analytical solutions). The describing-function method allows us to study the stability of non-linear systems from the viewpoint of the frequency domain., but it does not give the exact information about the characteristics of the time response.

Other methods to study non-linear systems with strong non-linearities are the phase-plane technique and other techniques based on the second Liapunov method. The phase-plane method gives information about the stability as well as about the time response, but it is limited to first and second-order systems [100]. The second Liapunov method can be applied to stability analysis of any non-linear system, but it can be difficult to find Liapunov functions for many non-linear systems [100].

As the kind of oscillators proposed herein consists of a non-linear part, its description is not generally trivial and some of the mathematical techniques discussed above could be required to study its oscillatory behavior. So, in the next chapter practical possibilities will be explored in order to examine the features of the test oscillators for different types of systems (filters, modulators,...). The aim of this analysis is to accurately determine the oscillation parameters used for testing the SUT.

1.3.6 Characterizing the test interpretation

In the OBT approach, a SUT is accepted as fault-free if the measured oscillation parameters lie close to their nominal values (i.e. within an acceptable range corresponding to the required specifications). Therefore, the goal is not only to obtain accurate values of the oscillation parameters to compare them with experimental measurements, but also to accurately define the place where all the good circuits must lie, that is, the acceptability region.

The ideal situation would be to devise a well-based procedure that attains two objectives:

1.- any component of the SUT must be closely related to at least one of the involved oscillation parameters in order to be sure that any fault can be observed (high fault coverage and high capability of identifying the fault location).

2.- the sensitivity of any oscillation parameter with respect to the deviations of any element must be high enough to guarantee that all parametric or catastrophic faults can be detectable. This is strongly linked to the fact that not only the test information given by the oscillation parameters but also the tolerance bands are accurately determined.

Finally, another crucial point is how to give support to frequency and amplitude measurements. Obviously, it would be preferable to encode them into a single digital signal. Unfortunately, although frequency information is easily coded to digital, it is not the same for amplitude. For example, to solve this problem, Roh and Abraham recently proposed the use of a Time-division Multiplexing comparator [69]. On the other hand, a solution based on over-sampling data converters has been proposed by our group in [85]-[90]. In this way, the generated waves are coded to digital and, thus, can be processed either internally or externally by a purely-digital tester. The work in [90] discusses the problem of accessing internal nodes and gives some ideas about possible procedures for on-chip decision mechanisms[12].

1.3.7 The test process

As can be seen in the literature [1]-[3], [5], [8]-[11], [13]-[15], [21], [25]-[41], [49]-[53], [96]-[98], in spite of the fact that some OBT methods have been proposed by several researchers for digital and mixed-signal systems, devising a completely general OBT strategy that can be employed to all the analog and mixed-signal circuits is a task still unfeasible. Nevertheless, this work is aimed at establishing a, more or less general, yet standard OBT technique applicable to many kinds of mixed-signal systems no matter how complex they are.

However, the way to convert a circuit in an oscillator and the choice of the oscillation parameters obviously depend on the involved fault-detection procedure. The general oscillator scheme in Fig. 1.21 (a harmonic oscillator with amplitude control by limitation) has been proposed as a general alternative to apply OBT. However, up to now, this approach has not considered a practical oscillator scheme oriented to the subsequent fault-detection.

Regarding fault-detection, this can be performed by a built-in self tester or in the frame of an external tester. In the former case, the original circuit is modified by inserting some test control logic which provides for the oscillation during the test mode. In the second case, the oscillation is achieved by an

---

[12]. All these issues will be studied in Chapter 6 where an application example of OBT is described.

external feedback loop network which is normally implemented as part of a dedicated tester. But, obviously, one of the objectives pursued in any OBT strategy must be that the main work of the fault-detection can be achieved on-chip. In fact, an intrinsic feature of the OBT concept is that the part of the fault-detection dedicated to obtain the test outputs (in this case, the oscillatory test outputs) can be made internally. It is so as long as an oscillator structure can be derived for the given BUTs and the impact of implementing it embedded in the system circuitry is minimum. In fact, the OBT strategy described in this thesis is basically intended for both, reconfiguring the BUT in the test mode and achieving the test outputs, all on-chip.

As was saw in the first section, Arabi and Kaminska proposed in [1] a general OBT-based test scheme at system level. Among other features, such a structure involved the use of many additional switches to provide the required programmability. Such switches are placed in the normal signal path. Obviously, all these switches may affect the normal functionality of the SUT and cause serious performance degradation. Therefore, many practical considerations must be made about how to manage such switches to be transparent in the SUT and not to degrade its performance [14].

Therefore, a goal in this thesis has been to devise an improved OBT-based test solution at system level which avoids many of the drawbacks of the Arabi and Kaminska's scheme, specially those involving the use of switches and other elements which could alter the normal performance of the system. See Chapter 6 where we explain how to implement our OBT approach into an embedded macrocell (OBIST scheme).

However, for the second part of the fault-detection which is devoted to measure, evaluate and interpret the oscillatory test outputs, different approaches have also been performed in Chapter 6 and Chapter 7. For instance, the oscillation experimental measurements were made with an oscilloscope, a counter, or any other designed procedure of measurement (on-chip or off-chip) such as a first order $\Sigma\Delta$ modulator. Apart from that, an evaluation and interpretation tester (internal or external) must be connected to the BUT output in order to process the oscillation features and thus determine the BUT malfunction. In this context, many considerations about the type of tester, the cost of the tester, the requirements of the tester, and so on, were made. Fig. 1.24 summarizes the main features of the general scheme proposed in Chapter 6 and Chapter 7 as a feasible OBIST solution.

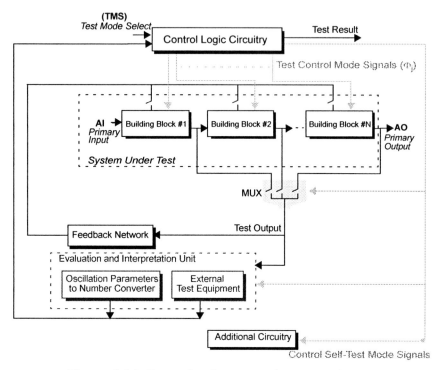

Figure 1.24: Example of an general OBIST scheme

1.4 SUMMARIZING THE NEW OBT CONCEPT

We are in a position to underline briefly the principles of the OBT strategy proposed in this Chapter. The basic idea is to implement the feedback loop shown in Fig. 1.25 for:

-converting the SUT into a ***Stable Self-starting Oscillator***

-relating the system specifications to the so-called indirect parameters (the oscillation frequency and the oscillation amplitude)[13].

-avoiding essential modifications in the normal signal path.

-ensuring that all system components are tested by the technique.

---

[13] The indirect parameters are not of interest in the normal mode of the SUT but are of vital importance in its test mode

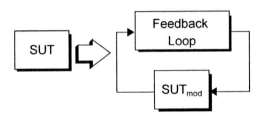

Figure 1.25: OBT strategy

But it is necessary to carefully think about the possible practical implementations as well. There are several alternatives to perform the approach shown in Fig. 1.25. We have to consider five points: granularity, local "manipulations", previous calculations, measurements and processing. In relation to the first point, we must select an implementation based on hierarchically splitting the entire SUT in lower-order subsystems because a meaningful/practical analysis to obtain the design relationships can not be handled for a complex/high-order system.

Another choice must be to add a "external" feedback loop instead of modifying the SUT transfer function. There are two main reasons, if we change the transfer functions, the actual circuit may not be tested in full. In any case, we must avoid to get anything out of the original SUT and only some essential extra test circuitry could be included if it were completely neccesary. Moreover, we reject the use of classical linear sinusoidal oscillator. Therefore, we need to guarantee a robust oscillation with a limiting mechanism (a nonlinear feedback block). This nonlinear feedback loop is preferred to be the same for any structure and its implementation must be as easy as possible. We will pursue to use, for any kind of SUT, only extra elements more or less "fixed" in advance.

On the other hand, both design equations and simulations have to be used as previous calculations because combining both is more flexible and helps to predict fault-free and faulty behaviours.

A test based only on the oscillation frequency measurement may not be practical when either a high precision is required or not all faults are related to the frequency. Other test parameters must be considered not only to improve the fault coverage but also to relax (if possible) the necessary accuracy of the measurements. In conclusion, we have to perform frequency and other measurements (i.e. amplitude) for achieving high fault coverage.

Finally, in the test process nothing is, in fact, determined. Reading digital signals should be preferred, just one test pin is ideal and, depending on the application, the measurement and its interpretation may be internal or external, but performing this on-chip is much more appealing.

As will be proven in next chapters, OBT seems a promising, yet effective test technique. However, using this strategy in practice shows there is no systematic way to apply it to complex circuits. Many decisions must be taken during the design process. In summary, for each practical application, we have two different problem levels:

1.-At Block level:

1.1.- Partitioning the SUT into components (BUTs) in an efficient decomposition.

1.2.- Providing a mechanism to isolate every BUT.

1.3.- Providing a (nonlinear) feedback path to convert every BUT in an self-starting and robust oscillator.

1.4.- Getting a valid and meaningful model for self-sustained oscillations (connected with the BUT design equations)[14].

2.-At System Level:

2.1.- Devising a straightforward method for reading the test outcome from every BUT.

2.2.- Interpreting the test results for the whole system.

These problems might be adequately solved for every specific example. Nevertheless, this thesis pursues to provide some guidelines useful to apply the basic ideas underlying the OBT/OBIST concept to any particular structure.

---

[14] The oscillator analysis may lead to quite accurate results, evaluating the oscillation conditions and the expected oscillation parameters.

Chapter 2

Mathematical Review of Non-linear Oscillators
Mathematical background

A GENERAL METHOD OF STUDYING systems containing nonlinear elements is perhaps impossible. However, a lot of work has been made [99], [104], [109]-[112], [114], [123] to develop mathematical techniques that can be applied to restricted classes of nonlinearities as well as to extend the application of known methods to a wider range of nonlinear systems.

In the first chapter we have revisited the basics of the Oscillation-Based Test (OBT) concept. The OBT approach, as introduced therein, requires to convert the System Under Test (SUT) in an oscillator structure by incorporating a nonlinear feedback loop. The requirement of this nonlinear block makes the study of such an oscillator very complex. Therefore, we are very interested in finding a practical model which can be straightforwardly used to design such an oscillator.

A possible solution would consist in employing the Describing-Function (DF) approach [99], [109]-[112], [114]. It is a simple and widely used method, easy to handle, intuitive, although restricted to systems fulfilling some conditions. The goal in this Chapter is to study under what conditions the DF approach provides an acceptable solution to model the oscillators used for OBT and when a more accurate mathematical approach is required in order to obtain a satisfactory oscillation solution.

This Chapter and its Appendix are included for the sake of completeness, to offer the reader a consistent, yet understandable presentation of the mathematical background required for dealing with the nonlinear oscillatiors handled for OBT. Most of this material is based on the work from Alistair I. Mees *et al.* (references [109]-[112])

2.1 FRAMEWORK

We are interested in closed-loop systems incorporating at least a nonlinear element. We intentionally disregard linear oscillators in order to avoid problems associated with the stabilization phenomenon[1]. In fact, we have opted for converting the SUT in a nonlinear oscillator. Particularly, an autonomous oscillator as shown in Fig. 2.1. This mechanism will be used as a standardized method to guarantee a self-maintaining, yet robust oscillator from any SUT.

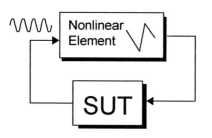

Figure 2.1: Basic OBT system

Studying this kind of nonlinear systems and its stability requires probably a lot of effort. Even more, sometimes an exhaustive and accurate analysis is impracticable. Only a rough calculation of the oscillation solution is possible. However, to successfully apply the OBT technique we need not only to detect whether the system may finally oscillate and to calculate the frequency and amplitude of the output signal, but also to check out its stability requirements (at least approximately). This Chapter will address the way to model the class of systems displayed in Fig. 2.1 in order to provide a valid analytical oscillation solution.

Many methods of analysis of this type of nonlinear systems have been proposed [99], [104], [109]-[112], [114], [123]. A lot of them are now widely used and there is a wealth of literature on them. Among these methods the DF technique emerges as one of the most important contributions. This method (or the simplest form of the method of harmonic balance) is a way for settling approximations to periodic solutions of non-linear systems by replacing the non-linear parts by a pseudo-linear representation of their effect on the fundamental harmonic. Our interest herein is to offer a straightforward methodology which provides insight and gives accurate information about the quantities defining the oscillations (amplitude, frequency, start-up, stability,...)

[1.] See Chapter 1 to go in depth in the stabilization phenomenon concept.

But such a methodology must be accompanied by an estimation on how accurate its solutions are since the DF approach is just approximated. In fact, when we started to develop the ideas and concepts of this work, we made use of the DF approach and the results were "good" or "acceptable". However, as the work advanced we spotted some contradictory results between the theoretical predictions given by the DF method and the outcomes from some more exact non-linear simulation tools.

In short, three different situations were found out:

1.- In most of the examples, the DF approach was **valid** not only to establish the oscillation conditions but also to calculate the oscillation parameters (such as amplitude and frequency) with very good accuracy.

2.- In some other examples, however, **discrepancies** were observed between the theoretical predictions and more accurate non-linear simulations[2].

3.- And finally, there were a few examples where (for some operation ranges) the DF approach gives results which **significantly deviated** from those obtained by simulations[3].

These last two cases (although corresponding to a relatively small number of practical cases) question whether the DF approach is an acceptable election. For them the proposed DF method sets up some limitations which force us to calculate the error bounds linked to this approach and, when this is not possible, to employ a more exact method of evaluation [99], [104], [109]-[112], [114], [123]. Therefore, two techniques for the estimation of error bounds in the results obtained by the DF approach are also included in this Chapter. Unfortunately, as will be seen, these techniques are cumbersome and do not often lead to intuitive solutions, requiring complex graphical methods.

2.2 THE DESCRIBING FUNCTION METHOD

The DF method is relatively practical in settling the stability of a nonlinear system but may not be directly applied to the optimization of the system design. The approach is usually classified as a frequency-response method

---

[2] In Chapter 3, we will go in depth in these cases.

[3] We will postpone until Chapter 4 more details about these cases.

rather than a time-domain technique and it is based on a study which neglects the effect of higher harmonics in the system. Therefore, this approach will be most successful in a system containing sufficient low-pass filtering.

The DF method is the first-order version of the method of harmonic balance, which pursues to find periodic solutions for nonlinear systems by adapting a truncated Fourier series. Basically, the DF method substitutes a Non-Linear Block (NLB) by a quasi-linear function, N, which represents the transfer function for the first Fourier component of the output generated by a pure sinusoidal input; one can then attempt to balance the first harmonic components in the system. If the linear part of the system is such that manages to attenuate higher harmonics, it seems reasonable that any balance found (in frequency and amplitude) will be "near" to an actual periodic solution of the system equation.

Let us assume the single-loop system with only a single nonlinearity as displayed in Fig. 2.2. The linear block has a transfer function, G, which is a frequency-sensitive function but it does not depend on the input signal amplitude. The nonlinear element has a transfer function, N which depends only on the input signal amplitude and is frequency-insensitive.

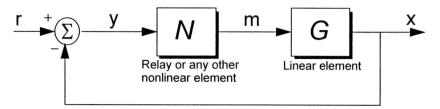

Figure 2.2: General nonlinear feedback system: division of the system into its linear and its nonlinear portions

In general, the nonlinear transfer function can be represented by the input-output equation

$$m = f(y) \tag{2.1}$$

However, for the sake of convenience, we can rewrite this expression as a new one consisting of two parts: a quasilinear gain and a distortion term

$$m \equiv N_{eq}y + f_d(y) \tag{2.2}$$

If the input, y, to the nonlinear block is sinusoidal ($y = a_1 \sin(\omega t)$), the first term on the right represents the fundamental component of the output, whereas the second term represents the distortion component due to high-order harmonics. The quantity N_{eq} is the equivalent linear gain and will be a function of the input-signal amplitude[4]. This is the ***describing-function (DF)*** associated to the nonlinearity $f(y)$.

It may be proved that with a proper selection of N_{eq}, and under certain conditions, the distortion terms may be quite often neglected. In short, if we desire to minimize $f_d(y)$ using a mean-square criterion, the right election of N_{eq} is the Fourier-series coefficient of the fundamental harmonic of the output waveform. Let us apply this approach, first to a relay system and then to other more complex forms of nonlinearities.

Depending on the specific relay characteristics, different hypotheses can be made. We can consider the ideal relay characteristic (Fig. 2.3-(a)) or a relay with a deadband, Δ (Fig. 2.3-(b)) or a more physically-realistic characteristic including also hysteresis, h (Fig. 2.3-(c)).

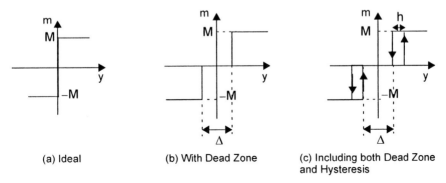

(a) Ideal   (b) With Dead Zone   (c) Including both Dead Zone and Hysteresis

Figure 2.3: Relay models

In general, the fundamental harmonic of the output of the nonlinear element N for a ***sinusoidal input*** will exhibit an amplitude and a phase shift which will depend on the amplitude of the input signal. That is, when the input is a sinusoidal wave

$$y = a_1 \sin(\omega t) = a_1 \sin(\theta) \qquad (2.3)$$

---

[4] Explicitly we should write $N_{eq}(a_1)$, being $y = a_1 \sin(\omega t)$.

The output can be represented by a Fourier series

$$m = f[a_1 \sin(\theta)] = h_1 \sin(\theta) + h_2 \cos(\theta) + \ldots \qquad (2.4)$$

We can normalize these coefficients

$$g(a_1) \equiv \frac{h_1}{a_1} \qquad b(a_1) \equiv \frac{h_2}{a_1} \qquad (2.5)$$

And these normalized Fourier coefficients can be found in the usual manner [114]

$$g(a_1) = \frac{1}{\pi a_1} \int_0^{2\pi} f[a_1 \sin(\theta)] \sin(\theta) d\theta$$

$$\qquad (2.6)$$

$$b(a_1) = \frac{1}{\pi a_1} \int_0^{2\pi} f[a_1 \sin(\theta)] \cos(\theta) d\theta$$

Note, therefore, that the describing-function N_{eq} is[5]

$$N_{eq} = g(a_1) + jb(a_1) \qquad (2.7)$$

As a simple and practical example, we can find the describing function for the ideal relay shown in Fig. 2.3-(a). Its output will be a square wave whose zero crossings occur exactly at the same instants than those of the input wave (see Fig. 2.4).

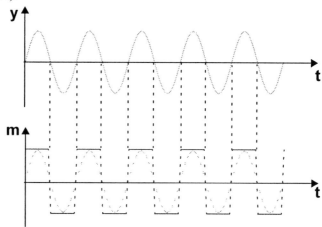

Figure 2.4: Input and output waves (ideal relay)

[5.] In complex number notation, $N_{eq} = |N_{eq}| e^{j\psi}$ where $|N_{eq}| = \frac{\sqrt{h_1^2 + h_2^2}}{a_1}$ and $\psi = \operatorname{atan}\left(\frac{h_1}{h_2}\right)$.

That is, there will be no phase shift between the fundamental component of the output and the input sine wave[6]. Then

$$b(a_1) = 0 \qquad (2.8)$$

The function $f(a_1 \sin(\theta))$ is, in this case, a symmetrical square wave[7], hence

$$g(a_1) = \frac{1}{\pi a_1} \int_0^{\pi} M \sin(\theta) d\theta - \frac{1}{\pi a_1} \int_{\pi}^{2\pi} M \sin(\theta) d\theta$$

$$g(a_1) = \frac{2}{\pi a_1} \int_0^{\pi} M \sin(\theta) d\theta = \frac{4M}{\pi a_1}$$

$$(2.9)$$

2.2.1 A General Describing-Function for Piecewise-linear Elements

Concerning the possible nonlinearities to be used in the feedback loop of Fig. 2.1, some restrictions must be made, otherwise the study of the oscillation mode by the DF methodology is unfeasible. We will exclusively focus on relays or Piecewise-Linear (PL) elements. A main motivation to consider only this kind of nonlinear elements when OBT is being applied comes from the fact that most of the nonlinearities we may find in practical circuits can be either represented or approximated by PL functions.

Then, we are mainly interested in a relay or at most a PL element. Therefore, it may be practical to analyze a general nonlinearity to obtain a describing-function that includes all those cases of interest in this thesis.

Let us study the general form of the nonlinearity shown in Fig. 2.5, [114]. In this general case several parameters are involved: a, b, c, d, e, F, D, M, n_1, n_2 and n_3. So, the general expression of the describing-function may seem complex. However, in all the cases of interest for this thesis, the resulting expression can be simplified because not all the parameters will be present.

---

[6] In fact, for any symmetric single-valued nonlinearity there will be no phase shift in the output fundamental component. That means that $b(a_1)$ will be zero. However, when a nonlinearity with memory is considered (that is, when a double-valued function is considered, see Fig. 2.3-(c)), there will be a phase shift associated with the describing function.

[7] This DF expression will be used during the thesis in many different contexts.

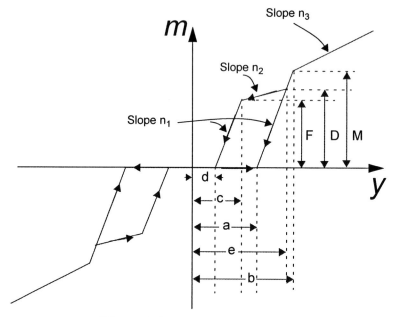

Figure 2.5: Generic nonlinearity

An equivalent form for (2.7) may be found by using the complex exponential form of (2.6)

$$N_{eq} = \frac{j}{\pi a_1} \int_0^{2\pi} f[a_1 \sin(\theta)] e^{-j\theta} d\theta \qquad (2.10)$$

From [114] the real and the imaginary parts of $N_{eq}(.)$ can be derived as

$$g(a_1) = \frac{n_1}{\pi a_1}[a_1(-\theta_1 + 2\theta_2 + \theta_3 - \theta_4 - \theta_5)]+$$

$$+\frac{n_1}{\pi a_1}\left[\frac{a_1}{2}(\sin 2\theta_1 - 2\sin 2\theta_2 - \sin 2\theta_3 + \sin 2\theta_4 + \sin 2\theta_5)\right]+$$

$$+\frac{n_1}{\pi a_1}[2a(-\cos\theta_1 + 2\cos\theta_2 - \cos\theta_5) + 2d(\cos\theta_3 - \cos\theta_4)]+ \qquad (2.11)$$

$$+\frac{n_2}{\pi a_1}\left[a_1(\theta_5 - \theta_3) + \frac{a_1}{2}(\sin 2\theta_3 - \sin 2\theta_5) + 2k_1(\cos\theta_5 - \cos\theta_3)\right]+$$

$$+\frac{n_3}{\pi a_1}[a_1(\pi - 2\theta_2) + a_1 \sin 2\theta_2 - 4k_2 \cos\theta_2]$$

and

$$b(a_1) = \frac{n_1}{\pi a_1}\left[\frac{a_1}{2}(\cos 2\theta_1 + \cos 2\theta_3 - \cos 2\theta_4 - \cos 2\theta_5)\right]+$$

$$+\frac{n_1}{\pi a_1}[2a(\sin\theta_1 - \sin\theta_5) + 2d(\sin\theta_3 - \sin\theta_4)]+ \qquad (2.12)$$

$$+\frac{n_2}{\pi a_1}\left[\frac{a_1}{2}(-\cos 2\theta_3 + \cos 2\theta_5) + 2k_1(\sin\theta_5 - \sin\theta_3)\right]$$

where

$$k_1 = \frac{Dc - Fe}{D - a_1} \qquad k_2 = b - \frac{B}{n_3}$$

$$\theta_1 = a\sin\frac{a}{a_1} \qquad \theta_2 = a\sin\frac{b}{a_1} \qquad \theta_3 = a\sin\frac{c}{a_1} \qquad (2.13)$$

$$\theta_4 = a\sin\frac{d}{a_1} \qquad \theta_5 = a\sin\frac{e}{a_1}$$

Therefore, (2.11) and (2.12) give the generic describing-function for the type of proposed nonlinearity. From this general result, we can derive every special case as needed in next chapters. These special cases fall into two general categories: single-valued nonlinearities and multivalued or memory-type nonlinearities. In Table 2.1 the equations of some of the most interesting describing-functions are given.

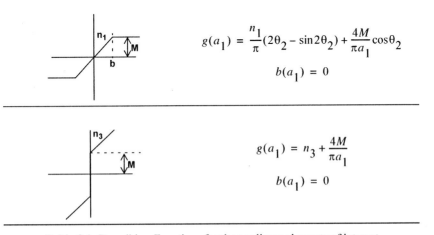

Non-Linear Characteristic	*Describing-Function*
	$g(a_1) = \frac{n_1}{\pi}(2\theta_2 - \sin 2\theta_2) + \frac{4M}{\pi a_1}\cos\theta_2$ $b(a_1) = 0$
	$g(a_1) = n_3 + \frac{4M}{\pi a_1}$ $b(a_1) = 0$

Table 2.1: Describing-Functions for the nonlinear elements of interest

Non-Linear Characteristic	*Describing-Function*
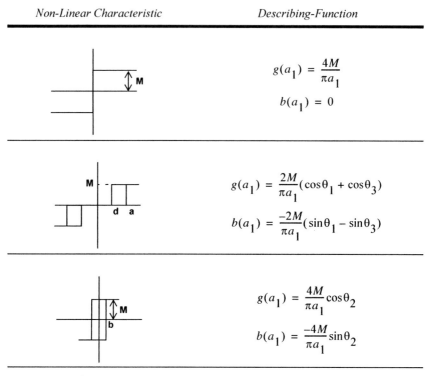	$$g(a_1) = \frac{4M}{\pi a_1}$$ $$b(a_1) = 0$$
	$$g(a_1) = \frac{2M}{\pi a_1}(\cos\theta_1 + \cos\theta_3)$$ $$b(a_1) = \frac{-2M}{\pi a_1}(\sin\theta_1 - \sin\theta_3)$$
	$$g(a_1) = \frac{4M}{\pi a_1}\cos\theta_2$$ $$b(a_1) = \frac{-4M}{\pi a_1}\sin\theta_2$$

Table 2.1: Describing-Functions for the nonlinear elements of interest

2.2.2 On the use of the DF method in oscillators

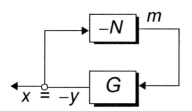

Figure 2.6: Generic oscillator

Let us now apply the DF approximation for the analysis of the existence of a self-sustained oscillation in the autonomous system displayed in Fig. 2.6 (a simple version of the general nonlinear feedback system in Fig. 2.2). As said above, we employ this type of closed-loop systems as instruments to apply the OBT strategy. Therefore, different versions of this closed-loop system will be object of study in the book.

Let us assume that the system in Fig. 2.6 oscillates, that is, a limit cycle exists (we will find later, in general terms, the sufficient conditions for this). Due to the nonlinear block, N, the periodic signal m is, in fact, nonsinusoidal. On the other hand, let us consider that the linear transfer function G has an attenuation which increases with frequency. Then, as a first approximation, we can assume that only the fundamental component of m is transferred to the output of G. Then the input y to N may be assumed to be sinusoidal. We can thus conclude that the output of N is related to its input through the describing-function as

$$m = -N_{eq}y \qquad (2.14)$$

that is, the distortion term is neglected.

Since exclusively the fundamental component of the limit cycle is retained in this first approximation, it will be enough to employ the steady-state form of the linear transfer function

$$\frac{x(j\omega)}{m(j\omega)} = G(j\omega) \qquad (2.15)$$

Therefore, a sufficient condition for the existence of sustained oscillations is the simultaneous fulfilment of these two requirements, (2.14) and (2.15), on the system. Solving both equations simultaneously yields the basic equation of the DF analysis[8]

$$-\frac{1}{G(j\omega)} = N_{eq} \qquad (2.16)$$

The derivation of the above equation is based on three main assumptions:
-the system must be autonomous (i.e. unforced and time invariant).
-the nonlinearity must be separable and frequency independent.
-the linear transfer function must perform enough low-pass filtering action to guarantee the exclusion of the higher harmonics in the input of N.

Doubtless, the most convenient way of analysing (2.16) is a polar plot of the two functions and a check for an intersection of the two curves (see Fig. 2.7-(a)). It could facilitate a more complete study of the stability requirements. In any case, this plot involves drawing in the complex plane the $-1/G(j\omega)$ locus as ω changes and the N_{eq} locus as a_1 changes (see Fig. 2.28-(a)).

[8.] Strictly we should write $N_{eq}(a_1)$, being $x = a_1 \sin(\omega t)$.

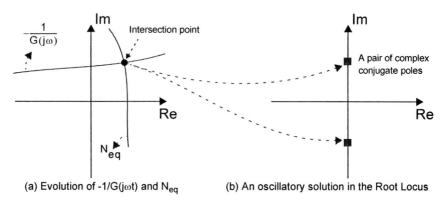

(a) Evolution of $-1/G(j\omega t)$ and N_{eq} (b) An oscillatory solution in the Root Locus

Figure 2.7: Two different representations of the oscillation solution for (2.16)

Let us suppose that, when plotting the $-1/G(j\omega)$ and the N_{eq} loci in the complex plane, an intersection point is obtained (this intersection point also corresponds to a pair of complex poles on the imaginary axis of the so-called *Root Locus*[9], see Fig. 2.7). Then, an oscillation may exist at the frequency and amplitude given by the intersection. But, however, it might happen that the oscillation corresponds to an unstable limit circle. If the system is in unstable equilibrium, the oscillation will break into either a destructive growth in magnitude or a decay to zero. We refer again to the stabilization phenomenon discussed in the previous chapter.

Let us now investigate the sinusoidal operation mode with a constant amplitude at an intersection of the amplitude locus and the frequency locus. We can define a ***convergent equilibrium*** as the oscillation mode when the system is stable under small perturbations, and a ***divergent equilibrium*** as the oscillation mode when the system is unstable. In a *convergent equilibrium,* a small positive or negative perturbation displaces the operation point in such a way that its own evolution forces the system to return to its equilibrium point (see Fig. 2.8). In a *divergent equilibrium,* a perturbation of the system causes it to reinforce the instability or discourage oscillation completely. So, a point of divergent equilibrium does not support sustained oscillations. Obviously we are particularly interested in building oscillators whose oscillation mode

---

[9.] The *Root Locus* of the closed-loop system in Fig. 2.6 is the collection of curves in the complex plane that show the possible positions of the roots of the characteristic equation, i.e, $1 - N_{eq}G(s) = 0$ for all possible values of N_{eq}.

be a convergent equilibrium. In short, we desire to establish the stability criteria in terms of the DF approach in order to be capable to determine when the OBT system (analysed by this linearized method) supports oscillations satisfactorily.

In broad terms, the criteria for convergent and divergent equilibrium conditions are [114]:

-A condition of convergent equilibrium exists if the amplitude of oscillation decreases as the operating point on the $-1/G(j\omega)$ locus moves within the N_{eq} locus.

-A condition of divergent equilibrium exists if the amplitude of oscillation increases as the operating point on the $-1/G(j\omega)$ locus moves within the N_{eq} locus.

These general criteria apply only to single-loop single-nonlinearity systems, it is, the class of systems we are interested in when OBT is applied. In next chapters we will shape and establish these conditions in a more explicit and intuitive way when we analyze several particular cases.

2.2.3 Convergent Equilibrium: *Steady Oscillation Mode*

As was already said any oscillation mode requires a mechanism to force a displacement of, at least, a pair of complex poles onto the imaginary axis. However, it is in practice unfeasible to locate exactly (and permanently) a pair of poles onto the imaginary axis. Instead, what can be done is to achieve a convergent equilibrium where a pair of system poles are moving periodically to the right and the left of the imaginary axis (observe Fig. 2.8).

In broad terms, the oscillation mechanism shown in Fig. 2.8 consists of displacing all poles, even the more remote poles to the left of the imaginary axis (*stable half-plane*) with the exception of a pair of complex conjugate poles. This pair, responsible of the oscillation, will be always placed in the neighbourhood of the imaginary axis (but in the unstable half-plane). Initially, these two poles move towards such imaginary axis, cross it and after a while reverse its movement crossing back to the unstable half-plane. This movement is perpetually repeated, forth and back, in the so-called *steady oscillation mode.*

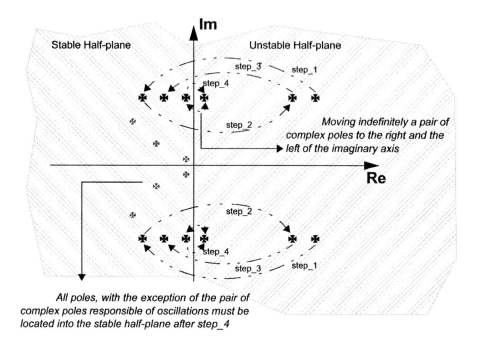

Figure 2.8: Oscillation Strategy

In short, this *steady oscillation mode* is a state which fulfils the condition of having a pair of complex conjugate poles moving continually to the right and the left of the imaginary axis, while the remainder poles are situated into the stable half-plane.

Obviously, there is a transient ending when the steady-state oscillation is reached. The transient duration is what is usually called the start-up time of the oscillator, and any mechanism involving the process of building the oscillation is called a start-up strategy.

2.3 APPLYING THE DF APPROACH

2.3.1 Determining the oscillation parameters

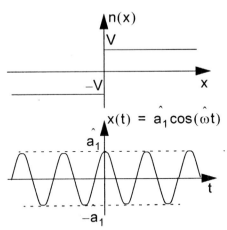

Figure 2.9: Nonlinearity of interest

Let us suppose that the first-order describing-function equation for the closed-loop system of Fig. 2.6

$$N(a_1) - \frac{1}{G(j\omega)} = 0 \qquad (2.17)$$

has a oscillation solution $(\hat{\omega}, \hat{a}_1)$ [10], being $\hat{\omega}$ the oscillation frequency and a_1 the oscillation amplitude.

On the other hand, let us study, for the sake of illustration, the nonlinearity shown in Fig. 2.9,

$n(x) = Vsign(x)$. The DF function will be in this case (see Table 2.1)

$$N(a_1) = \frac{4V}{\pi a_1} \qquad (2.18)$$

If the linear element in Fig. 2.6 has the following generic biquadratic transfer function

$$G(s) = \frac{k_2 s^2 + k_1 \dfrac{\omega_0}{Q} s + k_0 \omega_0^2}{s^2 + \dfrac{\omega_0}{Q} s + \omega_0^2} \qquad (2.19)$$

where ω_0 is the resonant frequency and Q is the quality factor of its poles, then we can rewrite the characteristic equation (2.17) as

$$s^2 + \left(\frac{\omega_0}{Q} \cdot \frac{1 - k_1 N(a_1)}{1 - k_2 N(a_1)} \right) s + \omega_0^2 \cdot \frac{1 - k_0 N(a_1)}{1 - k_2 N(a_1)} = 0 \qquad (2.20)$$

that can also be written as $s^2 + bs + \omega^2 = 0$

[10.] In what follows, $\hat{x}$ will be used for a solution point corresponding to a specific value of the parameter x.

where

$$b = \frac{\omega_0}{Q} \cdot \frac{1 - k_1 N(a_1)}{1 - k_2 N(a_1)} \qquad \omega^2 = \omega_0^2 \cdot \frac{1 - k_0 N(a_1)}{1 - k_2 N(a_1)} \qquad (2.21)$$

The two-pole solutions are given by

$$s_{1,2} = \frac{-b}{2} \pm \frac{\sqrt{b^2 - 4\omega^2}}{2} \qquad (2.22)$$

Graphically, we can see in Fig. 2.10 the two different possibilities of (2.22): a) a pair of complex conjugate poles or b) a pair of real poles.

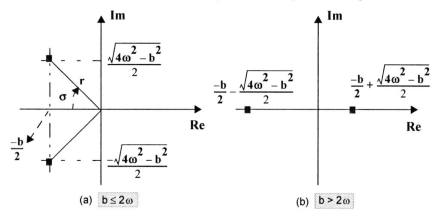

Figure 2.10: Poles Location of the closed-loop system shown in Fig. 2.6

As was explained in previous sections, in the so-called *Steady Oscillation Mode* the solution of (2.22) is a pair of complex conjugate poles placed just on the imaginary axis[11]. That means $b(\hat{a}_1) = 0$ (see Fig. 2.11-(a)). On the other hand, if the system suffers a small perturbation in such a way that $a_1 \neq \hat{a}_1$ (with $b(a_1) \neq 0$), then, the oscillation strategy is capable of returning the involved pair of complex poles to the imaginary axis (see Fig. 2.11-(b)). That is, reversing the trend of a_1. Accordingly, if at any time a_1 grows, the oscillation mechanism forces the complex poles to cross the imaginary axis and to turn back to the stable half-plane (Fig. 2.11-(b)). Otherwise, if at any time a_1 falls, the oscillation mechanism makes the poles to come back to the

---

[11]. Then, the output of the closed-loop system in Fig. 2.6, x, will be an oscillatory signal (with amplitude $\hat{a}_1$ and frequency $\hat{\omega}$).

unstable half-plane (Fig. 2.11-(b)). In fact, the final achieved state will be a pair of complex conjugate poles moving, forth and back, in the vicinity of the imaginary axis.

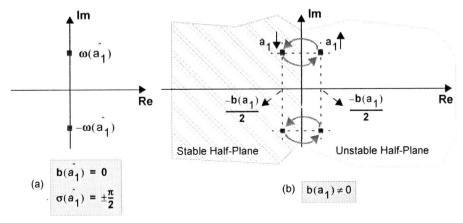

Figure 2.11: Poles in the oscillation mode of the closed-loop system shown in Fig. 2.6

To ensure that the system will oscillate the following requirements are needed:

1.- Start-up:

Initially, if no signal is present ($a_1 = 0$), the poles should be (safely) located in the right side of the plane (unstable system) to increase the value of the signal. Note that $N(0) \rightarrow \infty$ and the closed-loop system characteristic equation is reduced to the equation for $G(s)$ zero locations

$$k_2 s^2 + k_1 \frac{\omega_0}{Q} s + k_0 \omega_0^2 = 0 \qquad (2.23)$$

Then, the start-up condition is guaranteed if and only if G(s) has some zeros in the right side of the plane.

2.- Oscillation Frequency:

When the poles are on the imaginary axis, $b(\hat{a}_1) = 0$, the system will oscillate with a frequency given by

$$\hat{\omega} = \omega(\hat{a}_1) = \omega_0 \cdot \sqrt{\frac{1 - k_0 N(\hat{a}_1)}{1 - k_2 N(\hat{a}_1)}} = \omega_0 \cdot \sqrt{\frac{k_1 - k_0}{k_1 - k_2}} \qquad (2.24)$$

Then, in the case $k_0 = k_2$, the oscillation frequency, $\hat{\omega}$ is independent on the oscillation amplitude, $\hat{a}_1$.

3.- Oscillation Amplitude:

Once the steady state is achieved, the amplitude $\hat{a}_1$ can be determined because the poles-placement has to be just on the imaginary axis. That is

$$b(\hat{a}_1) = \frac{\omega_0}{Q} \cdot \frac{1 - k_1 N(\hat{a}_1)}{1 - k_2 N(\hat{a}_1)} = 0 \rightarrow N(\hat{a}_1) = \frac{1}{k_1} \tag{2.25}$$

Then, substituting $N(\hat{a}_1)$ (see Table 2.1)

$$\hat{a}_1 = \frac{4V}{\Pi} k_1 \tag{2.26}$$

Considering also that $\hat{a}_1$ has to be positive, we immediately have from (2.26) two new oscillation conditions. That is

$$k_1 \neq 0$$
$$sign(V) = sign(k_1) \tag{2.27}$$

4.- Amplitude Control Stability:

The amplitude control will be stable when [Ref]

$$\frac{\partial}{\partial a_1} b(a_1) > 0 \tag{2.28}$$

The above expression may be further developed

$$\frac{\partial}{\partial a_1} b(a_1) = \frac{\partial}{\partial a_1} \left(\left[\frac{1 - k_1 N(a_1)}{1 - k_2 N(a_1)} \right] \right)$$

$$= \frac{\partial}{\partial N(a_1)} \left(\left[\frac{1 - k_1 N(a_1)}{1 - k_2 N(a_1)} \right] \right) \frac{\partial}{\partial a_1} N(a_1) \tag{2.29}$$

$$= \frac{-4V}{\pi a_1^2} \cdot \frac{k_2 - k_1}{[1 - k_2 N(a_1)]^2} > 0$$

Consequently, an additional oscillation condition would be

$$V(k_2 - k_1) < 0 \rightarrow sign(V) \neq sign(k_2 - k_1) \tag{2.30}$$

Summarizing for the particular $N(a_1)$ function in Fig. 2.9, it can be shown that all the oscillation conditions are accomplished if

$$k_1 \neq 0$$
$$sign(V) = sign(k_1) \neq sign(k_2 - k_1) \tag{2.31}$$

and

$$\hat{\omega} = \omega_0 \cdot \sqrt{\frac{k_1 - k_0}{k_1 - k_2}} = f(\omega_0, k_0, k_1, k_2)$$

$$\hat{a}_1 = \frac{4V}{\pi} k_1 = f(V, k_1) \tag{2.32}$$

Consequently, when $k_0 = k_2$ the oscillation frequency found by the DF method is the resonant frequency, ω_0, and the oscillation amplitude is $\hat{a}_1 = \frac{4V}{\pi} k_1$.

2.3.2 Describing-Function limitations

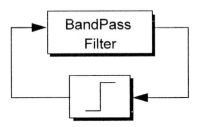

Figure 2.12: Sine-wave Oscillator

Let us suppose for this study the simple example of an oscillator consisting of a bandpass filter and a comparator as shown in Fig. 2.12. We are interested in applying the DF approach to the analysis of this oscillator, and then comparing the outcome with the results obtained by another analytical mathematical method [116]-[117]. The results derived from this section will help us to outline the weak and strong points of the DF method.

For the sake of simplicity, let us now set up the case of a second-order bandpass filter and a comparator with saturation levels 0 and 1 (Fig. 2.13). This closed-loop system verifies the premises postulated above: the system is autonomous, the nonlinearity is separable and frequency-independent, and the linear transfer function contains enough low-pass filtering to neglect the higher harmonics at the comparator output.

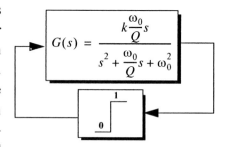

Figure 2.13: Using a second-order bandpass filter

Before studying the dynamic involved in the system of Fig. 2.13, we turn our attention to the s-plane location for poles of $G(s)$. Such poles are the values for which $s^2 + \dfrac{\omega_0}{Q}s + \omega_0^2 = 0$, where Q is the quality factor and ω_0 is the resonant frequency. Let their s-plane placement be $-\alpha \pm j\beta$, so that

$$s^2 + \frac{\omega_0}{Q}s + \omega_0^2 = s^2 + 2\alpha s + (\alpha^2 + \beta^2) \tag{2.33}$$

We find then (see Fig. 2.14) that

$$\alpha = \frac{\omega_0}{2Q} \qquad \beta = \omega_0\sqrt{1 - \frac{1}{4Q^2}} \tag{2.34}$$

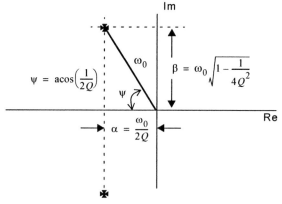

Figure 2.14: Definitions of parameters related to pole positions

Applying directly the DF method, (2.16), to solve the closed-loop system of Fig. 2.13, we obtain that the system produces a near-sinusoidal signal with high spectral purity whose oscillation frequency is the filter resonant frequency, $\omega_{osc} = \omega_0$ (regardless of the filter quality factor) and the oscillation amplitude is $A_{osc} = \frac{2}{\pi}k$.

An alternative method to study the circuit in Fig. 2.13 was given by Shanthi in [116]-[117]. Following this latter, the step response of the filter is calculated, resulting in an underdamped sine

$$s(t) = L^{-1}\left[\frac{G(s)}{s}\right] = \frac{2k}{\sqrt{4Q^2-1}}e^{-\frac{\omega_0}{2Q}t}\sin\left(\omega_0\sqrt{1-\frac{1}{4Q^2}}\,t\right)u(t) \qquad (2.35)$$

where $u(t)$ is the unit step function.

Then, using this result and examining in detail the Total Harmonic Distortion (THD) of the output waveform, the method obtains an exact oscillatory distorted waveform whose frequency, calculated through the zero crossings, depends also on the filter quality factor

$$\omega_{osc} = \omega_0\sqrt{1-\frac{1}{4Q^2}} \qquad (2.36)$$

This last expression reveals that the deviation between the resulting oscillation frequency given by the DF approach and the resulting oscillation frequency given by this last method is

$$\frac{\Delta\omega_{osc}}{\omega_{osc}} = \frac{\left[\sqrt{1-\frac{1}{4Q^2}}-1\right]}{\sqrt{1-\frac{1}{4Q^2}}} \qquad (2.37)$$

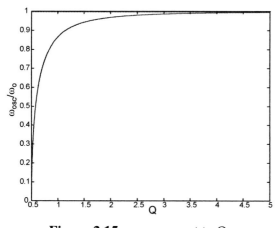

Figure 2.15: ω_{osc} respect to Q

Expression (2.36) is drawn in Fig. 2.15 as a function of Q. We can observe from this Figure that two asymptotes exist. One shows when the deviation of ω_{osc} with respect to ω_0 is high (for small values of Q) and the other one when this deviation is small (for high values of Q). As an example, if $Q < 1$ then $\omega_0 \gg \omega_{osc}$ whereas if $Q > 5$ then $\omega_{osc} \rightarrow \omega_0$.

The fact is that there is a dependency between the oscillation frequency ω_{osc} of the closed-loop system and the filter quality factor, Q. This dependency is not observed by the DF method, but may be completely negligible when Q exceeds the value of 10.

On the other hand, assuming steady-state response and calculating the amplitude peak by the method proposed in [116]-[117] for the special case when the filter quality factor is high, one obtains that it depends on k and Q according to

$$A_{osc} \approx \frac{2k}{\left[1 - \exp\left(-\frac{\pi}{\sqrt{4Q^2 - 1}}\right)\right]\sqrt{4Q^2 - 1}} \qquad Q \gg 1 \qquad (2.38)$$

As can be seen this expression practically coincides with the expression obtained by the DF approach $(A_{osc} = \frac{2}{\pi}k)$[12].

---

[12] If $Q \rightarrow \infty$, $A_{osc} \approx \dfrac{2k}{\left[1 - \exp\left(-\dfrac{\pi}{\sqrt{4Q^2 - 1}}\right)\right]\sqrt{4Q^2 - 1}} \approx \dfrac{2}{\pi}k$.

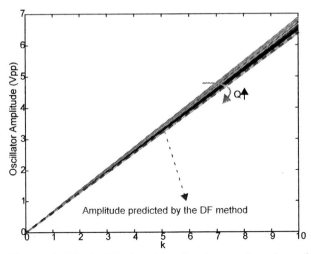

In Fig. 2.16 we have plotted the expression (2.38) as well as the amplitude predicted by the DF approach (dotted line). Observe from this Figure that when Q increases the oscillation amplitude curve matches up with the line predicted by the DF method. How-

Figure 2.16: Oscillation Amplitude as a function of k

ever, if Q is not large enough, as k increases both methods of analysis present significant disagreements between their predicted results. Nevertheless, it can be shown by simulation that this prediction of the amplitude given by (2.38), is even less accurate than the DF method for $k \gg 1$. Therefore, more accurate simulations show that for $k \gg 1$ the DF method is more appropriate compared to the method proposed in [116]-[117] (provided that $Q \gg 1$ as well).

Once reflected that the oscillation parameters predicted by other mathematical method [116]-[117], under some conditions, differ from the results given by the DF approach, let us think about our specific problems. In fact, we are mainly interested in:

1) determining a range of the quality factor, Q, and the parameter k where we can successfully apply the DF method.

2) establishing the limitations of this technique in order to understand why in some examples (found out in the context of this book) the DF approach is not entirely valid.

From an intuitive point of view the requirements to successfully apply the DF method to an oscillator such as the one in Fig. 2.12 are:

-A highly selective filter ($Q \gg 1$).

-Sufficient gain at the oscillation frequency ($k \gg 1$).

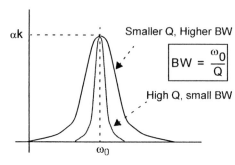

Figure 2.17: Illustration of a high-Q and a smaller-Q Banpass Function

The reason behind these conditions can be explained by considering the Bode diagram of the bandpass function (Fig. 2.17). It should be clear that as Q increases the transfer function is more selective around the filter resonant frequency, ω_0, avoiding thus the flow of higher harmonics. This effect is improved as $k \gg 1$. Therefore, the basic assumptions of the DF method are fulfilled as both Q and k increase.

However, let us now show in Fig. 2.18 the different results obtained by three different methods: a) the DF approach, b) the method by Shanthi ([116]-[117]) and c) simulations achieved by Matlab-Simulink. From this Figure a significant issue is reinforced by simulation: the results obtained by the three methods match up as both Q and k increase.

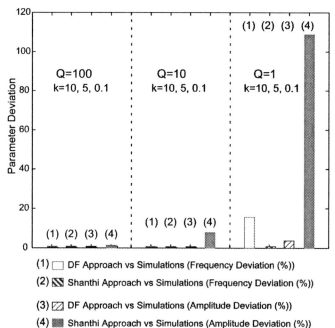

(1) ☐ DF Approach vs Simulations (Frequency Deviation (%))

(2) ☒ Shanthi Approach vs Simulations (Frequency Deviation (%))

(3) ▨ DF Approach vs Simulations (Amplitude Deviation (%))

(4) ▧ Shanthi Approach vs Simulations (Amplitude Deviation (%))

Figure 2.18: Summary of results obtained by the three different methods

In addition, it can be observed by simulation that as Q increases the start-up transient time also increases (we need more simulation time to establish completely the oscillations). In [116]-[117] it is demonstrated that as $Q \to \infty$, the time taken to reach the steady-state tends to ∞. However, this behavior is not shown explicitly if the DF method is used.

But, observe from Fig. 2.18 that, regardless of the value of Q and the value of k, the solution for the amplitude given by the DF approach is always closer to the solution given by the simulations than the solution obtained by Shanthi. Notice that even with $Q = 1$, the deviation between the amplitude given by simulation and the amplitude obtained by the DF approach is acceptable ($\sim 3.7\%$). Moreover, the restriction of high Q is not strict for the oscillation frequency either. See, for example, Fig. 2.15. This Figure shows that with $Q = 2,5$ (an intermediate value of Q), the frequency given by the DF approach differs from the solution given by Shanthi (close to the actual frequency obtained by simulation) less than a 2%.

In summary, a very interesting result comes out from Fig. 2.18: even for (relatively) small values of Q and k, the DF method provides quite a reasonable agreement with simulations.

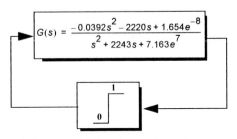

Figure 2.19: Example where the DF approach is acceptable

All this can be seen in the numerical example shown in Fig. 2.19. This is the case of a biquad which will be one of the main block-under-test in next chapters (actually is the circuit called HG #1 in the fifth chapter). In this example the filter quality factor Q is 3.77 and the corresponding k parameter is 0.99. Therefore, as we explained above, such values would be within the limits where the DF approach is almost valid. Notice, however, that all the above conclusions were extracted considering a bandpass function. But now, we have another type of function. Nevertheless, examining the accuracy of the model given by the DF method, we obtain that the predicted oscillations for HG #1 have a good agreement with more exact nonlinear simulations. Specifically, a neglected deviation in the frequency and a 1.20% of deviation in the amplitude.

On the other hand, all these assumptions seen above depend not only on the filter transfer function but also on the nonlinear element employed in the feedback loop (see Fig. 2.20). For instance, let us suppose, from a very simplified viewpoint, the dynamic of the closed-loop system proposed in Fig. 2.13. Notice from Fig. 2.20 that by changing slightly the nonlinear block, we could relax the characteristics of the bandpass function and the DF approach still is valid. It is because less harmonics are involved in the output of the nonlinear element.

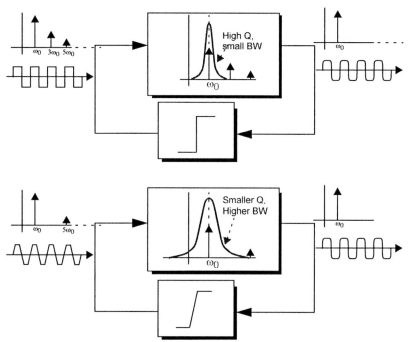

Figure 2.20: Dynamic of the closed-loop system of Fig. 2.13

The above described analysis is for the specific case of second-order bandpass filters and for very simple nonlinear feedback blocks. However, these results are illustrative for a more general case. In this sense, the DF approach could not always be accurate enough to model the OBT oscillator. It depends on both the linear transfer function and the nonlinear feedback element which determine the characteristics of the input and output signals. There may be cases in which we must correct the DF results with some kind of error analysis or with more exact non-linear simulations.

In next sections we will try to delimit from a more general point of view when the DF approach is completely valid and when, otherwise, their results must be corrected by a more useful method of analysis.

2.4 ERROR BOUND CALCULATION FOR THE DF APPROACH

2.4.1 First proposed method

Any error analysis is a point of vital importance in any approximate method as the DF approach. To verify that $(\omega, \hat{a}_1)$ (oscillation solution given

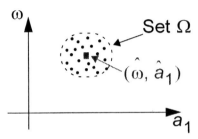

Figure 2.21: Confidence interval, Ω

by the DF approach) corresponds to a true oscillatory solution of the involved closed-loop system, we also have to find a *"confidence interval"* containing this fundamental frequency and amplitude (see Fig. 2.21). Such an interval, Ω, is formed by points in the space (ω, a_1) around $(\hat{\omega}, \hat{a}_1)$. Ω is found defining several error functions called p, q, and r (see Appendix 2.A extracted from [99] and [110]). Such error functions allow us to state a key inequality at all points in Ω. If, under certain conditions, the inequality is fulfilled, then, we can assert that an oscillatory solution exists and such a solution is into the set Ω, it is, near to the calculated solution point $(\hat{\omega}, \hat{a}_1)$.

The more we can confine Ω, the better we can settled the oscillation solution and the more accurate will be the result given by the DF method.

A revisited DF method is proposed and justified in Appendix 2.A ([99], [110]). In summary, this modified DF approach does not only regard the point of intersection given by the basic equation of the DF analysis (that is, (2.16)) but also takes into account some other features in the vicinity of such point and the implicated loci. This improved semi-graphical method involves solving various steps (see Appendix 2.A). The first step (*Step 0*) coincides with the first-order DF approach and the rest of the steps will correspond to the revised DF approach.

For the sake of clarity, let us summarize the different steps of the method presented in Appendix 2.A ([99], [110]):

Assuming that the non-linearity, n, is a single-valued odd function and is monotone increasing or decreasing; we only look for solutions composed of odd harmonics

Step 0: Find $(\omega, \hat{a}_1)$ satisfying $N(\hat{a}_1) + \dfrac{1}{G(j\omega)} = 0$. Check that the N and $-1/G$ loci are not parallel where they intersect at $(\hat{\omega}, a_1)$.

Step 1: Find $\rho(\omega)$. That is, the error function which estimates how well $G(j\omega)$ filters out high-order undesired harmonics. Small values of $\rho(\omega)$ are desirable; the smaller ρ is, the smaller the eventual error.

Step 2: Find $p(a_1)$. That is the DF output error or the error assuming that the output of n is sinusoidal when its input is sinusoidal.

Step 3: Find $q(a_1, \varepsilon)$. That is, the function that measures the error introduced by neglecting high harmonics at the input of n.

Step 4: Choose ε such that for all (ω, a_1) near $(\hat{\omega}, a_1)$, $\varepsilon \geq \rho(\omega)[p(a_1) + q(a_1, \varepsilon)]$.

Step 5: Find the set Ω of (ω, a_1) values near $(\omega, \hat{a}_1)$ such that $\left| N(a_1) + \dfrac{1}{G(j\omega)} \right| \leq \dfrac{q(a_1, \varepsilon)}{a_1}$. (This can be done graphically: see Appendix 2.A). Check that Ω is bounded.

Step 6: Check that Ω contains the DF solution, $(\hat{\omega}, a_1)$.

Step 7: Then, there is at least one true periodic solution with $(\omega, a_1) \in \Omega$ and $\|x^*\| \leq \varepsilon$.

2.4.1.1 Example #1: Oscillator with bandpass functions of different Q

Let us thus reintroduce the problem highlighted in previous sections where we can observe that the DF method on its own was incapable of determining a difference between the oscillator (given in Fig. 2.13) obtained with a bandpass of high Q and with a bandpass of small Q. Therefore, we will see in this section how the modified DF method described in Appendix 2.A is already capable of finding these discrepancies.

Suppose again the example of Fig. 2.13 but now considering two kind of bandpass functions (that is, $G(s) = k_1 \frac{\omega_0}{Q} s / s^2 + \frac{\omega_0}{Q} s + \omega_0^2$): one with a small Q (*Case #1*) and the other one with a high Q (*Case #2*), both cases with $k_1 > 1$ (see the specific values in Table 2.2 with $k_1 = 1.1$).

Case #1		Case #2	
$\frac{\omega_0}{Q} = 1000e^3$	$\omega_0^2 = 1000e^3$	$\frac{\omega_0}{Q} = 10e^3$	$\omega_0^2 = 1000e^3$

Table 2.2: Values for Example #1

As was discussed above, the result given by the simpler version of the DF method is only valid for the second case. Let us justify by means of the proposed semi-graphical method why it is so.

Step 0: That is, applying (2.32). Regardless of the case (considering $V = 1V$)[13]

$$\hat{a}_1 = \frac{2V}{\pi} k_1 = 1.4006V$$

$$\hat{\omega} = \omega_0 \cdot \sqrt{\frac{k_1 - k_0}{k_1 - k_2}} = 1000 \frac{rad}{s} (159.155 Hz)$$

(2.39)

Check that the $N(a_1)$ and $-1/G(j\omega)$ loci are not parallel where they intersect at $(\hat{\omega}, \hat{a}_1)$ (see Fig. 2.22).

Notice an important difference between both sets of loci in Fig. 2.22[14]: despite the fact that in both cases we have covered the same range of frequencies,

---

[13] Notice that for a bandpass function $k_0 = k_2 = 0$.
[14] Remark how different are the units in the vertical axis of both cases.

[100, 10000], in *Case #1* such a range takes up the range [-0.009, 0.009] in the imaginary axis whereas in *Case #2* it takes up [-0.9, 0.9].

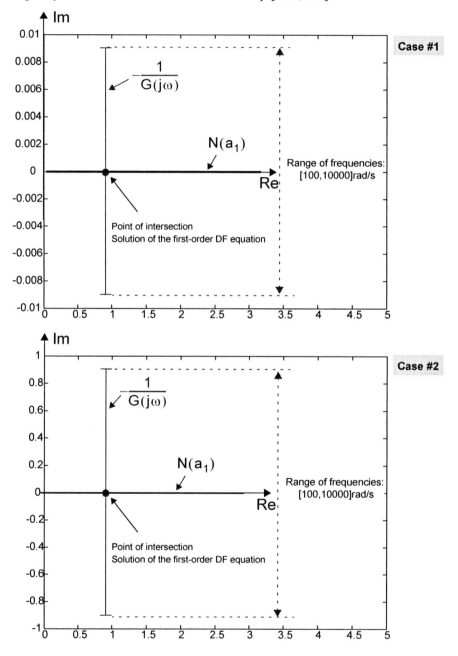

Figure 2.22: Loci of interest

Step 1: We have (see Appendix 2.A)

$$\rho(\omega) = \sqrt{\sum_{k = 3, 5, \ldots} \frac{\left(k_1^2 \frac{\omega_0^2}{Q^2}\right) k^2 \omega^2}{k^4 \omega^4 - k^2 \omega^2 \left(2\omega_0^2 - \frac{\omega_0^2}{Q^2}\right) + \omega_0^4}} \tag{2.40}$$

Observe that as $Q \to \infty$, then $\rho(\omega) \to 0$. This fact explains why the DF solution is more precise as $Q \to \infty$.

Step 2: We have (see Appendix 2.A)

$$p(a_1) = V \sqrt{2\left(\frac{\pi^2 - 8}{\pi^2}\right)} = 0,62 V \tag{2.41}$$

Step 3: We have (see Appendix 2.A)

$$q(a_1, \varepsilon)^2 = \frac{16 V^2}{\pi}\left[a sen\left(\frac{\varepsilon}{a_1}\right)\right] \approx \frac{16 V^2}{\pi} \frac{\varepsilon}{a_1} \tag{2.42}$$

Step 4: Let us take a 0.001% error, it is for instance, $\varepsilon = 0.001$.

Step 5: Find the set Ω of (ω, a_1) values near $(\hat{\omega}, a_1)$ such that

$$\left|N(a_1) + \frac{1}{G(j\omega)}\right| \le \sigma(\omega, a_1)$$

$$\sigma(\omega, a_1) = \frac{q(a_1, \varepsilon(\omega, a_1))}{a_1} \approx \left(\frac{16 V^2}{\pi} \frac{\varepsilon}{a_1}\right)^{\frac{1}{2}} \frac{1}{a_1} = \left(\frac{16 V^2}{\pi} \varepsilon\right)^{\frac{1}{2}} a_1^{-\frac{3}{2}}. \tag{2.43}$$

$$\left|N(a_1) + \frac{1}{G(j\omega)}\right| \le \left(\frac{16 V^2}{\pi} \varepsilon\right)^{\frac{1}{2}} a_1^{-\frac{3}{2}}$$

This can be done graphically in Fig. 2.23. Notice how different are the units in the vertical axis of both cases.

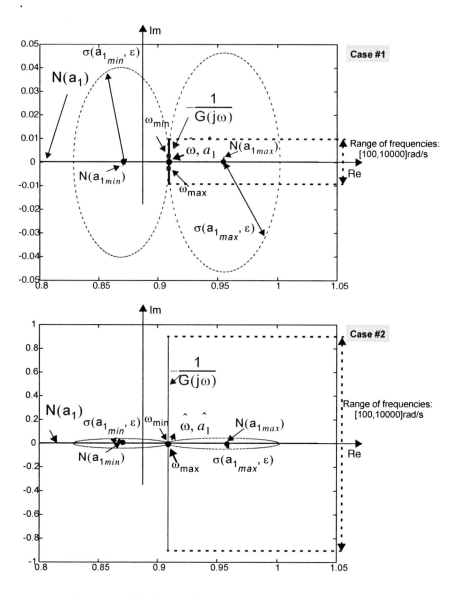

Figure 2.23: Error discs used in locating the set Ω, in which the exact solution lies

We have to check that Ω is bounded. Intuitively, it can be observed from Fig. 2.23 that in *Case #1* it is more difficult to establish the set of frequencies which form Ω because there is only a very short stretch in the imaginary axis which corresponds to the $[100, 10000]\dfrac{rad}{s}$ range of frequencies. Therefore, a

small deviation in defining the values of points where the discs are tangential

to the $-\dfrac{1}{G(j\omega)}$ locus increases significantly the frequency in the rectangle

$\Omega = [\omega_{min}, \omega_{max}] \times [a_{1min}, a_{1max}]$. Notice that this fact minimizes as $Q \gg 1$

(*Case #2* in Fig. 2.23).

If we consider, for example, in *Case #1* that the discs are tangential to the

$-1/G(j\omega)$ locus in the point on the imaginary axis ± 0.001, then,

$[\omega_{min}, \omega_{max}] = [1691.27, 591.27]\dfrac{rad}{s}$. However, simply supposing that

such a point is in ± 0.002, the rectangle $[\omega_{min}, \omega_{max}]$ enlarges a lot:

$[2586.61, 386.61]\dfrac{rad}{s}$. Alternatively, if in *Case #2* we assume that the discs

are tangential to the $-1/G(j\omega)$ locus in ± 0.001, then,

$[\omega_{min}, \omega_{max}] = [1005.52, 994.52]\dfrac{rad}{s}$ (very close to the DF solution, ω).

Similarly, if we assume that the point is in ± 0.002, the rectangle

$[\omega_{min}, \omega_{max}]$ hardly changes: $[1011.06, 989.06]\dfrac{rad}{s}$.

On the other hand, it must be clear from Fig. 2.23 that both cases (*Case #1*

and *Case #2*) give the same range of amplitude values. Doing the correspond-

ing calculations such a range is: $[1.3325, 1.4655]V$ or likewise $[\hat{a}_1 \pm 4.6\%]$.

It is consistent with the results obtained in previous sections where we studied

the example of Fig. 2.13. We conclude the DF approach was poor in deter-

mining the value of the frequency if Q was not large enough but, however, it

was not equal for the amplitude. The accuracy in the amplitude did not depend

on the value of Q.

Step 6: Check that Ω contains the DF solution

$(\hat{\omega}, \hat{a}_1) = \left(1000\dfrac{rad}{s}, 1.4006\,V\right)$. It can be shown that as $Q \gg 1$ we can fence

in the rectangle $[\omega_{min}, \omega_{max}] \times [a_{1min}, a_{1max}]$ to the DF solution.

Step 7: There is at least one true periodic solution with $(\omega, a_1) \in \Omega$ and $\|x*\| \leq \varepsilon$. It is the actual solution is in the close neighbourhood of the calculated one, $(\hat{\omega}, a_1)$ and the residual part of the solution due to the higher harmonics is smaller than ε.

2.4.1.2   Example #2: (Example of Fig. 2.19)

Let us again study the following linear transfer function[15]

$$G(s) = \frac{-0.0392s^2 - 2220.01s + 1.654e^{-8}}{s^2 + 2243s + 7.163e7} \tag{2.44}$$

Step 0: Applying (2.32), then we have

$$\hat{a}_1 = \frac{2V}{\pi}k_1 = 1.2602V$$

$$\hat{\omega} = \omega_0 \cdot \sqrt{\frac{k_1 - k_0}{k_1 - k_2}} = 8294.15\frac{rad}{s}(1320.0Hz) \tag{2.45}$$

Checking that the $N(a_1)$ and $-\dfrac{1}{G(j\omega)}$ loci are not parallel where they intersect at $(\hat{\omega}, a_1)$ (see Fig. 2.24).

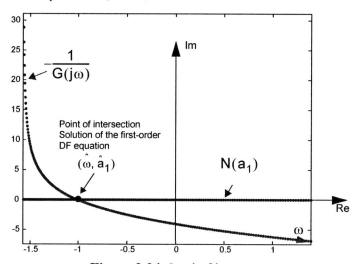

Figure 2.24: Loci of interest

---

[15] This example corresponds to a biquad which is one of main BUTs in next chapters (Fig. 2.19).

Step 1: We have (see Appendix 2.A)

$$\rho(\omega) = \sqrt{\sum_{k = 3, 5, \ldots} \frac{k_2^2 k^4 \omega^4 - \left(2k_2 k_0 \omega_0^2 - k_1^2 \frac{\omega_0^2}{Q^2}\right) k^2 \omega^2 + k_0^2 \omega_0^4}{k^4 \omega^4 - k^2 \omega^2 \left(2\omega_0^2 - \frac{\omega_0^2}{Q^2}\right) + \omega_0^4}}$$

(2.46)

Step 2: We have likewise: $p(a_1) = 0,62V$.

Step 3: We have likewise: $q(a_1, \varepsilon)^2 \approx \frac{16V^2}{\pi} \frac{\varepsilon}{a_1}$.

Step 4: Let us fix, for example, $\varepsilon = 0.001$.

Step 5: Find the set Ω of (ω, a_1) values near $(\hat{\omega}, a_1)$ such that

$$\left| N(a_1) + \frac{1}{G(j\omega)} \right| \leq \sigma(\omega, a_1) \approx \left(\frac{16V^2}{\pi} \varepsilon \right)^{\frac{1}{2}} a_1^{-\frac{3}{2}}.$$

(2.47)

This can be done graphically in Fig. 2.25.

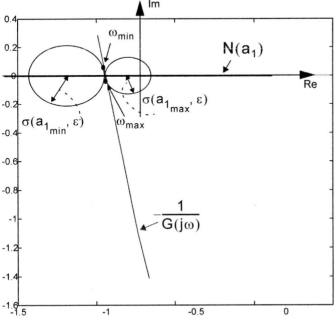

Figure 2.25: Error discs used in locating the set Ω, in which the exact solution lies

Check that Ω is bounded. That is right, $\omega \in [8289, 8309]\frac{rad}{s}$ and $a_1 \in [1.195, 1.322]$ V .

Step 6: Check that Ω contains the DF solution $(\hat{\omega}, a_1) = 8294.15, 1.2602$.

Step 7: There is at least one true periodic solution within $(\omega, a_1) \in \Omega$ and $\|x*\| \leq \varepsilon$.

Observe that now we can state that the actual frequency and amplitude of the oscillations are in the range $(\sim\tilde{\omega}, a_1 \pm 5\%)$. It agrees with the simulation results shown in previous sections, where the actual frequency and amplitude was in the range $(\sim\tilde{\omega}, a_1 \pm 1.2\%)$.

2.4.2 A graphical method for a particular type of nonlinearities

Let us recommend a modification of the revisited semi-graphical DF strategy (see the previous Section) in order to provide a reliable method for predicting whether or not certain types of closed-loop systems with certain kinds of nonlinear feedback loops (see [99], [104], [109]-[112], [114], [123]) can really oscillate. As will be seen, this alternative method is easy to apply and evaluates the usual intuitive ideas about the DF reliability. The goal is not to find a definite analytical oscillation solution, rather it is to find whether there is an oscillation in the system and to fence in the oscillation parameters (frequency and amplitude) within ranges. The idea is to draw a band which measures the amount of uncertainty introduced by the approximations inherent in the DF approach. Therefore, the method gives error bounds for oscillation predictions, as well as ranges of frequency and amplitude over which oscillation is possible. A particular restriction is that the nonlinear element must be single-valued and has bounded slope. Fortunately, most of our cases of interest are included in this kind of nonlinearities.

2.4.2.1   Proposed Strategy

Let us again study the autonomous feedback system shown in Fig. 2.6. As was said, this closed-loop system splits into a linear, time-invariant part, G

and a nonlinear part, n. Assume now that the characteristic of n has odd symmetry, is single-valued and has a slope between α and β, i.e.

$$\alpha(x_1 - x_2) \leq nx_1 - nx_2 \leq \beta(x_1 - x_2) \tag{2.48}$$

for all real numbers x_1 and $x_2 < x_1$.

We are mainly interested in determining **whether the system will oscillate**. As was explained in previous sections, the DF approach establishes that if $N(a_1) + 1/G(j\omega) = 0$ (2.17) has a solution $(\hat{\omega}, a_1)$, there is *"probably"* a π − symmetric oscillation (see Appendix 2.A) in the system with frequency and amplitude, at the input to the nonlinearity, close to $\hat{\omega}$ and $\hat{a}_1$. Contrarily, if the previous equation has no solutions, the system *"probably"* cannot sustain a π − symmetric oscillation. Let us explain how to replace *"**probably**"* with *"**certainly**"*.

As was discussed, (2.17) can be graphically solved plotting the loci in the complex plane of $N(a_1)$ as a_1 varies and of $-\dfrac{1}{G(j\omega)}$ as ω varies. Every intersection of the loci is a solution of (2.17).

On the other hand, in [99], [110], [112] it is shown that (2.17) is an approximate version of an exact equation

$$N(a_1) + \frac{1}{G(j\omega)} = F(\omega, a_1) \tag{2.49}$$

$F(-,-)$ cannot be found exactly but can often be bounded. If the errors introduced by neglecting this function are small enough, then solutions of (2.17) should be close to those of (2.49). The purpose of this section is to show that the bound on F can be used to define an uncertainly band around the $-1/G(j\omega)$ locus in such a way that the presence or absence of intersections between this band and the $N(a_1)$ locus guarantees the presence or absence of corresponding oscillations in the system of Fig. 2.6.

Let us determine the uncertainty band:

Let first $\rho(\omega)$ be again a quantity which estimates *how well $G(j\omega)$ filters out higher harmonics.*

On the other hand, let us proceed as follows: call the point on the $-1/G(j\omega)$ locus corresponding to ω, P_1; then, that corresponding to 3ω, P_3, and so on. Draw a circle with the interval $[\alpha, \beta]$ on the real axis as diameter[16]. Obviously, the $N(a_1)$ locus is placed inside this circle[17]. Now consider an ω such that every P_k ($k \neq 1$, k odd) lies outside this critical circle, as is shown in Fig. 2.26, and let P_{k_0} be the point closest to the circle. Then $\rho(\omega)$ is the distance from P_{k_0} to the center of the critical circle, i.e.

$$\rho(\omega) = min\left|\frac{\beta+\alpha}{2} + \frac{1}{G(jk\omega)}\right|\Bigg|_{\substack{k > 1 \\ k \text{ odd}}} \tag{2.50}$$

Observe that we have only define $\rho(\omega)$ for certain values of ω, specifically those in the set $\Gamma \equiv \left\{\omega: \rho(\omega) > \frac{(\beta-\alpha)}{2}\right\}$.

On the other hand, let us define a positive function $\sigma(\omega)$ on any subset of Γ

$$\sigma(\omega) = \frac{\left(\frac{\beta-\alpha}{2}\right)^2}{\left(\rho(\omega) - \frac{\beta-\alpha}{2}\right)} \tag{2.51}$$

We can obtain this value calculating straightforwardly, but also we can give it a geometrical interpretation which is useful when one is drawing the diagrams by hand. Referring to Fig. 2.26, draw the line segment from P_{k_0} to C [the length $\rho(\omega)$], and erect a perpendicular at C. Draw a square which determines the point C'. The point Q is defined in Fig. 2.26 as the positive quantity $\sigma(\omega)$. In our case we note that as ω increases $\sigma(\omega)$ decreases,

---

[16.] This circle will be named *critical circle* from this point forward.

[17.] It can be shown, in this case, $N(a_1)$ is a real-valued function for which $\alpha \leq N(a_1) \leq \beta$.

while as ω decreases $\sigma(\omega)$ increases until it diverges as P_{k_0} approaches the critical circle.

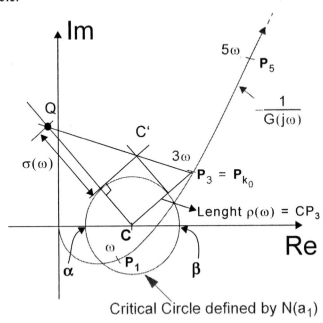

Figure 2.26: Finding $\rho(\omega)$ and $\sigma(\omega)$. Here P3 is the closest point to the circle, so $k_0=3$ and ρ in (2.50) is CP3

Now draw error circles centered on $-1/G(j\omega)$ with radius $\sigma(\omega)$ (see Fig. 2.27). The enveloped of all such circles over a connected subset Γ' of Γ is an uncertainty band. The reason for choosing a subset of Γ is that as P_{k_0} approaches the critical circle, the error circles become arbitrarily large and cease to give any useful information. The choice of subset Γ' is best made while the band is being drawn and is chosen with the objective of drawing a narrow band. If $G(j\omega)$ is low pass, the band can be quite narrow over Γ''. In any case let us assume that we have fixed Γ' and drawn the corresponding band.

We can highlight two cases for which we can make a definitive statements regarding the solution of (2.49):

Case a: No part of the band intersects the $N(a_1)$ locus.

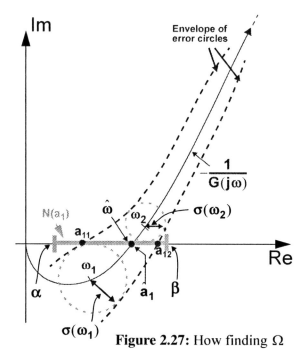

Figure 2.27: How finding Ω

Case b: The band intersects the locus completely as in Fig. 2.27.

Roughly speaking, in Case a there is no solution of (2.49) while in Case b there is one. This is in accordance with practical experience when using the DF method, since only a complete (nonglancing) intersection or noninter-section is treated with confidence, and then only when there is good reason to believe that higher harmonics are unimportant. The latter requirement is satisfied when the band is narrow, so all we are really doing is quantifying the low-pass hypothesis.

In Case b we can find error bounds by examining the intersection and reading off 1) the amplitudes a_{11} and a_{12} corresponding to the intersections of the boundary of the uncertainty band with the $N(a_1)$ locus, and 2) the frequencies ω_1 and ω_2 corresponding to the error circles (of radii $\sigma(\omega_1)$ and $\sigma(\omega_2)$) which are tangent to the $N(a_1)$ locus on either side of it.

On the basis of these numbers we can define a rectangle Ω in the (ω, a_1) plane, containing the point $(\hat{\omega}, a_1)$ for which the two loci intersect

$$\Omega = \{(\omega, a): \omega_1 < \omega < \omega_2, a_{11} < a_1 < a_{12}\} \qquad (2.52)$$

Therefore, a **complete intersection** between the uncertainty band and the $N(a_1)$ locus can be now defined as taking place when the $-1/G(j\omega)$ locus itself intersects the $N(a_1)$ locus and a finite Ω can be defined as above, on which $N(a_1)$ and $G(j\omega)$ are regular[18] and the loci are never parallel (see Ref for details).

Summarizing, Γ was the set on which $\rho(\omega)$ could be define and Γ' was that subset of Γ for which we chose to draw the uncertainty band. Let Γ'' be the subset of Γ for which all harmonics (including the first) have the corresponding $-1/G$ point outside the critical point. That is

$$\Gamma'' = \left\{ \omega : \left| \frac{\beta + \alpha}{2} + \frac{1}{G(j\omega)} \right| > \frac{\beta - \alpha}{2} \right\} \Bigg|_{k = 1, 3, 5, \ldots} \tag{2.53}$$

Therefore, we can affirm that the system in Fig. 2.6 (with a single-valued nonlinear element which has a bounded slope):

Theorem 1: can not have a π – symmetric oscillation of any fundamental frequency $\omega \in \Gamma''$.

Theorem 2: can not have a π – symmetric oscillation of any fundamental frequency $\omega \in \Gamma'$ if there is no intersection of any part of the uncertainty band with the $N(a_1)$ locus.

Theorem 3: can not have a π – symmetric oscillation of any fundamental frequency $\omega \in \Gamma$ and fundamental amplitude a_1 if the corresponding error circle does not contain the point $N(a_1)$.

Theorem 4: for each complete intersection of the uncertainty band with the $N(a_1)$ locus, there is at least one π – symmetric oscillation with frequency and amplitude contained in the corresponding Ω.

[18.] $N(a_1)$ is regular in Ω if $\dfrac{d}{da_1} N(a_1) \neq 0$ for all $a_1 \in \Omega$ and $-1/G(j\omega)$ is regular in Ω if $\dfrac{d}{d\omega} G(j\omega) \neq 0$ for all $\omega \in \Omega$.

2.4.2.2   Example #3: Non oscillatory solution

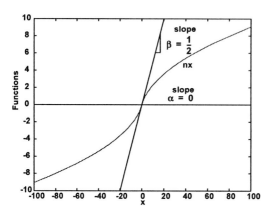

Figure 2.28: Example of Nonlinear Function

Consider the following non-linear function:

$$nx = (signx)(\sqrt{1 + |x|} - 1).$$ If we draw this expression we obtain two parabolic segments joined to give a smooth odd function (in red in Fig. 2.28). Then, the values of α and β are 0 and 1/2, respectively and the locus of $N(a)$ must be on the real axis between these limits. The diagram for the case where

$$G(s) = \frac{3}{2} \cdot \frac{(s-1)}{s^3(s+1)}$$ is shown

in Fig. 2.29.

In this case,

$$-\frac{1}{G(jk\omega)} = \frac{2}{3} \cdot \left[\frac{2(k\omega)^4}{1 + (k\omega)^2} + j\frac{((k\omega)^5 - (k\omega)^3)}{1 + (k\omega)^2} \right] \tag{2.54}$$

So, we have

$$-\frac{1}{G(j\omega)} = \frac{2}{3} \cdot \left[\frac{2\omega^4}{1 + \omega^2} + j\frac{(\omega^5 - \omega^3)}{1 + \omega^2} \right] \tag{2.55}$$

Then we define

$$\rho(\omega) = min\left| \frac{\beta + \alpha}{2} + \frac{1}{G(jk\omega)} \right|_{\substack{k > 1 \\ k \text{ odd}}}$$

$$= min\left| \frac{1}{4} - \frac{2}{3} \cdot \left[\frac{2(k\omega)^4}{1 + (k\omega)^2} + j\frac{((k\omega)^5 - (k\omega)^3)}{1 + (k\omega)^2} \right] \right|_{\substack{k > 1 \\ k \text{ odd}}} \tag{2.56}$$

An initial guess may be $k = 3$

$$\rho(\omega) = \left| \frac{1}{4} - \frac{2}{3} \cdot \left[\frac{2(3\omega)^4}{1+(3\omega)^2} + j\frac{((3\omega)^5 - (3\omega)^3)}{1+(3\omega)^2} \right] \right| \qquad (2.57)$$

And the set

$$\Gamma \equiv \left\{ \omega : \rho(\omega) > \frac{(\beta - \alpha)}{2} \right\} =$$

$$= \left\{ \omega : \sqrt{\left(\frac{1}{4} - \frac{2}{3} \cdot \frac{2(3\omega)^4}{1+(3\omega)^2} \right)^2 + \left(\frac{((3\omega)^5 - (3\omega)^3)}{1+(3\omega)^2} \right)^2} > \frac{1}{4} \right\} \qquad (2.58)$$

$$\Gamma \equiv \{ \omega : \omega > 0.30 \}$$

But if we choose $\Gamma' \equiv \{ \omega : \omega > 0.48 \}$, we have then the radius of the error

circle centered in $-\dfrac{1}{G(j\omega)}$ when $\omega = 0.48$

$$\sigma(0.48) = \frac{\left(\frac{1}{4}\right)^2}{\left(\rho(0.48) - \frac{1}{4}\right)}$$

$$\rho(0.48) = \left| \frac{1}{4} - \frac{2}{3} \cdot \left[\frac{2(3\omega)^4}{1+(3\omega)^2} + j\frac{((3\omega)^5 - (3\omega)^3)}{1+(3\omega)^2} \right] \right|_{\omega = 0.48} \qquad (2.59)$$

$$\rho(0.48) = 1.7586$$

$$\sigma(0.48) = 0.0414$$

If we define a set of error circles over $\Gamma' \equiv \{ \omega : \omega > 0.48 \}$, the envelope of all such circles will be an uncertainty band (in magenta Fig. 2.29). So, we can observe that this system cannot oscillate with an angular frequency greater than 0.48 because there is no intersection of any part of the uncertainty band with the $N(a_1)$ locus (*Theorem 2*).

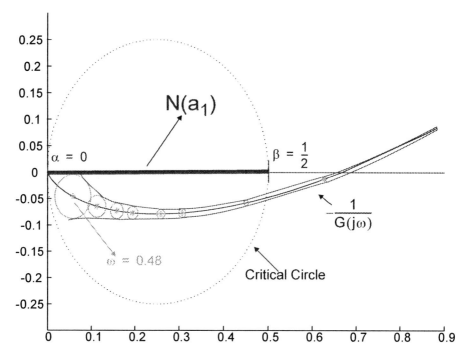

Figure 2.29: Band and locus for first example

2.4.2.3   Example #4: Existence of an oscillatory solution

Consider the following non-linear function (saturation nonlinearity):

$$nx = \begin{cases} x & |x| \le 1 \\ signx & |x| > 1 \end{cases}$$ so that $\alpha = 0$ and $\beta = 1$. The locus of $N(a)$ must

be on the real axis between these limits. The diagram for the case where

$G(s) = \dfrac{2(s-1)}{s^3(s+1)}$ is shown in Fig. 2.30.

In this case, $-\dfrac{1}{G(jk\omega)} = \dfrac{1}{2}\left[\dfrac{2(k\omega)^4}{1+(k\omega)^2} + j\dfrac{((k\omega)^5 - (k\omega)^3)}{1+(k\omega)^2}\right]$. So, we have

$$-\dfrac{1}{G(j\omega)} = \dfrac{1}{2}\left[\dfrac{2\omega^4}{1+\omega^2} + j\dfrac{(\omega^5 - \omega^3)}{1+\omega^2}\right].$$

Then we define

$$\rho(\omega) = min\left|\frac{\beta+\alpha}{2}+\frac{1}{G(jk\omega)}\right|\Bigg|_{\substack{k>1 \\ k\ odd}} =$$

$$=min\left|\frac{1}{2}-\frac{1}{2}\left[\frac{2(k\omega)^4}{1+(k\omega)^2}+j\frac{((k\omega)^5-(k\omega)^3)}{1+(k\omega)^2}\right]\right|\Bigg|_{\substack{k>1 \\ k\ odd}}$$

If we impose, for example, $k = 3$:

$$\rho(\omega) = \left|\frac{1}{2}-\frac{1}{2}\left[\frac{2(3\omega)^4}{1+(3\omega)^2}+j\frac{((3\omega)^5-(3\omega)^3)}{1+(3\omega)^2}\right]\right|$$

The situation is as shown in Fig. 2.30, with $[\omega_1, \omega_2] = [0.94, 1.03]$ and $[a_{11}, a_{12}] = [2.25, 2.90]$. We have then $\Omega = \{(\omega, a):0.94 < \omega < 1.03, 2.25 < a_1 < 2.90\}$. We can assert that there is an oscillation solution whose frequency and first harmonic amplitude, $(\omega, a_1) \in \Omega$.

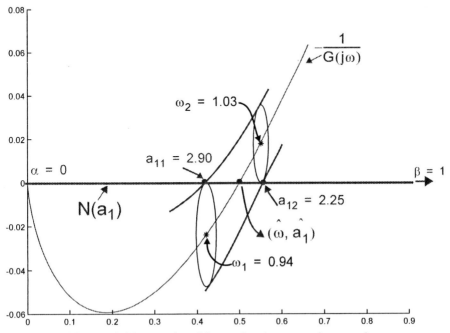

Figure 2.30: Band and locus for the second example

These last examples revealed two important issues:

-when the nonlinear element is single-valued and has bounded slope, there is a way to graphically find if an oscillation solution exists without having to calculate the specific describing-function. This method also allow us to define error bounds of the solution.

- the existence of an oscillation solution in the scheme of Fig. 2.1 depends not only on the involved linear block but also on the features of the non-linear element. Notice, for instance, that when we change the linear block of the Example #4 by the one from the Example #3 (although they look like very similar) the resulting system has no an oscillatory solution. Therefore, we can find very different results with systems composed of the same non-linear element and very similar transfer functions and vice versa.

2.5 SUMMARY

Experience coined from the study of many cases allows us to assert that the DF approach is the simplest and intuitive method to study the proposed scheme of OBT (Fig. 2.1). However, we can find some particular examples where the predictions extracted by the most basic version of this method are not completely satisfactory. Therefore, we have discussed in this Chapter how to get the best out of the DF approach introducing two modified versions of it. Both versions allow us not only to find an analytical oscillation solution and error bounds for this solution, but also to measure the amount of uncertainty introduced by the approximations in the DF approach. The value of this uncertainty allow us to determine if the result will be a "good" or "acceptable" prediction.

But, problems concerning the accuracy of the oscillation solution obtained by the DF approach appear in different contexts throughout the thesis. Most of them can be studied by employing the modified versions of the DF strategy presented in this Chapter. These methods do not give a solution as only one point. Instead, a set of points is obtained. Therefore, we will work with a set of solutions and the actual oscillation will be within this set. Smaller we can define this set, more accurate our results will be.

However, as will be shown in later chapters, we can observe that there are some cases where even working with a set of DF approach solutions does not guarantee a "valid" result. These examples force us to examine our linearized models (DF approaches) and even, sometimes, to define a new way to go on: that is, complementing the DF results with simulations[19].

[19.] In Chapter 4, we will go in depth in this case.

Chapter 3

OBT Methodology for Discrete-Time Filters
Principles and architecture

IN ORDER TO EVALUATE the feasibility and applicability of the new general OBT approach proposed and studied mathematically in previous chapters, the particular case of discrete-time filters will be herein considered. In practice, the Arabi and Kaminska's OBT concept [1]-[15], was especially tailored for active filters that may be divided into second-order sections. In fact, its application to continuous-time filters implemented by off-the-shelf components was reported elsewhere [14]. Now, considering many other references of our own [36]-[37], it may be reinforced that this methodology seem especially appealing for active filters. In all these cases, [36]-[37], discrete-time filters were studied, although the technique is not limited to them.

This Chapter aims to exhaustively review the essential points of the previously presented OBT technique for discrete-time filters, discussing the practical modifications needed for obtaining economical and viable test results. In particular, we will describe a pragmatical approach for applying OBT to discrete-time active filters. A rather general, yet simple to implement, mechanism will be proposed to guarantee oscillations without almost modifying the SUT. As was seen in Chapter 1, such a mechanism has to be nonlinear in order to force robust oscillations. However, it will be shown that the resulting oscillator is quite predictable in terms of both frequency and amplitude, making it feasible to use these two parameters to obtain very high fault coverages when typical faults are considered.

3.1 FEASIBLE OBT STRATEGY IN DISCRETE-TIME FILTERS

Converting a discrete-time system under test (and more specifically a discrete-time filter) into an oscillator requires a mechanism to force a displacement of, at least, a pair of poles onto the unit circle. In fact, the

97

oscillation mechanism illustrated in Fig. 3.1 consists in reaching, after a transient state, a steady state where a pair of complex conjugate poles is continually moving in and out the unit circle, while the remaining poles are placed into the unit circle to avoid the system to become unstable.

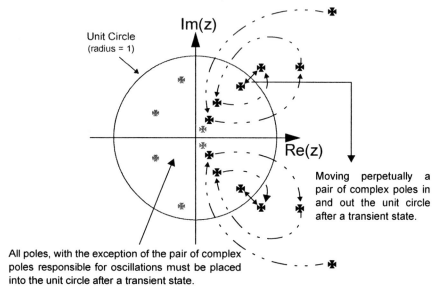

Figure 3.1: Robust oscillation strategy in a discrete-time SUT

As was demonstrated elsewhere [100] and postulated in previous chapters, a basic means to enforce this proposed oscillation strategy lies in closing a loop around the involved SUT. The introduced feedback must be a non-linear block with a series of required features which allow to govern the magnitude of the pole

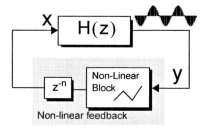

Figure 3.2: Oscillator

radius. Accordingly, this approach builds up an oscillator with amplitude control by limitation. A simplified z -domain scheme serving this proposal is displayed in Fig. 3.2, where $H(z)$ represents the filter transfer function (of the filter under test) in the z-domain and (z^{-n}) stands for the possible extra delays that a real circuitry may insert in the system.

A linearized model for this closed-loop system formed by a generic transfer function and a non-linear block can be derived using the DF approach (see Chapter 2). Provided that the discrete-time linear part of the system filters out the highest harmonics of its input, and the nonlinear block accomplishes some conditions, the system will have a pair of complex poles on the unit circle and a sinusoidal discrete-time signal with general form $y(t) \approx A\cos(\theta t)$ will be obtained[1], where $\theta = 2\pi f_{signal}/f_{sampling}$ (being $f_{sampling}$ the sampling frequency of the system and f_{signal} the frequency of the output signal).

The closed-loop system in Fig. 3.2 is characterized in the z-domain by

$$1 - z^{-n}N(A)H(z) = 0 \qquad (3.1)$$

where $N(A)$ is the describing-function of the non-linear block as a function of the amplitude (A) of the first harmonic of its input[2] (signal called y in Fig. 3.2).

Consequently, the pole placements is determined by the roots of the equation (3.1). But, obviously, depending on the value of n (number of delays in the feedback loop) and on the order of the filter transfer function, several types of oscillation solutions can be found.

Expression (3.1) formulates the mathematical basis of the proposed OBT strategy. Next sections aim to solve this mathematical problem but in such a way that feasible oscillation solutions can be easily obtained for every particular case.

3.1.1 Oscillation solutions for a generic filter

The general mathematical procedure to extract viable oscillation solutions, efficient for the OBT approach, involves: **1)** establish (3.1) (that is, determine the pole placements of the SUT reconfigured as an oscillator) in a suitable way, **2)** check the conditions for robust and sustained oscillations and **3)** express the oscillation parameters and the oscillation conditions as a function of the coefficients of the system transfer function.

---

[1] Considering the relation between continuous and discrete-time frequency domains

$z = re^{sT_s} = re^{j\omega T_s}$. A pair of complex poles in the unit circle ($r = 1$) will be placed in

$z = e^{j\omega T_s} = e^{j2\pi\frac{f_{signal}}{f_{sampling}}}$.

[2] See Chapter 2 for details.

The main goal of this mathematical procedure is to find out the oscillation parameters. Such oscillation parameters have to be related to all the elements of the SUT as well as the restrictions of the steady oscillation mode have to be established.

Consider the most general case when the filter transfer function in Fig. 3.2 presents the following expression

$$H(z) = \frac{k_i z^i + k_{i-1} z^{i-1} + \ldots + k_0}{z^j + b_{j-1} z^{j-1} + \ldots + b_0} \qquad j \geq i \qquad (3.2)$$

where indexes i and j would be, respectively, the numerator and the denominator order of the filter transfer function.

Thus, the characteristic equation (3.1), would now turn into

$$Q(z) - N(A)P(z) = 0$$

$$P(z) = k_i z^i + k_{i-1} z^{i-1} + \ldots + k_0 \qquad (3.3)$$

$$Q(z) = z^{j+n} + b_{j-1} z^{j+n-1} + \ldots + b_0 z^n$$

Depending on the order of (3.3), the number of the poles vary, but if we are interested in an oscillatory solution, we have to impose, as minimum, a pair of complex conjugate poles (obviously, such poles have to be located, after a transient mode, onto the unit circle). For the sake of clarity, (3.3) may be rewritten as

$$(z^2 - 2r\cos\theta z + r^2)[\text{Remaining Poles}] = 0 \qquad (3.4)$$

where those poles responsible for an oscillatory result[3] have been separated from the remaining poles. Notice that r and θ are functions of $N(A)$ and the coefficients of $H(z)$. Observe, on the other hand, that the solutions of the remaining poles must be located into the unit circle in order to guarantee the consistent stability of the closed-loop system.

Handling an expression as (3.4) from an abstract viewpoint and deriving practical conclusions can be an arduous and most of times an impossible task. In fact, the roots of (3.4) can be either generic pairs of complex conjugate poles, or real poles or/and pairs of pure imaginary poles. But, regardless of the other existing poles, let us focus our attention on the pairs of complex

---

[3] For convenience, the polar form has been used to express the pair of complex poles of interest.

conjugate poles which control the oscillations. As was said, the expression of such poles is, in polar form, $z^2 - 2r\cos\theta + r^2$, where, as is shown in Fig. 3.3, r represents the corresponding radius and θ the respective angle for an instantaneous amplitude condition (that is an instantaneous A^4). Such an angle is related to the frequency of the non-linear block input signal (signal y in Fig. 3.2) as well as with the sampling ratio of the discrete-time system in such a way that $\theta = 2\pi f_{signal}/f_{sampling}$.

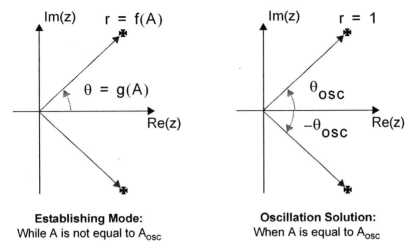

Establishing Mode:
While A is not equal to A_{osc}

Oscillation Solution:
When A is equal to A_{osc}

Figure 3.3: Pair of complex poles ruling oscillations

On the other hand, let us illustrate in Fig. 3.4 the pole reorganization (the evolution of the complex poles responsible for oscillations as well as the behaviour of the remaining poles) in the so-called steady oscillation mode.

---

4. Observe again (from (3.3) and (3.4)) that both, r and θ, depend on A, that is, the amplitude of the first harmonic of the non-linear block input signal.

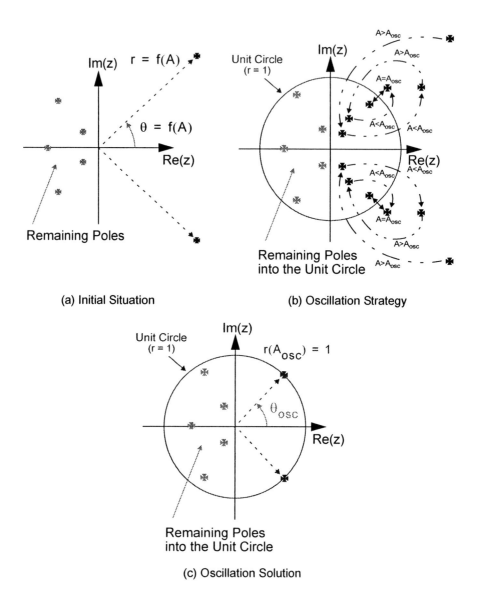

(a) Initial Situation   (b) Oscillation Strategy

(c) Oscillation Solution

Figure 3.4: Description of the closed-loop system pole reorganization

The problem will have an oscillation solution as long as A_{osc} and θ_{osc} can be found, satisfying the following oscillation conditions:

C1.- Start-up condition. Initially, if no signal is present ($A = 0$), at least one pole should be moved out of the unit circle in order to convert the system to an unstable system and thus, increasing the value of the signal.

C2.- The pair of complex conjugate poles commanding the oscillations must be placed just in the unit circle. That is, $r^2(A_{osc}) = 1$.

C3.- Amplitude control stability. That is, at the end of the oscillation strategy process, a pair of complex poles have to be continually moving out and into the unit circle as illustrated in Fig. 3.4-(b). To fulfil it, if A increases, r must decrease and viceversa. That means: $\dfrac{\partial r}{\partial A} = \dfrac{\partial r}{\partial N(A)} \cdot \dfrac{\partial}{\partial A} N(A) < 0$.

C4.- The solution must be one and only one. That is, only a value of A_{osc} and θ_{osc} can be found. Then, A_{osc} and $f_{osc}(\theta_{osc})$ are respectively the amplitude and the frequency of the oscillations.

C5.- The remaining poles must be placed into the unit circle in order to guarantee the stability of the system as shown in Fig. 3.4-(c).

According to this, some oscillation conditions can be derived for all the cases proposed in Table 3.1 (only the types of poles shown in Table 3.1 will be considered because more complex solutions are a combination of them).

For instance, let us assume the case of having in (3.4) a pair of complex poles plus a real pole (named p_1) (first row of Table 3.1). Then, (3.4) is reduced to

$$(z^2 - 2r\cos\theta z + r^2)[z + p_1] = z^3 + a_2 z^2 - a_1 z + a_0 = 0 \qquad (3.5)$$

where the coefficients a_i are related to $N(A)$ and consequently to A (first-harmonic amplitude) as well as the coefficients of the filter transfer function. Then, when we impose the oscillation mode ($r(A_{osc}) = 1$), the derived oscillation condition is $p_1 = a_0[A_{osc}] < 1$.

Existing Poles	Main part of the characteristic equation ((3.4))		Oscillation Solution($r = 1$)
	Polar Form	General Form	
Pair complex poles	$z^2 - 2r\cos\theta z + r^2$	$z^2 - a_1 z + a_0$	$\cos\theta_{osc} = \dfrac{a_1}{2}$ $a_0 = 1$
Pair complex poles + Real pole	$(z^2 - 2r\cos\theta z + r^2)(z + p_1)$	$z^3 + a_2 z^2 - a_1 z + a_0$	$\cos\theta_{osc} = \dfrac{a_2 - a_0}{2}$ $1 + 2p_1\cos\theta_{osc} = a_1$ $p_1 = a_0$
Pair complex poles + Pair pure imaginary poles	$(z^2 - 2r\cos\theta z + r^2)(z^2 + |p_2|)$	$z^4 + a_3 z^3 + a_2 z^2 - a_1 z + a_0$	$2\cos\theta = a_3$ $1 + |p_2| = a_2$ $2\cos\theta|p_2| = a_1$ $|p_2| = a_0$

Table 3.1: Summary of the simplest pole placements for oscillations

In the same way, if a pair of complex poles plus a pair of pure imaginary poles $(\pm j|p_2|)$ is considered, the oscillation condition would be $|p_2| = a_0[A_{osc}] < 1$.

On the other hand, another evident oscillation condition would be $-1 \le \cos(\theta_{osc}) \le 1$. Even more, the result $\cos(\theta_{osc}) \to \pm 1$ is not desirable either due to the fact that, regardless of the filter coefficient values, θ_{osc} tends to π or 2π (poles on the real axis) and consequently, f_{osc} tends always to $f_{sampling}$ or $f_{sampling}/2$, respectively. In such a case no information for test purpose could be obtained from the frequency parameter. Clearly, this fact halves the efficiency of the test strategy. Moreover, when this solution is obtained it would be needed a faster sampling frequency to find out and evaluate the oscillation frequency.

But, in general, another hypothetical situations must be, most of the times, avoided not only those associated with the result $\cos(\theta_{osc}) \to \pm 1$. Also, the result $\theta_{osc} \to \pm(\pi/2)$ (when poles are on the imaginary axis) must be rejected for test purposes because under this condition the oscillation frequency cosine is virtually zero and, likewise, f_{osc} is $f_{sampling}/4$. Consequently, in most cases, the closed-loop system would not behave as an efficient oscillator for testing (see Chapter 4 for details). Therefore, the four rejected situations associated with θ_{osc} are those where that angle is placed near to the axes: either the imaginary or the real one.

To go right away, we must aim to achieve the following oscillation condition for finding accurate values with the DF approach as well as for test purposes: $f_{osc} \ll f_{sampling}$. We are interested in $f_{osc} = f_{sampling}/M$ with $M \gg 1$. That means that $\theta_{osc} = \left. \dfrac{2\pi}{M} \right|_{M \gg 1} \to 0$.

Consequently, a trade-off between this condition and the aforementioned condition where we had to circumvent that $\cos(\theta_{osc}) \to \pm 1$ is required. This condition allows us to use the system sampling frequency to measure and evaluate the oscillation frequency. However, this point is important mainly when we are interested in on-chip interpretation. In an off-chip strategy we could have a tester with a faster sampling frequency and, then, this condition may not be essential. All these issues will be separately considered for specific examples in next sections.

To end this section, let us regard the case of a generic filter. Normally, under this circumstance, a more complex combination of oscillation results is found. However, although a more complicated case is being considered, the critical points to be taken into account are essentially the same than those above-mentioned.

3.1.2 Oscillation solutions for the biquadratic case

As can be deduced from the previous section, splitting the whole SUT into simpler components is necessary. Otherwise predicting the fulfilment of the oscillation conditions and the main oscillation parameters, is no trivial and most of the times an impracticable task. So, a first approach to test a high-order filter should be to convert the complete system into an oscillator and then to solve the general case, (3.3), where the transfer function, $H(z)$, has the order of the complete SUT. But in this case, as can be seen from (3.3), a lot of poles have to be handled and thus it would be very difficult to control the specific pair originating the oscillation, especially when all these poles differ by a relatively small value and lie very close to each other. Furthermore, relating the oscillatory behaviour to specific faults is much more difficult, making test interpretation very complex as well. Since each oscillatory mode is associated with a pair of complex poles, the minimum order to achieve an oscillation is two. For this reason, we focus on decomposing the SUT into **biquadratic components.**

Let us consider the case of generic second-order structures (biquads) whose transfer function in the z-domain can then be expressed by

$$H(z) = \frac{x(z)}{y(z)} = \frac{k_2 z^2 + k_1 z + k_0}{z^2 + b_1 z + b_0} \tag{3.6}$$

where the coefficients k_i and b_i are related to the particular circuit implementation.

For this particular case, the characteristic equation, (3.1), can be rewritten as

$$z^{n+2} + b_1 z^{n+1} + b_0 z^n - N(A)[k_2 z^2 + k_1 z + k_0] = 0 \tag{3.7}$$

being n the number of delays existing in the closed-loop oscillator.

Let us, likewise, isolate the pair of complex poles responsible for oscillations. That means, (3.7) is rewritten as

$$(z^2 - 2r\cos\theta z + r^2)[\text{Remaining Poles}] = 0 \tag{3.8}$$

being r and θ are, respectively, the radius and the angle of the complex poles of interest. Notice that the number of remaining poles depends on the number of delays ('n').

Several types of oscillation solutions for (3.7) will be described in next sections. All these solutions will depend on both the number of existing delays, n, and the specific non-linear block selected to close the feedback loop. Thus, let us, first, sort the solutions for different values of 'n'.

3.1.2.1 Type a: Delay-free loop oscillator (n=0)

Let us consider that there are no delays in the feedback loop. Expanding (3.7) the following equation is obtained

$$z^2 + b_1 z + b_0 - N(A) \cdot [k_2 z^2 + k_1 z + k_0] = 0 \tag{3.9}$$

whose result is only a pair of complex poles. (3.9) can also be rewritten as (polar form)

$$z^2 - 2r\cos\theta z + r^2 = 0 \tag{3.10}$$

where $r^2 = \dfrac{b_0 - k_0 N(A)}{1 - k_2 N(A)}$ and $-2r\cos\theta = -\dfrac{b_1 - k_1 N(A)}{1 - k_2 N(A)}$.

Notice that the placement of the poles (the radius and the angle in polar form) will depend on the particular form of the non-linear block describing-function.

In this case, considering the oscillation conditions **C2** and **C3**, a simple system of equations is obtained

$$1 = \frac{b_0 - k_0 N(A_{osc})}{1 - k_2 N(A_{osc})}$$

$$\cos\theta_{osc} = \frac{1}{2}\left[\frac{b_1 - k_1 N(A_{osc})}{1 - k_2 N(A_{osc})}\right] \tag{3.11}$$

$$\frac{dr^2}{dN(A)} = 2r\frac{dr}{dN(A)} = \frac{k_2 b_0 - k_0}{[1 - k_2 N(A)]^2} < 0$$

Then, the oscillation parameters are related to the biquadratic function coefficients by

$$N(A_{osc}) = \frac{b_0 - 1}{k_0 - k_2}$$

$$\cos\theta_{osc} = \frac{1}{2}\left[\frac{(k_0 - k_2)b_1 - k_1(b_0 - 1)}{k_0 - k_2 b_0}\right] \tag{3.12}$$

fulfilling, at least, the following conditions

$$sign(A_{osc}) > 0$$

$$-1 < \cos\theta_{osc} < 1 \Rightarrow -2 < \frac{(k_0 - k_2)b_1 - k_1(b_0 - 1)}{k_0 - k_2 b_0} < 2 \tag{3.13}$$

On the other hand, as was explained earlier, the result $\cos(\theta_{osc}) = \pm 1$ has to be avoided. This is the reason why the second obtained oscillation condition is a strict inequality. Otherwise, no effective information from the oscillation frequency can be extracted since it would be simply the sampling frequency or half of the sampling frequency. Even the case in that $\cos\theta_{osc} \rightarrow \pm 1$ must be avoided in order to have perfectly characterized the oscillation frequency and be able to faultlessly distinguish this one from the sampling frequency.

3.1.2.2   Type b: Single-delay loop oscillator (n=1)

Let us consider that there is a unit delay in the feedback loop. Expanding (3.7) the following relation is obtained

$$z^3 + z^2[b_1 - k_2N(A)] + z[b_0 - k_1N(A)] - k_0N(A) = 0 \qquad (3.14)$$

whose result is a pair of complex poles and also a real pole.
(3.14) can also be rewritten as

$$(z + p_1)\langle z^2 - 2r\cos\theta z + r^2 \rangle = 0 \qquad (3.15)$$

where p_1 represents the real pole.

The solution is given by the following set of equations

$$r^2 p_1 = -k_0N(A)$$

$$-2p_1 r\cos\theta + r^2 = b_0 - k_1N(A) \qquad (3.16)$$

$$-2r\cos\theta + p_1 = b_1 - k_2N(A)$$

In this case, considering the oscillation conditions (**C2**), we have

$$p_1 = -k_0N(A_{osc})$$

$$-2p_1\cos\theta_{osc} + 1 = b_0 - k_1N(A_{osc}) \qquad (3.17)$$

$$-2\cos\theta_{osc} + p_1 = b_1 - k_2N(A_{osc})$$

Depending on the explicit expression of the non-linear block describing-function, $N(A)$, this set of equations will have one solution or another and consequently, the oscillation conditions which satisfy $sign(A_{osc}) > 0$ and $-1 < \cos\theta_{osc} < 1$ will adopt a particular form. But, independently of the specific $N(A)$, a requirement has to be guaranteed in order to finally attain an equilibrium state: the real pole, p_1, must be placed into the unit circle when the steady-state is reached. That means, then, that we have to impose: $|k_0 N(A_{osc})| < 1$ (**C5**).

3.1.2.3 Type c: Two-delay loop oscillator (n=2)

Let us consider that there are two unit delays in the feedback loop. Expanding (3.7) the following relation is obtained

$$z^4 + z^3 b_1 + z^2 [b_0 - k_2 N(A)] - z[k_1 N(A)] - k_0 N(A) = 0 \qquad (3.18)$$

that can also be expressed as follows

$$(z^2 + az + b)\langle z^2 - 2r\cos\theta z + r^2 \rangle = 0 \qquad (3.19)$$

where

$$br^2 = -k_0 N(A)$$
$$-2ra\cos\theta + r^2 + b = b_0 - k_2 N(A)$$
$$-2r\cos\theta b + ar^2 = -k_1 N(A) \qquad (3.20)$$
$$-2r\cos\theta + a = b_1$$

In this case, considering the oscillation conditions (**C2**), we have

$$b = -k_0 N(A_{osc})$$
$$-2a\cos\theta + 1 + b = b_0 - k_2 N(A_{osc})$$
$$-2\cos\theta b + a = -k_1 N(A_{osc}) \qquad (3.21)$$
$$-2\cos\theta + a = b_1$$

Again, depending on the explicit expression of the non-linear block describing-function, $N(A)$, this set of equations will have one solution or another and the oscillation conditions which satisfy $sign(A_{osc}) > 0$ and $-1 < \cos\theta_{osc} < 1$ will adopt a determined form. But, in addition, a requirement has to be forced in order to attain an equilibrium state: the pole which is not responsible for the oscillations whose expression is $z = -\dfrac{a}{2} \pm \sqrt{a^2 - 4b}$ must be located into the unit circle. That is, the values of $a(A_{osc})$ and $b(A_{osc})$

must be such that their poles associated are into the unit circle. That condition is also determined by the involved $N(A)$.

3.1.3 A simple Non-Linear Block

As can be seen from all the above considerations, it would be convenient to select a specific non-linear feedback element capable of generating robust oscillation for any biquadratic section. Although many non-linear blocks are possible it is important from a practical viewpoint to ease it as much as possible taking into account not only the required hardware for its implementation (area&power) but also its simplicity (robustness) as well as the particular involved analytical expressions obtained from (3.7).

The first issue to be considered when the OBT strategy is applied to a system in general (discrete-time filter or whatever other system) is the circuitry available into the core in order to reuse any part of it to implement the OBT technique. It was one of the main reasons why, in some papers, [26], [36], a low-resolution quantizer was used as the non-linear element to feed the system under test input when it was reconfigured as an oscillator.

For the sake of simplicity, let us consider the non-linear function shown in Fig. 3.5 which implements a saturation function. This non-linear functionality was studied in many preliminary works [25]-[28], [30]-[37]. As was mentioned in previous chapters, this is very important to perfectly characterize the test oscillator and to achieve general expressions giving accurate nominal values for fault-free oscillations. So, the main advantage of using such a nonlinearity is that it allows to simplify the theoretical analysis and easily establish, in most of the cases, closed-form expressions for the estimation of the oscillation features by analytical computations. That means, closed-form oscillation conditions can be effortlessly derived as well as good approximations for the frequency and the amplitude of the resulting oscillation. Moreover, this non-linear block is very straightforward and the additional required hardware for its implementation is very reduced.

In a first approach, the feedback element is formed by an analog comparator providing one of two voltage levels, V_A or V_B to connect or disconnect one of two switches, as depicted in Fig. 3.5-(b). Obviously, more complex versions of this non-linear block can be regarded. However, in this scheme, the nonlinear block can be formally described by a 1-bit ADC followed by a 1-bit DAC, and implemented by an analog comparator and some switches.

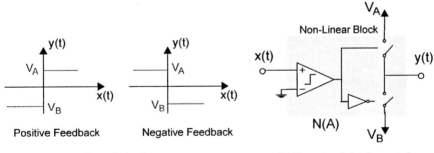

Figure 3.5: A very simple Non-Linear Block

It was shown in Chapter 2 that the describing- function, $N(A)$, for the case of a non-linearity given by a saturation function is

$$N(A) = \frac{2V_{ref}}{\pi A} \qquad (3.22)$$

being $V_{ref} = V_A - V_B$.

An important fact derived of (3.22) is that the describing-function does not only depend on the value of the first harmonic amplitude but also can be controlled by the voltage V_{ref} (whose exact value must be fixed by the designer). So, this voltage can further be exploited as a valuable parameter to select the best set of values for the amplitude and the frequency of the oscillations. That is, V_{ref} is a practical parameter to place the oscillation features in the zone of the space, amplitude versus frequency, where the detection of faults could be more clear. This issue will be studied in detail in next section.

3.1.4 Oscillation Conditions

Once the non-linear block has been selected, $N(A)$ is known, and the oscillation equations can further be developed to obtain practical oscillation conditions. In what follows, these conditions will be derived for the so-called delay-free oscillator (Type a, n=0), using the simple nonlinear block proposed in the previous section.[5]

---

[5] For the other types of oscillator (Type b and Type c), only a symbolic set of equations is found and no important conclusions can be extracted from the general case.

1.- Start-up conditions:

As was seen, initially, if no signal is present ($A = 0$), at least one pole should be out of the unit circle. Note that in this case $N(0) \rightarrow \infty$ and then, the closed-loop system characteristic equation (3.7) is reduced to the equation for $H(z)$ zero locations: $k_2 z^2 + k_1 z + k_0 = 0$.

In view of above, we have that the start-up condition is guaranteed *if and only if* $H(z)$ *has some zeros out of the unit circle.*

2.- Oscillation Frequency:

When the poles are on the unit circle, the system oscillates with a frequency given by (3.12) (but now replacing $\theta_{osc} = 2\pi f_{osc}$)

$$
f_{osc} = \frac{1}{2\pi T_s} \cdot acos\left[\frac{-1}{2} \cdot \frac{b_1 - k_1 N(A_{osc})}{1 - k_2 N(A_{osc})}\right] =
$$

$$
= \frac{1}{2\pi T_s} \cdot acos\left[\frac{1}{2} \cdot \frac{b_1(k_2 - k_0) + k_1(b_0 - 1)}{k_0 - b_0 k_2}\right]
$$

(3.23)

being T_s the sampling period of the discrete-time system.

3.- Oscillation Amplitude:

Once the steady state is reached, the amplitude A_{osc} will be determined by imposing the pole-placements to be just onto the unit circle. That is, substituting A_{osc} into (3.12)

$$
A_{osc} = \frac{2V_{ref}}{\pi} \cdot \frac{k_2 - k_0}{1 - b_0}
$$

(3.24)

Considering also that A_{osc} has to be positive, it can immediately be deduced from (3.24) two new oscillation conditions (supposing, without loss of generality, that $1 > b_0 \geq 0$)

$$
k_2 \neq k_0
$$

$$
sign(V_{ref}) = sign(k_2 - k_0)
$$

(3.25)

(3.25) shows the influence of V_{ref} on the oscillation conditions and the flexibility that this factor can add to our OBT approach.

4.- Amplitude Control Stability:

As was discussed, the amplitude control mechanism imposes

$$\frac{\partial}{\partial A} r(A) < 0 \qquad (3.26)$$

As $r(A) > 0$, the above expression may be further written by

$$\frac{\partial}{\partial A} r^2(A) = 2r(A)\frac{\partial}{\partial A} r(A) < 0 \qquad (3.27)$$

Then, using (3.11) and (3.22), we obtain

$$\frac{\partial}{\partial A} r^2(A) = \frac{\partial}{\partial A}\left[\frac{b_0 - k_0 N(A)}{1 - k_2 N(A)}\right] = \frac{\partial}{\partial N(A)}\left[\frac{b_0 - k_0 N(A)}{1 - k_2 N(A)}\right]\frac{\partial}{\partial A}N(A)$$

$$\frac{\partial}{\partial A} r^2(A) = \frac{-2V_{ref}}{\pi A^2} \cdot \frac{k_2 b_0 - k_0}{[1 - k_2 N(A)]^2} < 0 \qquad (3.28)$$

Consequently

$$V_{ref}(k_2 b_0 - k_0) > 0 \rightarrow sign(V_{ref}) = sign(k_2 b_0 - k_0) \qquad (3.29)$$

This last expression points out an important fact, the factor V_{ref} is vital to carry out the required strategy of establishing the oscillation amplitude. Since k_0, k_2 and b_0 are coefficients fixed by the biquad characteristic, the coefficient V_{ref} has to be modified in order to attain the amplitude control stability. The importance of this parameter, V_{ref}, relies mainly on the case where the existing relationship between the biquad coefficients is not suitable to support oscillations. In this case, an additional limitation element is required in the feedback loop.

Concluding for this particular $N(A)$ function and the Type a structure[6], it can be shown that oscillation conditions are all accomplished if the biquad coefficients fulfil

---

6. Assuming that $1 > b_0 \geq 0$ without loss of generality.

$$k_2 \neq k_0$$

$$sign(V_{ref}) = sign(k_2 - k_0) = sign(k_2 b_0 - k_0) \qquad (3.30)$$

and then the oscillation parameters according to such coefficients will be

$$A_{osc} = \frac{2V_{ref}}{\pi} \cdot \frac{k_2 - k_0}{1 - b_0}$$

$$f_{osc} = \frac{1}{2\pi T_s} \cdot acos\left[\frac{1}{2} \cdot \frac{b_1(k_2 - k_0) + k_1(b_0 - 1)}{k_0 - b_0 k_2} \right] \qquad (3.31)$$

For the general case, it can be seen from (3.31) that the oscillation amplitude only depends on the coefficients k_2, k_0 and b_0 whereas the oscillation
frequency depends on all coefficients. Obviously, at a first glance, it may
seem that the frequency is enough to discriminate any deviation of any coefficient. Accordingly, frequency may serve as the only test parameter to detect
faults and the amplitude may be just a complement. But, observing in detail
(3.31) two considerations can be made:

a) The frequency expression contains all the needed information, whereas
the amplitude expression only cover a part of the needed information (the
information associated with three of the biquad coefficients). Consequently,
the amplitude measurement would be not enough to achieve a high fault coverage. However, to be sure that the frequency measurement would be enough
to obtain a high fault coverage, (3.31) must be carefully examined in order to
study also the sensitivity of the frequency with respect to the variations of
every coefficient. It could occur that though all coefficients appear in the frequency expression, the observability of a fault due to a coefficient deviation
can be insufficient because the sensitivity of the frequency respect to that
deviation is too small. So, an exhaustive analysis of (3.31) leads to advance
that both oscillation parameters are required if a high fault coverage is
pursued.

In view of (3.31), if the case $k_2 = k_1 = 0$ is carefully examined, f_{osc}
becomes independent of the oscillation amplitude, A_{osc}. This is a very illustrative case because reveals a situation where the expressions of the
oscillation frequency and the oscillation amplitude are mathematically disconnected. In this example, at least from a theoretical perspective, both
oscillation parameters are strictly required to cover all the biquad coefficients.

This mathematical result is a proof of that, **both parameters, frequency and amplitude, must be evaluated to obtain a high fault coverage.**

b) On the other hand, another point to be taken into account in (3.31) is the usefulness of the parameter V_{ref} which is only present in the amplitude expression. This parameter helps to adjust the amplitude value. Thus, the amplitude can be chosen to achieve the best testing conditions. This is a way to control the sensitivity of our measurements and, thus, the test quality. Furthermore, (3.31) paves the way for the potential use of distortion measurements, which, in turn, can be controlled by the actual value of V_{ref}.

5.- Amplitude Sensitivities and Frequency Cosine Sensitivities:

Once we have defined the oscillation parameters (frequency and amplitude) in relation to the biquad coefficients, we are in a position to calculate the corresponding sensitivities. Table 3.2 shows the respective expressions as functions of the filter coefficients (k_2, k_1, k_0, b_1, b_0).

From Table 3.2, we can again observe that the oscillation amplitude does not depend on the coefficients k_1 and b_1 of the biquadratic transfer function. Therefore, if these coefficients are present in the biquad structure, we can never use the oscillation amplitude to detect possible deviations (or faults) in them. However, the study of the oscillation amplitude sensitivity with respect to the coefficients k_2 and k_0 reveals that they are correlated and, depending on their actual values, the amplitude value deviations will be higher or lesser. Fig. 3.6-(a), for example, shows the case when $k_0 = 0$. The smaller k_2, the higher the amplitude sensitivity. However, the main result in this sense is displayed in Fig. 3.6-(b) where we can observe that such sensitivity changes depending on the particular value of k_0. Finally, we can also note from Table 3.2 that the value of oscillation amplitude sensitivity with respect to the coefficient b_0 is exclusively subject to the own value of b_0.

On the other hand, from the general expressions of the frequency cosine sensitivities shown in Table 3.2, we can not extract useful conclusions because they involve all the biquad coefficients. To study the importance of the coefficient deviations we must particularize for each specific case. Next sections will deal with this point, but starting from simpler expressions of the sensitivities.

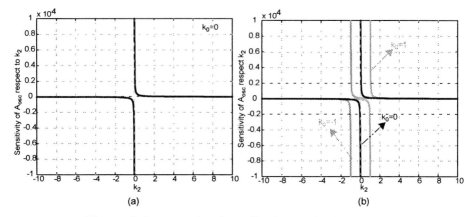

Figure 3.6: Example of amplitude sensitivity

$$S_{p_i}^{A_{osc}} = 100 \cdot \dfrac{\dfrac{dA_{osc}}{dp_i}}{A_{osc}}$$	$$S_{p_i}^{\cos(\theta_{osc})} = 100 \cdot \dfrac{\dfrac{d}{dp_i}\cos(\theta_{osc})}{\cos(\theta_{osc})}$$
$$S_{k_2}^{A_{osc}} = \dfrac{100}{k_2 - k_0}$$	$$S_{k_2}^{\cos(\theta_{osc})} = 100\dfrac{\left[b_1 + \dfrac{[b_1(k_2 - k_0) + k_1(b_0 - 1)]b_0}{k_0 - b_0 k_2}\right]}{b_1(k_2 - k_0) + k_1(b_0 - 1)}$$
$$S_{k_1}^{A_{osc}} = 0$$	$$S_{k_1}^{\cos(\theta_{osc})} = 100\dfrac{[b_0 - 1]}{b_1(k_2 - k_0) + k_1(b_0 - 1)}$$
$$S_{k_0}^{A_{osc}} = -\dfrac{100}{k_2 - k_0}$$	$$S_{k_0}^{\cos(\theta_{osc})} = 100\dfrac{\left[-b_1 - \dfrac{b_1(k_2 - k_0) + k_1(b_0 - 1)}{k_0 - b_0 k_2}\right]}{b_1(k_2 - k_0) + k_1(b_0 - 1)}$$
$$S_{b_0}^{A_{osc}} = \dfrac{100}{1 - b_0}$$	$$S_{b_0}^{\cos(\theta_{osc})} = 100\dfrac{\left[k_1 + \dfrac{[b_1(k_2 - k_0) + k_1(b_0 - 1)]k_2}{k_0 - b_0 k_2}\right]}{b_1(k_2 - k_0) + k_1(b_0 - 1)}$$
$$S_{b_1}^{A_{osc}} = 0$$	$$S_{b_1}^{\cos(\theta_{osc})} = 100\dfrac{[k_2 - k_0]}{b_1(k_2 - k_0) + k_1(b_0 - 1)}$$

Table 3.2: Sensitivities as functions of biquad coefficients

3.2 APPLICATION TO A PARTICULAR BIQUAD STRUCTURE

The OBT methodology considering the Type a (n=0) structure and the suggested $N(A)$ will be applied to a family of active switched capacitor biquads [101], [103] proposed by P.E. Fleischer and K.R. Laker (1979, FL-biquad). They presented two active switched capacitor filter topologies. Each of these circuits comprises two operational amplifiers and at most nine capacitors. Most commonly used transfer functions can be made with any of these topologies.

The general SC FL topology is shown in Fig. 3.7. The circuit consists of two integrators, the first stage being inverting while the second stage is noninverting. Damping is provided by the capacitors E and F. In any particular application, only one of these capacitors need to be present, leaving a total of nine capacitors, but for analysis purposes it is convenient to handle the two cases together.

When the input signal is held over the full clock period, the equivalent circuit given in Fig. 3.7-(b) is obtained for the circuit in Fig. 3.7-(a). By means of this z-domain equivalent circuit, the transfer functions can be drawn out by directly using straightforward nodal analysis

$$H_{o1}(z) = \frac{V_{o1}(z)}{Vin(z)} =$$

$$= \frac{(IC + IE - GF - GB)z^2 + (FH + BH + BG - JC - JE - IE)z + (EJ - BH)}{D(F + B)z^2 + (AC + AE - DF - 2DB)z + (DB - AE)}$$

(3.32)

$$H_{o2}(z) = \frac{V_{o2}(z)}{Vin(z)} =$$

$$= -\frac{DIz^2 + (AG - DI - DJ)z + (DJ - AH)}{D(F + B)z^2 + (AC + AE - DF - 2DB)z + (DB - AE)}$$

Some conflicting degrees of freedom can be eliminated if, it is arbitrarily chosen $B = 1$ and $D = 1$ (each one of the two groups of capacitors (C, D, E, G, H) and (A, B, F, I, J) may be arbitrarily and independently scaled without changing the transfer functions).

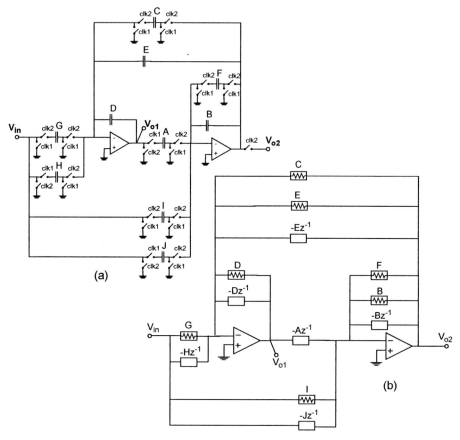

Figure 3.7: General SC FL-topology

In view of the above, $B = D = 1$ is taken and therefore, (3.32) can be reduced to the following simpler expressions

$$H_{o1}(z) = \frac{(IC + IE - GF - G)z^2 + (FH + H + G - JC - JE - IE)z + (EJ - H)}{(F+1)z^2 + (AC + AE - F - 2)z + (1 - AE)}$$

$$H_{o2}(z) = -\frac{Iz^2 + (AG - I - J)z + (J - AH)}{(F+1)z^2 + (AC + AE - F - 2)z + (1 - AE)}$$

(3.33)

These expressions have to be related to the generic transfer function of second-order structures (3.6). The relationships are shown in Table 3.3.

	k_2	k_1	k_0	b_1	b_0
$H_{o1}(z)$	$\dfrac{IC+IE-GF-G}{1+F}$	$\dfrac{FH+H+G-JC-JE-IE}{1+F}$	$\dfrac{EJ-H}{1+F}$	$\dfrac{AC+AE-F-2}{1+F}$	$\dfrac{1-AE}{1+F}$
$H_{o2}(z)$	$\dfrac{-I}{1+F}$	$\dfrac{1+J-AG}{1+F}$	$\dfrac{AH-J}{1+F}$	$\dfrac{AC+AE-F-2}{1+F}$	$\dfrac{1-AE}{1+F}$

Table 3.3: Transfer function coefficients in relation to Fleischer and Laker elements

3.2.1 Properties of the FL-Biquad

3.2.1.1   The E- and F-circuits

Let us introduce, for the sake of simplicity, another representation of the FL-biquad given by Fig. 3.8 where it is possible to see that the two available outputs, V_{o1} and V_{o2}, are always correlated[7].

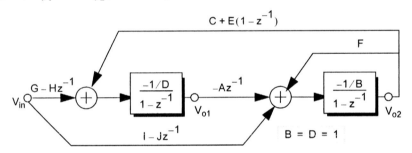

Figure 3.8: FL-biquad implementation

One final simplification that can be made to the general biquad in Fig. 3.8, involves the elements E and F. These elements are redundant in the sense that they both provide damping. It is, therefore, convenient to define an "E-circuit" in which $E \neq 0$ and $F = 0$, and an "F-circuit" in which $F \neq 0$ and $E = 0$.

Accordingly, the transfer functions for these two types of circuits in relation to the elements of the generic biquad structure are summarized in Table 3.4.

---

[7] This detail will be very important hereinafter because this feature will allow us to replace, if needed, an output by the other one to close the required OBT feedback loop.

	k_2	k_1	k_0	b_1	b_0
$H_{o1}(z)\big|_E$	$IC + IE - G$	$H + G - JC - JE - IE$	$EJ - H$	$AC + AE - 2$	$1 - AE$
$H_{o2}(z)\big|_E$	$-I$	$I + J - AG$	$AH - J$	$AC + AE - 2$	$1 - AE$
$H_{o1}(z)\big|_F$	$\dfrac{IC - GF - G}{1 + F}$	$\dfrac{FH + H + G - JC}{1 + F}$	$\dfrac{-H}{1 + F}$	$\dfrac{AC - F - 2}{1 + F}$	$\dfrac{1}{1 + F}$
$H_{o2}(z)\big|_F$	$\dfrac{-I}{1 + F}$	$\dfrac{I + J - AG}{1 + F}$	$\dfrac{AH - J}{1 + F}$	$\dfrac{AC - F - 2}{1 + F}$	$\dfrac{1}{1 + F}$

Table 3.4: Transfer function coefficients for the FL-biquad

3.2.1.2   Pole placement

For any pair of complex conjugate poles in the z-domain, one can write the denominator as

$$D(z) = z^2 + b_1 z + b_0 \qquad (3.34)$$

At this point, it would be appropriate to consider simultaneously the oscillation conditions of the proposed OBT configuration with the stability and the realizability of the proposed E-circuit and F-circuit in order to establish some guidelines and criteria of DfT. It is mandatory to realize all stable pole positions. Stability for a biquad can be conveniently expressed [101], [103] in the b_1, b_0 parameter space as the area within the shaded triangle shown in Fig. 3.9. The upper parabolic area of the triangle (in red) represents the b_1, b_0 values for stable complex poles. On the other hand, the zone under the parabola where $b_0 > 0$, corresponds to the real pole pairs which lie to the left or right of $z = 0$, whereas the lower triangular portion where $b_0 < 0$, corresponds to the real poles which lie on alternate sides of $z = 0$. Observe that, clearly, the upper area of the triangle (where $b_0 > 0$) depicts most of the useful pole locations for frequency selective filters.

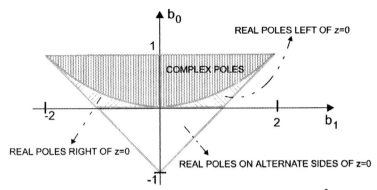

Figure 3.9: Triangle of stable pole positions for $D(z) = z^2 + b_1 z + b_0$

Let us consider firstly the E-circuit realizability properties. From Table 3.4, we have $b_1 = AC + AE - 2$, $b_0 = 1 - AE$ and, thus $b_1 + b_0 = AC - 1$.

Since $A > 0$, $E > 0$ and $C > 0$, it would be $b_1 > -2$; $b_0 < 1$ and $b_1 + b_0 > -1$.

Consequently, the b_1, b_0 values realizable with the E-circuit are confined within the wedge-like area shown in Fig. 3.9. This area includes the whole stable region as well as a portion of the remaining unstable area. E-circuits which are unstable must have real poles.

Similarly, it can be derived the F-circuit realizability conditions from the Table 3.4: $b_1 = \dfrac{AC - F - 2}{1 + F}$; $b_0 = \dfrac{1}{1 + F}$.

Since $A > 0$, $F > 0$ and $C > 0$, it would immediately be $0 < b_0 < 1$ and $b_1 + b_0 > -1$.

Graphically, the regions of the realizable poles are as shown in Fig. 3.10.

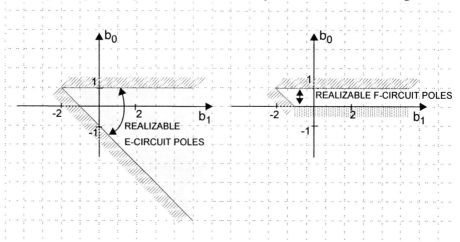

Figure 3.10: Realizable Circuit Poles

3.2.1.3   Zero Placement

Table 3.5 and Table 3.6 list all possible zero placement functions that can be implemented by the FL-structure. They are named, low-pass (LP), high-pass (HP), bandpass (BP), low-pass notch (LPN), high-pass notch (HPN), and all-pass (AP). Such a table will allow us, in next sections, to establish in which FL-structure it is possible to implement the OBT strategy from a theoretical viewpoint. The method is to replace the values of the biquad coefficients (k_2, k_1 and k_0 responsible for the zero placements)[8] into the oscillation conditions given by (3.30) as well as into the expressions of the oscillation parameters (3.31).

[8.]   Note we have replaced in Table 3.5 the biquad coefficients by the simplest solution associated to each possible generic form.

Generic Form	Numerator $N(z)$	$H(z)$	k_2	k_1	k_0
			Simple Solution		
LP 20 (bilinear transform)	$K(1+z^{-1})^2$	$\dfrac{K(z^2+2z+1)}{z^2+b_1 z+b_0}$	K	2K	K
LP11	$Kz^{-1}(1+z^{-1})$	$\dfrac{K(z+1)}{z^2+b_1 z+b_0}$	0	K	K
LP10	$K(1+z^{-1})$	$\dfrac{K(z+1)z}{z^2+b_1 z+b_0}$	K	K	0
LP02	Kz^{-2}	$\dfrac{K}{z^2+b_1 z+b_0}$	0	0	K
LP01	Kz^{-1}	$\dfrac{Kz}{z^2+b_1 z+b_0}$	0	K	0
LP00	K	$\dfrac{Kz^2}{z^2+b_1 z+b_0}$	K	0	0
BP10 (bilinear transform)	$K(1+z^{-1})(1-z^{-1})$	$\dfrac{K(z^2-1)}{z^2+b_1 z+b_0}$	K	0	$-K$

Table 3.5: Generic transfer functions in the z-domain (I)

Generic Form	Numerator $N(z)$	$H(z)$	k_2	k_1	k_0
				Simple Solution	
BP01	$Kz^{-1}(1-z^{-1})$	$\dfrac{K(z-1)}{z^2+b_1z+b_0}$	0	K	$-K$
BP00	$K(1-z^{-1})$	$\dfrac{K(z-1)z}{z^2+b_1z+b_0}$	K	$-K$	0
HP	$K(1-z^{-1})^2$	$\dfrac{K(z^2-2z+1)}{z^2+b_1z+b_0}$	K	$-2K$	K
LPN	$K(1+\varepsilon z^{-1}+z^{-2})$ $\varepsilon > \dfrac{b_1}{\sqrt{b_0}},\ b_0>0$	$\dfrac{K(z^2+\varepsilon z+1)}{z^2+b_1z+b_0}$	K	εK	K
HPN	$K(1+\varepsilon z^{-1}+z^{-2})$ $\varepsilon < \dfrac{b_1}{\sqrt{b_0}},\ b_0>0$	$\dfrac{K(z^2+\varepsilon z+1)}{z^2+b_1z+b_0}$	K	εK	K
AP	$K(\beta+\alpha z^{-1}+z^{-2})$	$\dfrac{K(\beta z^2+\alpha z+1)}{z^2+b_1z+b_0}$	βK	αK	K
GENERAL	$\gamma+\varepsilon z^{-1}+\delta z^{-2}$	$\dfrac{\gamma z^2+\varepsilon z+\delta}{z^2+b_1z+b_0}$	γ	ε	δ

Table 3.6: Generic transfer functions in the z-domain (II)

3.2.1.4 Design Equations

For the sake of clarity, let us summarize in Table 3.7 all the results obtained in the last sections. This Table will be used in next sections to set up some DfT rules.

Pole Placement Formulas				Zero Placement Formulas for H_{oE} and H_{oF}
Realizability Conditions		Synthesis Equations		Depending on the Generic Biquadratic Transfer Function
H_{oE}	H_{oF}	H_{oE}	H_{oF}	
$b_1 > -2$ $b_0 < 1$ $b_1 + b_0 > -1$	$0 < b_0 < 1$ $b_1 + b_0 > -1$	$AE = 1 - b_0$ $AC = 1 + b_0 + b_1$	$F = \dfrac{1 - b_0}{b_0}$ $AC = \dfrac{1 + b_1 + b_0}{b_0}$	(LP20, LP11, LP10, LP02, LP01, LP00, BP10, BP01, BP00, HP, LPN, AP, General) as shown in Table 3.5

Table 3.7: Design Equations for the FL-biquad

3.2.2 Applying the OBT technique to the FL-biquad

The objective herein is to implement, starting from the FL-biquad, the oscillator with amplitude control by limitation shown in Fig. 3.11.

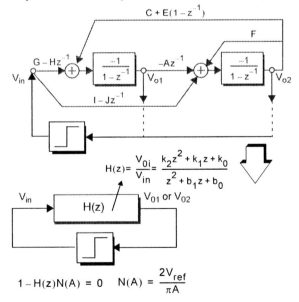

$$H(z) = \frac{V_{oi}}{V_{in}} = \frac{k_2 z^2 + k_1 z + k_0}{z^2 + b_1 z + b_0}$$

$$1 - H(z)N(A) = 0 \qquad N(A) = \frac{2V_{ref}}{\pi A}$$

Figure 3.11: SC Oscillator from the FL-biquad

If we choose the scheme displayed in Fig. 3.11, the oscillation equations will be those in (3.30) and (3.31) with k_2, k_1, k_0, b_1, b_0 given by Table 3.4. Any FL-biquad must verify these oscillation equations to be tested by this OBT technique. For the sake of clarity, let us summarize the expressions in Table 3.8.

Amplitude Oscillation Conditions	Oscillation Amplitude
$k_2 \neq k_0$ $sign(V_{ref}) = sign(k_2 b_0 - k_0) = sign(k_2 - k_0)$	$A_{osc} = \dfrac{2V_{ref}}{\pi} \cdot \dfrac{k_2 - k_0}{1 - b_0}$
Frequency Oscillation Condition	Oscillation Frequency
$-2 < \dfrac{b_1(k_2 - k_0) + k_1(b_0 - 1)}{k_0 - b_0 k_2} < 2$	$f_{osc} = \dfrac{1}{2\pi T_s} acos\left[\dfrac{1}{2} \cdot \dfrac{b_1(k_2 - k_0) + k_1(b_0 - 1)}{k_0 - b_0 k_2}\right]$

Table 3.8: Oscillation equations for the FL-biquad reconfigured as an oscillator (k_2, k_1, k_0, b_1 and b_0 are given in Table 3.4)

We can distinguish from Table 3.8 two kinds of oscillation conditions. One is related to the oscillation amplitude and the other one to the oscillation frequency. Observe that the ***amplitude oscillation conditions*** are ***only*** related to k_2 and k_o (parameters which define the zero placement) whereas the ***frequency oscillation condition*** is related to both the parameters which define the zero placement (k_2, k_1, k_0) and the parameters which define the pole placement (b_0, b_1).

On the other hand, if we particularize for each specific generic second-order function which can be obtained from the FL-structure (Table 3.5), we obtain Table 3.9 containing the particular oscillation conditions for each case as well as whether a start-up strategy is needed or not.

Generic Form	Oscillation Conditions		Zeros H(z) (Start-up)
	Amplitude Oscillation Condition	Frequency Oscillation Condition	
LP 20	$k_2 = k_0 \Rightarrow$ **NO**	------------	------------
LP 11	IFF $\text{sign}(V_{ref}) \neq \text{sign}(K)$	IFF $-1 < \dfrac{b_0 - b_1 - 1}{2} < 1$	$z_1 = -1 \in (r^2 = 1)$ $z_2 = \infty \subset (r^2 > 1)$ NO NEEDED
LP 10	IFF $\text{sign}(V_{ref}) = \text{sign}(K)$	IFF $-1 < \dfrac{-b_0 - b_1 + 1}{2b_0} < 1$	$z_1 = -1 \in (r^2 = 1)$ $z_2 = 0 \subset (r^2 < 1)$ NEEDED
LP 02	IFF $\text{sign}(V_{ref}) \neq \text{sign}(K)$	IFF $-1 < \dfrac{-b_1}{2} < 1$	$z_{1,2} = \infty \subset (r^2 > 1)$ NO NEEDED
LP 01	$k_2 = k_0 \Rightarrow$ **NO**	------------	----------------
LP 00	$\text{sign}(V_{ref}) = \text{sign}(K)$	IFF $-1 < \dfrac{-b_1}{2b_0} < 1$	$z_{1,2} = 0 \subset (r^2 < 1)$ NEEDED
BP 10	$\text{sign}(V_{ref}) = \text{sign}(K)$	IFF $-1 < \dfrac{-b_1}{1 + b_0} < 1$	$z_{1,2} = \pm 1 \in (r^2 = 1)$ NEEDED
BP 01	$\text{sign}(V_{ref}) = \text{sign}(K)$	IFF $-1 < \dfrac{-b_0 - b_1 + 1}{2} < 1$	$z_1 = 1 \in (r^2 = 1)$ $z_2 = \infty \subset (r^2 > 1)$ NO NEEDED
BP 00	$\text{sign}(V_{ref}) = \text{sign}(K)$	IFF $-1 < \dfrac{b_0 - b_1 - 1}{2b_0} < 1$	$z_{1,2} = 1 \in (r^2 = 1)$ $z_2 = 0 \subset (r^2 < 1)$ NEEDED
HP	$k_2 = k_0 \Rightarrow$ **NO**	------------	----------------
LPN	$k_2 = k_0 \Rightarrow$ **NO**	------------	----------------
HPN	$k_2 = k_0 \Rightarrow$ **NO**	------------	----------------
GENERAL	$\text{sign}(V_{ref}) = \text{sign}(\gamma - \delta) =$ $= \text{sign}(\gamma b_0 - \delta)$	IFF $-1 < \dfrac{b_1(\gamma - \delta) + \varepsilon(b_0 - 1)}{2(\delta - b_0 \gamma)} < 1$	IFF $z_{1,2} \subset (r^2 > 1)$

Table 3.9: Structures which can be converted in oscillators with the nonlinear mechanism

Observe from Table 3.9 that there are some functions where it is not possible to implement an oscillator. It is because the *oscillation amplitude condition* is not fulfilled. The remaining functions must additionally satisfy some particular oscillation conditions to perform an oscillator.

Notice as well, that in Table 3.9 the *frequency oscillation condition* has been uncorrelated (following the simple solution given in Table 3.4) from the parameters k_0, k_1 and k_2 responsible for the zero placement.

Summing up, Table 3.10 contains the sign of the feedback loop required to achieve the oscillations for the different types of the biquad.

IFF $\text{sign}(V_{ref}) = \text{sign}(\gamma - \delta) = \text{sign}(\gamma b_0 - \delta)$	IFF $\text{sign}(V_{ref}) \neq \text{sign}(K)$	IFF $\text{sign}(V_{ref}) = \text{sign}(K)$
GENERAL	LP11	LP10
	LP02	LP00
		BP10
		BP01
		BP00

Table 3.10: Feedback Sign for the types of biquads convertible in an oscillator

Finally, the oscillation parameters and the oscillation conditions for each type of biquad configuration convertible in an oscillator are shown in Table 3.11.

Generic Form	Oscillation Parameters	Oscillation Conditions
LP11	$$A_{osc} = \frac{2V_{ref}}{\pi} \cdot \frac{-K}{1-b_0}$$ $$f_{osc} = \frac{1}{2\pi T_s} acos\left[\frac{b_0 - b_1 - 1}{2}\right]$$	$sign(V_{ref}) \neq sign(K)$ $$-1 < \frac{b_0 - b_1 - 1}{2} < 1$$
LP10	$$A_{osc} = \frac{2V_{ref}}{\pi} \cdot \frac{K}{1-b_0}$$ $$f_{osc} = \frac{1}{2\pi T_s} acos\left[\frac{-b_0 - b_1 + 1}{2b_0}\right]$$	$sign(V_{ref}) = sign(K)$ $$-1 < \frac{-b_0 - b_1 + 1}{2b_0} < 1$$
LP02	$$A_0 = \frac{2V_{ref}}{\pi} \cdot \frac{-K}{1-b_0}$$ $$f_{osc} = \frac{1}{2\pi T_s} acos\left[\frac{-b_1}{2}\right]$$	$sign(V_{ref}) \neq sign(K)$ $$-1 < \frac{-b_1}{2} < 1$$
LP00	$$A_0 = \frac{2V_{ref}}{\pi} \cdot \frac{K}{1-b_0}$$ $$f_{osc} = \frac{1}{2\pi T_s} acos\left[\frac{-b_1}{2b_0}\right]$$	$sign(V_{ref}) = sign(K)$ $$-1 < \frac{-b_1}{2b_0} < 1$$
BP10	$$A_0 = \frac{4V_{ref}}{\pi} \cdot \frac{K}{1-b_0}$$ $$f_{osc} = \frac{1}{2\pi T_s} acos\left[\frac{-b_1}{1+b_0}\right]$$	$sign(V_{ref}) = sign(K)$ $$-1 < \frac{-b_1}{1+b_0} < 1$$
BP01	$$A_0 = \frac{2V_{ref}}{\pi} \cdot \frac{K}{1-b_0}$$ $$f_{osc} = \frac{1}{2\pi T_s} acos\left[\frac{-b_1 - b_0 + 1}{2}\right]$$	$sign(V_{ref}) = sign(K)$ $$-1 < \frac{-b_0 - b_1 + 1}{2} < 1$$
BP00	$$A_0 = \frac{2V_{ref}}{\pi} \cdot \frac{K}{1-b_0}$$ $$f_{osc} = \frac{1}{2\pi T_s} acos\left[\frac{b_0 - b_1 - 1}{2b_0}\right]$$	$sign(V_{ref}) = sign(K)$ $$-1 < \frac{b_0 - b_1 - 1}{2b_0} < 1$$
GENERAL	$$A_0 = \frac{2V_{ref}}{\pi} \cdot \frac{\gamma - \delta}{1-b_0}$$ $$f_{osc} = \frac{1}{2\pi T_s} acos\left[\frac{1}{2} p \frac{b_1(\gamma - \delta) + \varepsilon(b_0 - 1)}{\delta - b_0\gamma}\right]$$	$sign(V_{rf}) = sign(\gamma - \delta) = sign(\gamma b_0 - \delta)$ $$-1 < \frac{b_1(\gamma - \delta) + \varepsilon(b_0 - 1)}{2(\delta - b_0\gamma)} < 1$$

Table 3.11: Oscillation equations for the types of biquads convertible in an oscillator

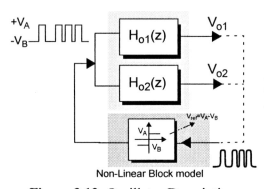

Non-Linear Block model

Figure 3.12: Oscillator Description

It can be deduced from Table 3.9 that the proposed OBT non-linear technique is valid in principle for all cases except for the LP20, LP01, HP and NOTCH ones. However, it does not mean that the proposed non-linear block can not be applied in such cases. To see this, it has to be considered the real implementation of the involved oscillators. Since the FL-circuit has two workable outputs that can be used interchangeably depending on the transfer function, the general model for forcing oscillations is given in Fig. 3.12, where any of the biquad outputs is assumed to be connected once to the nonlinear feedback block.

Table 3.12 shows the existing correlation between V_{o1} and V_{o2} depending on the implemented function for the cases LP20, LP01, HP and NOTCH where oscillations can not be achieved. Fortunately, as can be seen, always one of the two available outputs belong to the valid set of functions. So it demonstrates that the oscillator can normally be built just selecting properly the output used as the input of the proposed non-linear block.

E-TYPE		F-TYPE	
Vo1	Vo2	Vo1	Vo2
LP20	GENERAL	LP20	BP01
GENERAL	LP20	GENERAL	LP20
LP01	GENERAL	LP01	BP01
BP00	LP01	GENERAL	LP01
HP	BP01	HP	BP01
GENERAL	HP	GENERAL	HP
NOTCH	GENERAL	NOTCH	GENERAL
GENERAL	NOTCH	GENERAL	NOTCH

Table 3.12: Correlation between Vo1 and Vo2

3.2.2.1   Regions of interest in the plane b_0, b_1

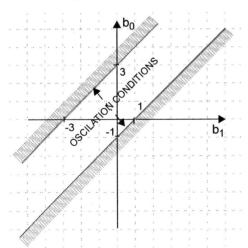

Figure 3.13: b_0, b_1 region satisfying the frequency oscillation condition (case LP11)

On the other hand, let us study in detail Table 3.8 in order to define the different b_0, b_1 regions associated with the FL topologies where it is feasible to apply the OBT strategy (obviously, we refer exclusively to the types LP11, LP10, LP02, LP00, BP10, BP01 and BP00). We have to consider separately each case. If, for example, we bear in mind the case LP11, we can specify the region for b_0 and b_1 that fulfils the *frequency oscillation condition*[9] by simply solving these corresponding implicit inequalities. Graphically this region is as displayed in Fig. 3.13.

We can combine all requirements in order to compose the acceptable region satisfying both the realizability conditions as well as the frequency oscillation condition. Then, a set of graphics are obtained as shown in Fig. 3.14.

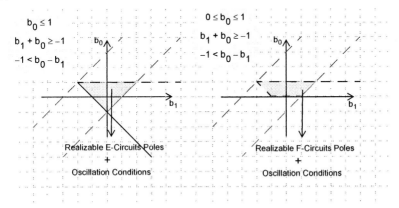

Figure 3.14: b_0, b_1 regions satisfying both the frequency oscillation conditions and the realizable pole conditions (case LP11)

---

9. $-1 < \dfrac{b_0 - b_1 - 1}{2} < 1$

In the same way, we can define graphically all b_0, b_1 regions satisfying the frequency oscillation conditions for every case. The results are displayed in Fig. 3.15.

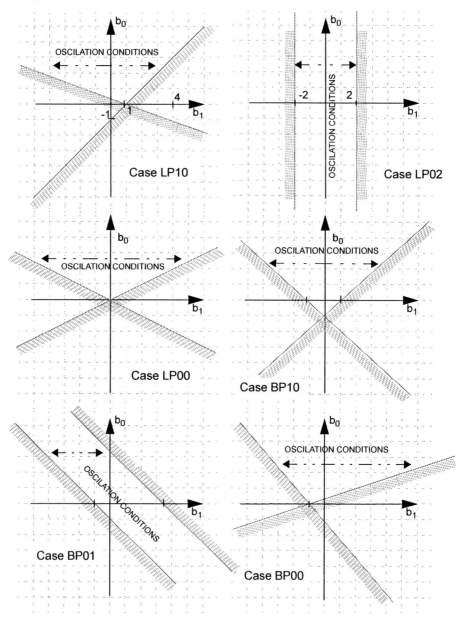

Figure 3.15: b_0, b_1 regions satisfying the frequency oscillation conditions

In summary, all regions where oscillations are possible from a mathematical point of view are represented in Fig. 3.16 and Fig. 3.17.

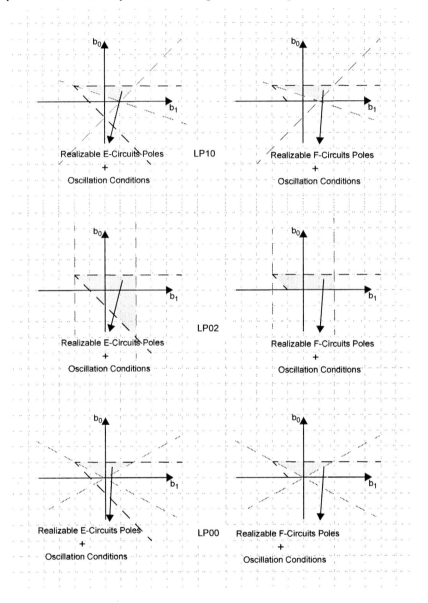

Figure 3.16: b_0 and b_1 regions satisfying both the frequency oscillation conditions and the realizable pole conditions

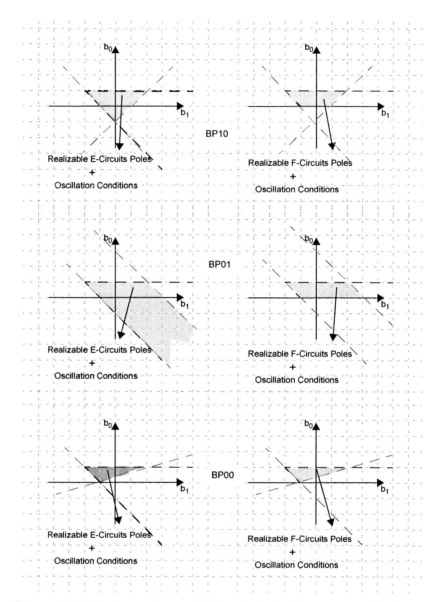

Figure 3.17: b_0, b_1 regions satisfying both the frequency oscillation
conditions and the realizable pole conditions

We can gather in Table 3.13 the results of all these previous graphics.

Generic Form	E_CIRCUIT	F_CIRCUIT
LP11	$b_0 < 1$ $b_1 + b_0 > -1$ $-1 < b_0 - b_1$	$0 < b_0 < 1$ $b_1 + b_0 > -1$ $-1 < b_0 - b_1$
LP10	$b_0 < 1$ $b_1 - b_0 < 1$ $3b_0 + b_1 > 1$	
LP02	$b_0 < 1$ $b_1 + b_0 > -1$ $b_1 < 2$	$0 < b_0 < 1$ $b_1 + b_0 > -1$ $b_1 < 2$
LP00	$b_0 < 1$ $2b_0 - b_1 > 0$ $2b_0 + b_1 > 0$	
BP10	$b_0 < 1$ $b_1 + b_0 > -1$ $b_0 - b_1 > 1$	$0 < b_0 < 1$ $b_1 + b_0 > -1$ $b_0 - b_1 > 1$
BP01	$b_0 < 1$ $-1 < b_1 + b_0 < 3$	$0 < b_0 < 1$ $-1 < b_1 + b_0 < 3$
BP00	$b_0 < 1$ $b_1 + b_0 > -1$ $3b_0 - b_1 > 1$	

Table 3.13: b_0, b_1 regions of interest

Comparing all the b_0, b_1 regions, we can conclude that the BP01 configuration has a b_0, b_1 region (where both, the *realizable pole condition* and the *frequency oscillation condition*, are fulfilled) much larger than the rest of FL-topologies. Accordingly, since b_0 and b_1 depends on capacitors A, C, E, F (responsible for the pole placement), this BP01configuration makes it possible to expand the range of values of the A, C, E, F parameters where we can apply the OBT strategy.

But, obviously, we have also to take into account the elements G, H, I, J (the elements which define the *oscillation amplitude condition*) to determine which FL-topologies we are dealt with and whether the OBT method can be applied to it.

3.2.2.2   OBT routine

In this point, we are in a position to establish a set of steps which, one designer interested in applying OBT to a particular biquad, can use as a sort of guide. The steps are the following:

-**Step_1**: Determining the biquad type for V_{o1} and V_{o2}. If one of the outputs belongs to the groups HP, AP, LPN, HPN, LP01 or LP20, you must immediately reject it to close the corresponding OBT feedback loop (see Table 3.9). But, as was seen in Table 3.12, the other output will belong to the group LP11, LP10, LP02, LP00, BP00, BP10, BP01 or GENERAL where, otherwise, it can be feasible to close the OBT feedback loop (see Table 3.9).

-**Step_2**: Using Table 3.13 to decide if oscillations are really possible for such a particular case. The way is to carefully check the involved b_0, b_1 region (where both the realizable pole condition and the frequency oscillation condition are fulfilled).

-**Step_3**: Using Table 3.11 to determine the oscillation parameters (amplitude, A_{osc}, and frequency, f_{osc}) by substituting the specific values of the biquad coefficients (k_0, k_1, k_2) in the corresponding expressions.

There is, however, an underlying question: *what happen if the chosen biquad does not oscillate using either of its outputs?*. In this case, there are two possible paths to follow. One is focused on employing extra delays in the feedback block (*type b* or *type c*) as was said above. Then, the oscillation expressions become more complicated and there is not a general way to proceed as well (that is, no general tables can be deduced). But, in this case, however, you can solve the oscillation equations by replacing the particular biquad elements. Some examples will be studied in next chapters. The other way is to slightly transform, during the test mode, the internal biquad structure, building a modified biquad structure where it is feasible to achieve the required oscillations (as shown in Fig. 3.18). Therefore, the goal in the next section will be to define the aforementioned modified biquad structure which allows us to design a generic oscillator scheme useful for OBT.

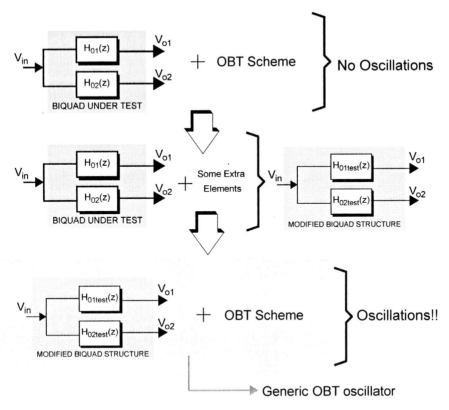

Figure 3.18: Means to satisfactorily apply OBT

3.3 A GENERIC OBT OSCILLATOR

In the last section, we set up the possibility of finding a general oscillator scheme to apply OBT, starting from any generic biquad structure. The idea in this Section is to study every specific FL-biquad in order to define such a generic oscillator. Several requirements have to be considered:

1.- The number of involved members of the generic oscillator structure has to be high (the objective is to include in the generic OBT biquad structure the maximum number of FL-topologies).

2.- The involved b_0, b_1 region (zone satisfying both the realizability conditions as well as the frequency oscillation condition) has to be large enough (to be sure to accommodate a wide range of applications).

3.- A good fault coverage is required. It means that the sensitivities of the oscillation parameters in relation to the biquad elements must allow it.

4.- It is convenient that our linearized model (DF approach, see Chapter 2) can be exact and/or at least "acceptable".

For the sake of clarity, some other conflicting degrees of freedom can also be eliminated in the FL-structure if it is arbitrarily assumed $A = 1$ (it may be shown that the net effect of this choice is to remove our ability to control the gain constants associated with H_{01} and H_{02} simultaneously). Then, Table 3.4 is automatically converted into Table 3.14.

	k_2	k_1	k_0	b_1	b_0
$H_{01}(z)\vert_E$	$IC + IE - G$	$H + G - JC - JE - IE$	$EJ - H$	$C + E - 2$	$1 - E$
$H_{02}(z)\vert_E$	$-I$	$I + J - G$	$H - J$		
$H_{01}(z)\vert_F$	$\dfrac{IC - GF - G}{1 + F}$	$\dfrac{FH + H + G - JC}{1 + F}$	$\dfrac{-H}{1 + F}$	$\dfrac{C - F - 2}{1 + F}$	$\dfrac{1}{1 + F}$
$H_{02}(z)\vert_F$	$\dfrac{-I}{1 + F}$	$\dfrac{I + J - G}{1 + F}$	$\dfrac{H - J}{1 + F}$		

Table 3.14: Transfer function coefficients

In general terms, the proposed OBT structure for the FL biquad is as shown in Fig. 3.19. Observe that any hypothetical fault in the elements (G, H, I, J, C, E, F) (design elements of the FL-device) has to be covered. Therefore, the oscillation parameters have to be defined as a function of such coefficients in order to determine not only if a certain fault can be detected but also the sensitivity of oscillations to such a fault.

To extract reasonable conclusions, let us study, from this point forward, the proposed OBT implementation for every FL-topology.

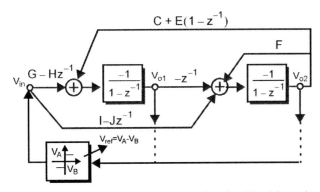

Figure 3.19: OBT structure for the FL- biquad

3.3.1 Conclusions extracted by the simplified results

Let us, firstly, obtain the simplified versions of the oscillation parameters for every FL-topology (Table 3.15). In these simplified expressions the involved biquad elements, (G, H, I, J, C, E, F), are reduced to four: $|K|$, C, E and F for all the cases. As can be seen from Table 3.15, the oscillation amplitude only depends on $|K|$, and E or F whereas the oscillation frequency only depends on C, and E or F. Therefore, both oscillation parameters will be necessary to obtain a good fault coverage.

From a simplified viewpoint, Table 3.15 shows the simplest expressions of the oscillation parameters for all the FL-biquads of interest. The simplest form of the zero placement for every FL-configuration has been employed (see Table 3.5). However, these expressions are exclusively valid when the elements responsible for the zero placement (I, J, G and H) are built in such a way that their values are correlated. That basically means that a deviation or fault in one of them is perceived by all them in the same direction and magnitude. Then and only then, we can assert that these simplified oscillation parameters can be employed to extract conclusions. And then and only then, Table 3.15 can be considered as valuable.

Generic Form	Oscillation Parameters			
	E-Circuit		F-Circuit	
	Vo2	Vo1	Vo2	Vo1
LP11	$A_{osc} = \frac{2\left|V_{ref}\right|}{\pi} \cdot \frac{\left|K\right|}{E}$		$A_{osc} = \frac{2\left|V_{ref}\right|}{\pi} \cdot \frac{(1+F)\left|K\right|}{F}$	$A_{osc} = \frac{2\left|V_{ref}\right|}{\pi} \cdot \frac{\left|K\right|}{F}$
	$f_{osc} = \frac{1}{2\pi T_s} acos\left[\frac{2-C-2E}{2}\right]$		$f_{osc} = \frac{1}{2\pi T_s} acos\left[\frac{2-C}{2(F+1)}\right]$	
LP10	$A_{osc} = \frac{2\left|V_{ref}\right|}{\pi} \cdot \frac{\left|K\right|}{E}$		$A_{osc} = \frac{2\left|V_{ref}\right|}{\pi} \cdot \frac{(1+F)\left|K\right|}{F}$	$A_{osc} = \frac{2\left|V_{ref}\right|}{\pi} \cdot \frac{\left|K\right|}{F}$
	$f_{osc} = \frac{1}{2\pi T_s} acos\left[\frac{2-C}{2-2E}\right]$		$f_{osc} = \frac{1}{2\pi T_s} acos\left[\frac{2-C+2F}{2}\right]$	
LP02	$A_{osc} = \frac{2\left|V_{ref}\right|}{\pi} \cdot \frac{\left|K\right|}{E}$		$A_{osc} = \frac{2\left|V_{ref}\right|}{\pi} \cdot \frac{(1+F)\left|K\right|}{F}$	$A_{osc} = \frac{2\left|V_{ref}\right|}{\pi} \cdot \frac{\left|K\right|}{F}$
	$f_{osc} = \frac{1}{2\pi T_s} acos\left[\frac{2-C-E}{2}\right]$		$f_{osc} = \frac{1}{2\pi T_s} acos\left[\frac{2-C+F}{2(F+1)}\right]$	
LP00	$A_{osc} = \frac{2\left|V_{ref}\right|}{\pi} \cdot \frac{\left|K\right|}{E}$		$A_{osc} = \frac{2\left|V_{ref}\right|}{\pi} \cdot \frac{(1+F)\left|K\right|}{F}$	$A_{osc} = \frac{2\left|V_{ref}\right|}{\pi} \cdot \frac{\left|K\right|}{F}$
	$f_{osc} = \frac{1}{2\pi T_s} acos\left[\frac{2-C-E}{2(1-E)}\right]$		$f_{osc} = \frac{1}{2\pi T_s} acos\left[\frac{2-C+F}{2}\right]$	
BP10	$A_{osc} = \frac{2\left|V_{ref}\right|}{\pi} \cdot \frac{2\left|K\right|}{E}$		$A_{osc} = \frac{2\left|V_{ref}\right|}{\pi} \cdot \frac{2\left|K\right|}{F}$	
	$f_{osc} = \frac{1}{2\pi T_s} acos\left[\frac{2-C-E}{2-E}\right]$		$f_{osc} = \frac{1}{2\pi T_s} acos\left[\frac{2-C+F}{F+2}\right]$	
BP01	$A_{osc} = \frac{2\left|V_{ref}\right|}{\pi} \cdot \frac{\left|K\right|}{E}$		$A_{osc} = \frac{2\left|V_{ref}\right|}{\pi} \cdot \frac{(1+F)\left|K\right|}{F}$	$A_{osc} = \frac{2\left|V_{ref}\right|}{\pi} \cdot \frac{\left|K\right|}{F}$
	$f_{osc} = \frac{1}{2\pi T_s} acos\left[\frac{4-C-2E}{2}\right]$		$f_{osc} = \frac{1}{2\pi T_s} acos\left[\frac{4-C+2F}{2(F+1)}\right]$	
BP00	$A_{osc} = \frac{2\left|V_{ref}\right|}{\pi} \cdot \frac{\left|K\right|}{E}$		$A_{osc} = \frac{2\left|V_{ref}\right|}{\pi} \cdot \frac{(1+F)\left|K\right|}{F}$	$A_{osc} = \frac{2\left|V_{ref}\right|}{\pi} \cdot \frac{\left|K\right|}{F}$
	$f_{osc} = \frac{1}{2\pi T_s} acos\left[\frac{2-C-2E}{2(1-E)}\right]$		$f_{osc} = \frac{1}{2\pi T_s} acos\left[\frac{2-C}{2}\right]$	

Table 3.15: Simplified oscillation parameters for each kind of oscillator

From this last table, there would be no difference between implementing one or other biquad structure. The resulting oscillators seem equivalent. The oscillation parameters depend on the involved elements (K, C, E and F) in a similar way. However, with the simplified expressions of the oscillation parameters shown in Table 3.15 a lot of basic information is hidden. In fact, we have to study each FL structure but for a more general zero placement solution and then establish which structure is preferable to be used within the generic OBT scheme.

3.3.2 Conclusions extracted by the no-simplified results

One of the key points which must satisfy a generic OBT scheme is that it has to involve all the particular FL structures or, at least, the maximum number of them. From Table 3.16, (table including all the elements required in each kind of FL-biquad), we can deduce that the more general case is the type BP10. If we use the second output to build an oscillator we can test the elements I, G, H, C, E or F whereas if we use the first output we can test the elements I, J, G, C, E or F. Therefore, the BP10 structure seems to be the best candidate to be employed within the generic oscillator scheme.

Generic Form	Involved Elements with the more general solution (Seven Elements: I, J, G, H, C, E, F)			
	E-Circuit		F-Circuit	
	Vo2	Vo1	Vo2	Vo1
LP11	J, G, C, E	I, J, G, C, E	J, G, C, F	$\hat{J}, \hat{H}, C, F$
	(4 Elements)	(5 Elements)	(4Elements)	(4 Elements)
LP10	I, G, C, E	I, G, C, E	I, G, C, F	$\hat{J}, \hat{G}, C, F$
	(4 Elements)	(4 Elements)	(4 Elements)	(4 Elements)
LP02	H, C, E	I, J, G, C, E	H, C, F	$\hat{J}, \hat{H}, C, F$
	(3 Elements)	(5 Elements)	(3 Elements)	(4 Elements)
LP00	I, G, C, E	I, G, C, E	I, G, C, F	$\hat{I}, C, F$
	(4 Elements)	(4 Elements)	(4 Elements)	(3 Elements)
BP10	I, G, H, C, E	I, J, G, C, E	I, G, H, C, F	$\hat{I}, \hat{J}, \hat{H}, C, F$
	(5 Elements)	(5 Elements)	(5 Elements)	(5 Elements)
BP01	J, C, E	H, C, E	J, C, F	$\hat{J}, \hat{H}, C, F$
	(3 Elements)	(3 Elements)	(3 Elements)	(4 Elements)
BP00	I, C, E	G, C, E	I, C, F	$\hat{I}, \hat{J}, C, F$
	(3 Elements)	(3 Elements)	(3 Elements)	(4 Elements)

Table 3.16: Number of elements involved in each type of oscillator

3.3.3 Selected Generic Oscillator: Case BP10

As demonstrated, the biquadratic type called BP10 allows us to cover the maximum number of FL structures. And even more, it must be clear from Table 3.16, that the two potential oscillators obtained from this configuration allow us to test all the possible involved elements.

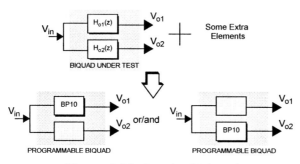

Figure 3.20: Standard Biquad

The idea is, starting from the biquad under test and always adding (never taking away) the corresponding extra elements, to build a programmable biquad (type BP10). Then, the biquad can be converted in one of the two possible oscillators in the test phase (Fig. 3.20). One or the other output will be employed depending on the existing elements in the original biquad under test. If I, G, H, C, E or F are involved, then, we must employ the second output (see Table 3.16). On the contrary, if I, J, G, C, E or F are involved, then, we must use the first output. Obviously, this strategy supposes an additional effort of design. However, in next section, we will give some guidelines to make this work easier.

3.3.4 Guidelines to implement a generic OBT scheme

3.3.4.1 Conclusions related to K, b_0 and b_1

Observe again Table 3.11 where the simplified oscillation equations for every FL-structure are shown. Let us pay attention in the expressions of the BP10 configuration. These expressions allowed us to define the acceptable b_0, b_1 region satisfying both the oscillation frequency condition and the realizable pole condition. The acceptable regions are related to the elements A, C, E, F and, due to the nature of the case BP10, such regions include a wide range of values for such elements. In fact, the first requirement when designing the generic BP10 oscillator is that the particular values of A, C, E, F make b_0, b_1 values belong to the acceptable region.

3.3.4.2 Conclusions related to the zero placement formulas (I,J,G,H)

Table 3.17 shows again the zero placement formulas in the case BP10. Keeping this table in mind, we will study each particular point in order to build the adequate OBT oscillator.

Transfer Function	H_{o2E} H_{o2F}	H_{o1E}	H_{o1F}
Zero Placement Formulas	$I = \lvert K \rvert$ $G - I - J = 0$ $J - H = -\lvert K \rvert$	$I(E + C) - G = \pm\lvert K \rvert$ $H + G - IE - J(E + C) = 0$ $EJ - H = \mp\lvert K \rvert$	$\hat{G}F + \hat{G} - \hat{I}\hat{C} = -\lvert K \rvert(1 + F)$ $\hat{J}\hat{C} - F\hat{H} - \hat{H} - \hat{G} = 0$ $\hat{H} = \lvert K \rvert(1 + F)$
Simple Solution	$I = \lvert K \rvert$ $J = 0$ $G = \lvert K \rvert$ $H = \lvert K \rvert$	$I = \dfrac{\lvert K \rvert}{E}$ $J = \dfrac{\lvert K \rvert}{E}$ $G = \dfrac{\lvert K \rvert(2E + C)}{E}$ $H = 0$	$\hat{I} = \dfrac{\lvert K \rvert(1 + F)}{C}$ $\hat{J} = \dfrac{\lvert K \rvert(1 + F)^2}{C}$ $\hat{G} = 0$ $\hat{H} = \lvert K \rvert(1 + F)$

Table 3.17: General Design Equations for BP10

Four oscillators are derived from the case BP10 (Table 3.17). Only two branches must be adapted to transform the original biquad under test in the BP10 structure. These branches are displayed in dotted lines in Fig. 3.21. On the other hand, the E and C elements would have to be also modified when their associated b_0, b_1 regions do not satisfy the required conditions.

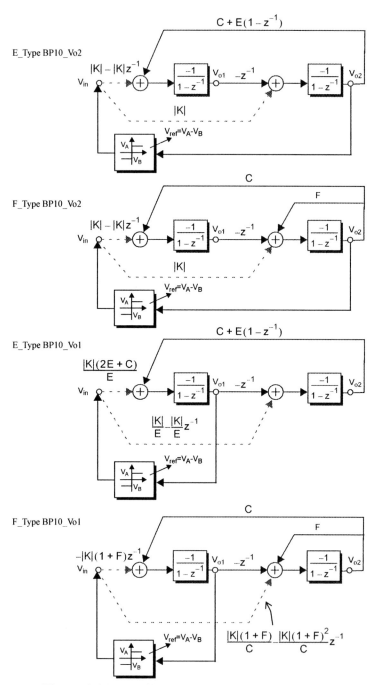

Figure 3.21: Possible oscillators from the case BP10

However, we have to consider the generic form BP10 (Table 3.18) to obtain no-simplified oscillation parameters where a higher number of elements are involved and more complex terms are obtained containing a more exhaustive test information.

H_{o2E}	H_{o2F}	H_{o1E}	H_{o1F}
BP10_Vo2		BP10_Vo1	
$k_2 = -I$ $k_1 = I - G$ $k_0 = H$ $b_1 = C + E - 2$ $b_0 = 1 - E$	$k_2 = \dfrac{-I}{1+F}$ $k_1 = \dfrac{I-G}{1+F}$ $k_0 = \dfrac{H}{1+F}$ $b_1 = \dfrac{C-F-2}{F+1}$ $b_0 = \dfrac{1}{F+1}$	$k_2 = IC + IE - G$ $k_1 = G - JC - JE - IE$ $k_0 = EJ - H$ $b_1 = C + E - 2$ $b_0 = 1 - E$	$k_2 = \dfrac{IC}{1+F}$ $k_1 = \dfrac{FH + H - JC}{1+F}$ $k_0 = \dfrac{-H}{1+F}$ $b_1 = \dfrac{C-F-2}{F+1}$ $b_0 = \dfrac{1}{F+1}$

Table 3.18: Generic Form BP10. No-simplified Versions

Let us compare then the no-simplified oscillation parameters with the simplified ones (Table 3.19 and Table 3.20, case BP10_Vo2 for the sake of clarity).

Generic Form	Oscillation Parameters
	E-Circuit
	Vo2
BP10 Simplified	$A_{osc} = \dfrac{2V_{ref}}{\pi} \cdot \dfrac{-2|K|}{E}$
	$f_{osc} = \dfrac{1}{2\pi T_s} acos\left[\dfrac{-(C + E - 2)}{2 - E}\right]$
BP10 No Simplified	$A_{osc} = \dfrac{-2V_{ref}}{\pi} \cdot \dfrac{1+H}{E}$
	$f_{osc} = \dfrac{1}{2\pi T_s} acos\left[\dfrac{I(2 - C - 2E) - H(C + E - 2) + GE}{2(H + I(1 - E))}\right]$

Table 3.19: Oscillation Equations (E-Circuit)

Generic Form	Oscillation Parameters		
	F-Circuit		
	Vo2		
BP10 Simplified	$A_{osc} = \dfrac{2V_{ref}}{\pi} \cdot \dfrac{-2\lvert K \rvert}{F}$
	$f_{osc} = \dfrac{1}{2\pi T_s}\, acos\left[\dfrac{-(C-F-2)}{2+F}\right]$
BP10 No Simplified	$A_{osc} = \dfrac{-2V_{ref}}{\pi} \cdot \dfrac{I+H}{F}$
	$f_{osc} = \dfrac{1}{2\pi T_s}\, acos\left[\dfrac{I(2-C)-H(C-F-2)+GF}{2(H(1+F)+I)}\right]$

Table 3.20: Oscillation Equations (F_Circuit)

Notice that when I, H and G (which have to be designed as $\lvert K \rvert$ in order to carry out the BP10 biquad, see Table 3.17) are actually implemented in such a way that a deviation of ΔK in, for example, the I element, is observed by H and G in the same manner (that means, $H = \lvert K \rvert + \Delta K$ and similarly $G = \lvert K \rvert + \Delta K$). Then, the oscillation frequency will be dissociated from these elements and only will be useful to detect faults in C and E. In this case the oscillation amplitude will cover the coefficient $\lvert K \rvert$ (apart from E as predicted the simplified expressions given in Table 3.19).

Note, however, that if no-simplified expressions for the oscillation parameters are studied, different considerations can be made. In short, if deviations in one of the elements (I, H or G) are uncorrelated with deviations in the rest of the elements, then, the role of the oscillation frequency is crucial to identify a fault in I, H or G because in this case, the expression of the oscillation frequency contains all the biquad elements.

If no-simplified expressions of BP10_Vo2 are considered, then the sensitivities also change (see Table 3.21 and Table 3.22).

E-Circuit (%)	
$S_I^{A_{osc}} = \dfrac{100}{I+H}$	$S_I^{\cos(\Theta_{osc})} = 100\dfrac{E[H(C+E-1)+G(1-E)]}{[(C-2+2E)I+H(C+E-2)-GE][H+I(1-E)]}$
$S_H^{A_{osc}} = \dfrac{100}{I+H}$	$S_H^{\cos(\Theta_{osc})} = -100\dfrac{EI(C+E-1)-GE}{[(C-2+2E)I+H(C+E-2)-GE][H+I(1-E)]}$
$S_E^{A_{osc}} = -\dfrac{100}{E}$	$S_E^{\cos(\Theta_{osc})} = 100\dfrac{H(I+H-G+IC)+I(IC-G)}{[(C-2+2E)I+H(C+E-2)-GE][H+I(1-E)]}$
$S_G^{A_{osc}} = 0$	$S_G^{\cos(\Theta_{osc})} = 100\dfrac{E}{[(C-2+2E)I+H(C+E-2)-GE]}$
$S_C^{A_{osc}} = 0$	$S_C^{\cos(\Theta_{osc})} = 100\dfrac{I+H}{[(C-2+2E)I+H(C+E-2)-GE]}$

Table 3.21: Sensitivities in frequency and amplitude (case BP10_Vo2)

F-Circuit (%)	
$S_I^{A_{osc}} = \dfrac{100}{I+H}$	$S_I^{\cos(\Theta_{osc})} = 100\dfrac{F[HC-H+G]}{[(C-2)I+H(C-F-2)-GF][H(1+F)+I]}$
$S_H^{A_{osc}} = \dfrac{100}{I+H}$	$S_H^{\cos(\Theta_{osc})} = -100\dfrac{F[I(C-1)-G(1+F)]}{[(C-2)I+H(C-F-2)-GF][H(1+F)+I]}$
$S_F^{A_{osc}} = -\dfrac{100}{F}$	$S_F^{\cos(\Theta_{osc})} = -100\dfrac{H[-H+CH-I+G+CI]+GI}{[(C-2)I+H(C-F-2)-GF][H(1+F)+I]}$
$S_G^{A_{osc}} = 0$	$S_G^{\cos(\Theta_{osc})} = -100\dfrac{F}{[(C-2)I+H(C-F-2)-GF]}$
$S_C^{A_{osc}} = 0$	$S_C^{\cos(\Theta_{osc})} = 100\dfrac{I+H}{[(C-2)I+H(C-F-2)-GF]}$

Table 3.22: Sensitivities in frequency and amplitude (case BP10_Vo2)

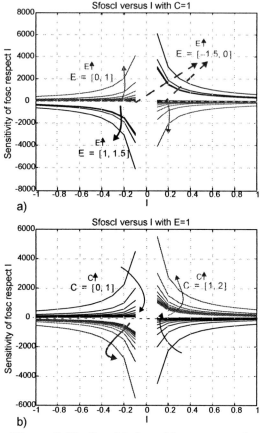

a)

b)

Figure 3.22: Sensitivity with respect to I but varying C and E

As an example suppose the case $S_I^{\cos(\Theta_{osc})}$. This sensitivity depends not only on the coefficients E and C but also on the rest of the coefficients I, H and G in such a way that the graphics in Fig. 3.22 can be drawn. In these graphics it is reflected that when I changes its value (for a fixed value of H and G), its deviation is observed reasonably well by the oscillation frequency depending on the particular values of E or C. When $C = 1$ and $E \in [0, 1.5]$ (Fig. 3.22-(a)), the sensitivity is higher as E increases. However, if $E \in [-1.5, 0]$, as E increases the sensitivity is smaller. Something similar occurs when $E = 1$ and C is varying (Fig. 3.22-(b)).

Another way to see this is to study the simple solution $I = |K|$, $J = 0$, $G = |K|$, $H = |K|$ and then define the oscillation parameters when a deviation of one of these elements occurs (Table 3.23 and Table 3.24).

Deviations	Oscillation Parameters
	E-Circuit
	Vo2
$I = |K| + \Delta I$	$A_{osc} = -\frac{2V_{ref}}{\pi} \cdot \frac{2|K|}{E} - \frac{2V_{ref}}{\pi} \cdot \frac{\Delta I}{E}$
	$f_{osc} = \frac{1}{2\pi T_s} acos\left[\frac{-2|K|(C+E-2)+\Delta I(2-C-2E)}{2(|K|(2-E)+\Delta I(1-E))}\right]$
$H = |K| + \Delta H$	$A_{osc} = -\frac{2V_{ref}}{\pi} \cdot \frac{2|K|}{E} - \frac{2V_{ref}}{\pi} \cdot \frac{\Delta H}{E}$
	$f_{osc} = \frac{1}{2\pi T_s} acos\left[\frac{-2|K|(C+E-2)+\Delta H(C+E-2)}{2(|K|(2-E)+\Delta H)}\right]$
$G = |K| + \Delta G$	$A_{osc} = \frac{-2V_{ref}}{\pi} \cdot \frac{2|K|}{E}$
	$f_{osc} = \frac{1}{2\pi T_s} acos\left[\frac{-2|K|(C+E-2)+\Delta GE}{2|K|(2-E)}\right]$

Table 3.23: Oscillation parameters in function of a deviation of the *I, G, H* elements

Deviations	Oscillation Parameters
	F-Circuit
	Vo2
$I = |K| + \Delta I$	$A_{osc} = \frac{-2V_{ref}}{\pi} \cdot \frac{2|K|}{F} - \frac{2V_{ref}}{\pi} \cdot \frac{\Delta I}{F}$
	$f_{osc} = \frac{1}{2\pi T_s} acos\left[\frac{-2|K|(C-F-2)+\Delta I(2-C)}{2(|K|(2+F)+\Delta I)}\right]$
$H = |K| + \Delta H$	$A_{osc} = \frac{-2V_{ref}}{\pi} \cdot \frac{2|K|}{F} - \frac{2V_{ref}}{\pi} \cdot \frac{\Delta H}{F}$
	$f_{osc} = \frac{1}{2\pi T_s} acos\left[\frac{-2|K|(C-F-2)+\Delta H(C-F-2)}{2(|K|(2+F)+\Delta H(F+1))}\right]$
$G = |K| + \Delta G$	$A_{osc} = \frac{-2V_{ref}}{\pi} \cdot \frac{2|K|}{E}$
	$f_{osc} = \frac{1}{2\pi T_s} acos\left[\frac{-2|K|(C+E-2)+\Delta GE}{2|K|(2-E)}\right]$

Table 3.24: Oscillation parameters in function of a deviation of the *I, G, H* elements

Let us compare as well the no-simplified oscillation parameters with the simplified ones for the case BP10_Vo1 (Table 3.25).

Generic Form	Oscillation Parameters
	E-Circuit
	Vo1
BP10 Simplified	$A_{osc} = \dfrac{2V_{ref}}{\pi} \cdot \dfrac{-2|K|}{E}$
	$f_{osc} = \dfrac{1}{2\pi T_s} acos\left[\dfrac{-(C + E - 2)}{2 - E}\right]$
BP10 No-Simplified	$A_{osc} = \dfrac{-2V_{ref}}{\pi} \cdot \dfrac{I(E + C) - G - EJ}{E}$
	$f_{osc} = \dfrac{1}{2\pi T_s} acos\left[\dfrac{(G - IE)(2 - C - 2E) + IC(C + E - 2) + 2EJ(1 - E)}{EJ + [G - I(E + C)](1 - E)}\right]$

Table 3.25: Oscillation Equations

Generic Form	Oscillation Parameters
	F-Circuit
	Vo1
BP10 Simplified	$A_{osc} = \dfrac{2V_{ref}}{\pi} \cdot \dfrac{-2|K|}{F}$
	$f_{osc} = \dfrac{1}{2\pi T_s} acos\left[\dfrac{-(C - F - 2)}{2 + F}\right]$
BP10 No-Simplified	$A_{osc} = \dfrac{-2V_{ref}}{\pi} \cdot \dfrac{IC + H}{F}$
	$f_{osc} = \dfrac{1}{2\pi T_s} acos\left[\dfrac{-(H + IC)(C - F - 2) + F(H(1 + F) - JC)}{H(1 + F) + IC}\right]$

Table 3.26: Oscillation Equations

Then, a similar reasoning as the case BP10_Vo2 can be made. Because of its similitude, we do not go through this case in detail.

3.3.4.3   Applying the generic OBT scheme

Suppose a simple example: the case LP01 E_circuit obtained by the second output. In this case, we would have Table 3.27 and the corresponding Table 3.28.

Transfer Function	Zero Placement Formulas	Simple Solution
H_{o2E}	$I = 0$ $G - I - J = \pm \lvert K \rvert$ $J - H = 0$	$I = 0$ $J = 0$ $G = \lvert K \rvert$ $H = 0$

Table 3.27: General Design Equations for LP01

LP01_Vo2	BP00_Vo1
$k_2 = 0$	$k_2 = -\lvert K \rvert$
$k_1 = -\lvert K \rvert$	$k_1 = \lvert K \rvert$
$k_0 = 0$	$k_0 = 0$
$b_1 = C + E - 2$	$b_1 = C + E - 2$
$b_0 = 1 - E$	$b_0 = 1 - E$

Table 3.28: Generic Form LP01 E_Circuit

We saw that it is not possible to fulfil the amplitude oscillation condition if we feed the second output back. But imagine now that for the first output (with BP00 structure, see Table 3.12) the oscillation conditions are not fulfilled either. Then as an option, we can turn to the proposed generic OBT scheme.

In fact, we have here two possibilities. Let us describe both of them in what follows.

-First approach: the simplest option will be to modify the BP00 structure also adding some extra elements (E_{test} and C_{test}) until achieving the frequency oscillation condition as shown in Fig. 3.23.

In this case, you have a very simple oscillator with the following oscillation parameters (see Table 3.15): $A_{osc} = \dfrac{2|V_{ref}|}{\pi} \cdot \dfrac{|K|}{E + E_{test}}$ and $f_{osc} = \dfrac{1}{2\pi T_s} \mathrm{acos}\left[\dfrac{4 - C - C_{test} - 2(E + E_{test})}{2} \right]$.

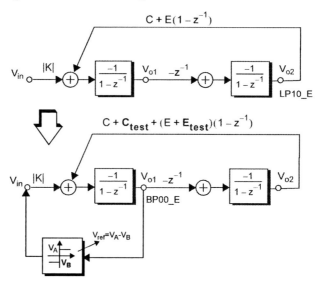

Figure 3.23: First manner to transform the LP10_E circuit to a feasible OBT scheme

To design this oscillator for test purposes, you would have to consider, apart from the feedback sign condition $\left(sign(V_{ref}) = sign\left(\dfrac{1}{E + E_{test}} \right) \right)$ and the start-up requirements, other three important points:

1.- The b_0, b_1 region. That is:

$$\begin{matrix} b_0 + b_1 > -1 \\ 3b_0 - b_1 > 1 \end{matrix} \quad \longrightarrow \quad \begin{matrix} C + C_{test} > 0 \\ 4(E + E_{test}) + C + C_{test} < 4 \end{matrix} \qquad (3.35)$$

2.- The corresponding sensitivities of the oscillation parameters in function of E_{test} and C_{test}. It means to resort again to Table 3.18 but now replacing E with $E' = E + E_{test}$ and C with $C' = C + C_{test}$.

3.- The accuracy of our linearized model (DF approach). It involves to simulate, with a simulation tool, the oscillator proposed in Fig. 3.23-(b) for different ranges of values of E' and C', and then, compare with the theoretical results obtained by the describing-function method. Use always these analytical results as a starting point and choose always an area of E' and C values where the linearized model matches up with accurate simulations. Such an area of E' and C values must be compatible with the choice of the b_0, b_1 region and the conclusions about the sensitivities made above.

This last point may seem very complicated. However, we will study, in next chapters, particular examples and through them, it will be shown that it is feasible to simultaneously satisfy all the mentioned requirements. For instance, in Chapter 4, a similar problem is set out. In that chapter, a generic oscillator is proposed for testing sigma- delta modulators and some guidelines are presented to optimize the design of the obtained OBT scheme.

-Second approach: a second mean to solve the problem of testing the LP01 E_circuit would be to employ the generic OBT scheme. In fact, we could use the previous proposed OBT scheme. However, we are interested in characterizing other cases where the available output performs as a GEN-ERAL biquad. It could happen that not only the oscillation condition associated with the frequency is not satisfied but also the oscillation condition associated with the amplitude. Then we would have to turn out to the generic OBT scheme using the BP10 structure. But as it would be very difficult to obtain conclusions for a general case, let us, for the sake of clarity, consider the following simple example.

3.3.4.4   Designing the oscillator

Several points must be taken into account:

1.- Checking the I, J, G, H, C, E or F elements. That is, how they must be changed in the original structure in order to design one of the oscillators associated with the BP10 structure (Fig. 3.21).

As $H = 0$ and $J = 0$ in LP01_E, we can choose both configurations, BP10_Vo1 E_circuit and BP10_Vo2 E_circuit to build the required oscillator (see Table 3.16). If we observe Fig. 3.21, we can use the first or the third oscillator shown in this Figure. Obviously, in principle, we need only to modify two branches of the LP10_E circuit until finding the BP10_E feasible structures (see Fig. 3.24).

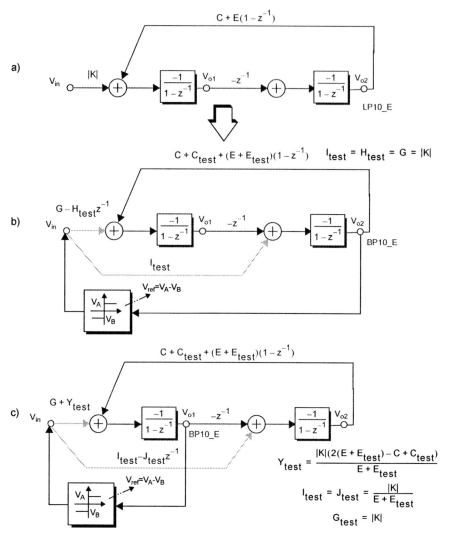

Figure 3.24: Second manner to transform the LP10_E circuit in
a feasible OBT scheme

Choosing, because of its simplicity, the oscillator structure shown in
Fig. 3.24-(b), we can repeat all the points explained in the previous approach.
Obviously, now I_{test} and H_{test} have a different role than I and H.

2.- Adding E_{test} and C_{test} to E and C respectively in order to guarantee the oscillation frequency condition.

3.- Proposing a set of values of E_{test} and C_{test} considering the corresponding b_0, b_1 region. That is, checking the b_0, b_1 region to choose the conditions for the E_{test} and C_{test} values:

$$
\begin{aligned}
b_0 &< 1 & E + E_{test} &> 0 \\
b_1 + b_0 &> -1 \longrightarrow & C + C_{test} &> -2 \\
b_0 - b_1 &> 1 & 2(E + E_{test}) + C_{test} &< 3
\end{aligned}
\tag{3.36}
$$

4.- Delimiting the set of values of E_{test} and C_{test} considering the sensitivities of the oscillation parameters in function of the values of such elements. It is, to examine the sensitivities in order to choose another conditions for the E_{test} and C_{test} values. Use again Table 3.22 but now replacing E with $E' = E + E_{test}$ and C with $C' = C + C_{test}$.

5.- Defining exactly the set of values of E_{test} and C_{test} considering the accuracy of our linearized model (DF approach). It is, to check the accuracy to choose the specific values of the E_{test} and C_{test} parameters.

3.4 SUMMARY

This Chapter presents an exhaustive study showing how to systematically apply the OBT strategy to active discrete-time filters. The discussion begins analysing a generic filter structure and, finishes concluding many analytical details about an efficient implementation of the OBT approach to a particular filter topology, the so-called Fleischer and Laker biquad. However, this filter topology covers most filter configurations. Therefore, the inferred practical conclusions can be easily translated into fundamental DfT guidelines, at low level, useful to apply OBT to any kind of discrete-time filters.

Many tables are obtained throughout the Chapter in order to establish steps or rules to follow when applying the OBT technique to a specific discrete-time filter. All these tables may be helpful to scan them when one reads the

Chapter and to consult them later when one is working out his/her own examples. In general lines, this Chapter has been conceived as a kind of cook-book to be considered when a designer is employing the OBT method to discrete-time filters.

Chapter 4

OBT Methodology for discrete-time ΣΔ Modulators
Principles and architectures

IN THE LAST FEW YEARS, attention to testing ADCs and DACs circuits has been paid for, leading to new methods for increasing test application speed and/or test quality, as well as for Design for Test (DfT) and Built-In-Self Test (BIST) strategies [6], [16], [22], [42], [47]-[48], [64]-[65], [73], [76], [91], [130]. In broad terms, testing data converters is nowadays recognized as a hot research topic. This is so because of their almost ubiquitous presence in mixed-signal systems, for which they can be considered a cornerstone.

A main feature for ADCs is the variety of circuit architectures and techniques in use. The need to manage a wide range of applications, from low to high frequencies and from low to high resolution, has encouraged engineers to develop a constellation of design techniques and has forced to consider their test methodologies. An important problem when dealing with data converters -especially with high-accuracy ones- is the long time required to characterize these circuits. Traditional test procedures rely on the acquisition of large amounts of data to determine parameters such as INL, DNL, and SNR. This effort is unavoidable for prototype testing or even for production test, where functional characterization is requested.

Traditionally, the so-called specification-based test techniques have been used when testing data converters. However, structural testing is emerging as an alternative since it can be less costly and the introduction of test methodologies based on this concept can be of great importance for efficient, yet fast testing of ADCs, specially those embedded in SoCs. Therefore, structural testing is very appealing since a fault-driven approach promises a high reduction in test cost. It is in this context where we consider OBT is worth attention.

There is a strong difference regarding the application of the OBT concept to filters and to converters. In the former case, the associated test technique is truly structural and only a couple of measurements (amplitude and frequency of a quasi-sinusoidal signal) is enough to gather information on defects. On the other hand, concerning data converters, OBT and OBIST have been

157

scarcely applied in practice for testing actual converters; and when done, it was as a variation of the classical servo approach to characterize data converters [2]. In this case, oscillations are forced for every converter code, in order to obtain an accurate analog measurement for it. Then, the test process is very time-consuming.

However, an alternative concept can be developed to force oscillations in just one shot. In particular, the so-called oversampling converters are good candidates to extend the OBT concept coming up from filters, since they are essentially a quasi-stable nonlinear filter. This kind of converters are widely used in many applications and are difficult to test by conventional methods [106]-[107], [115], [118], [127].

This Chapter aims to illustrate how the OBT/OBIST technique can be extended to ΣΔ modulators, forming the analog core of oversampling converters. The new approach is based on the previous methodology developed for discrete-time filters and is intended for a fast validation of the converter. Instead of using the OBT implementation of the servo method, a single oscillator is formed, from whose frequency and amplitude can be discriminated whether the modulator is or not fault-free. The forthcoming sections will first describe a method to transform a second-order modulator into an oscillator. Later, the validity of the introduced method will be discussed and modified to increase its efficiency. Afterwards, to provide an idea of the existing possibilities, a few examples based on well-known modulator architectures are presented, just to illustrate how they can be split and how the basic method can be extended.

4.1 OBT CONCEPT IN LOW-PASS DISCRETE-TIME ΣΔ MODULATORS

4.1.1 Basic approach: forcing oscillations using local extra feedback loops

ΣΔ modulators are used for analog-to-digital and digital-to-analog conversion in a wide and increasing range of applications [118], [127]. The robustness and simplicity of this conversion method make it the preferred choice in many contexts. A typical discrete-time 2nd-order lowpass ΣΔ modulator is shown in Fig. 4.1 where parameters δ_0 and δ_1 are chosen to optimize the (normal) modulator operation without endangering it because of unwanted oscillations. Mainly, this second-order modulator structure consists of two integrators and a feedback loop which is composed of a comparator and diverse local gains responsible for completing the desired system operation.

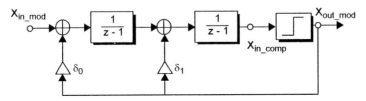

Figure 4.1: A second-order low-pass ΣΔ modulator

The work in previous chapters inspires us to devise a simple way to apply OBT to the modulator. It is based on disconnecting the ordinary modulator input (X_{in_mod} in Fig. 4.1) and then, adding a new feedback loop to force system oscillations in an almost sinusoidal regime. This added loop is depicted in Fig. 4.2 by dotted red lines. It must be clear to recognize that the proposed system structure is formally similar to the oscillator scheme shown in Fig. 4.2-(b): there is a linear network (in the case of this figure, a second-order function) and a nonlinear block in a closed-form feedback.

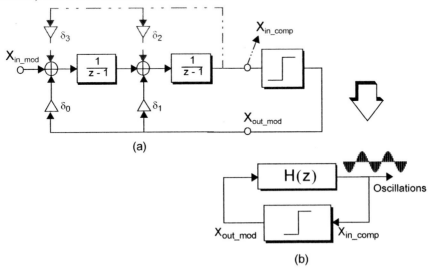

Figure 4.2: Oscillator scheme

If we calculate the resultant transfer function for the linear part, $H(z)$, we obtain

$$H(z) = \frac{X_{in\_comp}}{X_{out\_mod}} = \frac{k_2 z^2 + k_1 z + k_0}{z^2 + b_1 z + b_0} \qquad (4.1)$$

That is

$$H(z) = \frac{G^2(z)\delta_0 + G(z)\delta_1}{1 - G^2(z)\delta_3 - G(z)\delta_2} \qquad (4.2)$$

where $G(z) = \dfrac{1}{z-1}$.

Then, the global transfer function for the closed-loop system displayed in Fig. 4.2-(b) can be expressed as

$$\frac{H(z)}{1 - N(A) \cdot H(z)} \qquad (4.3)$$

or

$$\frac{G^2(z)\delta_0 + G(z)\delta_1}{1 - G^2(z)\delta_3 - G(z)\delta_2 - N(A) \cdot [G^2(z)\delta_0 + G(z)\delta_1]} \qquad (4.4)$$

where $N(A)$ represents again the comparator describing function which, as was explained in the previous chapter, depends on the first harmonic amplitude, A, of the signal called X_{in_comp} (the output signal of the $H(z)$ block or the input signal to the comparator).

Determining the oscillation condition as predicted by the DF approach can be performed following the concepts given in previous chapters for filters, i.e., equating the denominator of (4.3) to zero

$$1 - G^2(z)\delta_3 - G(z)\delta_2 - N(A) \cdot [G^2(z)\delta_0 + G(z)\delta_1] = 0 \qquad (4.5)$$

or, equivalently

$$z^2 - z(2 + \delta_2 + N(A)\delta_1) + 1 + \delta_2 - \delta_3 - N(A)[\delta_0 - \delta_1] = 0 \qquad (4.6)$$

Notice that the poles placement will depend on the particular comparator describing function. Different options can be considered depending on how the actual circuit is implemented: the particular model of the non-linear feedback element and the different values of the parameters δ_j. The procedure to solve the corresponding oscillation mode (when the pole placement is on the unit circle) is similar to the procedure established in previous chapters when we dealt with discrete filters.

For example, when a simple comparator is studied, clearly a pair of complex poles are the roots of the characteristic function (4.6) when oscillation exists. For the sake of simplicity, we can rewrite (4.6) in polar form

$$z^2 - 2r\cos\theta z + r^2 = 0 \tag{4.7}$$

where r and θ symbolize the instantaneous pole position.

Comparing (4.6) and (4.7), we have

$$2r\cos\theta = 2 + \delta_2 + N(A)\delta_1$$
$$r^2 = 1 + \delta_2 - \delta_3 - N(A)[\delta_0 - \delta_1] \tag{4.8}$$

Replacing the specific expression[1] $N(A_{osc}) = \dfrac{2V_{ref}}{\pi A_{osc}}$, the oscillation

mode equations will be (forcing $r = 1$ and solving for θ_{osc}, A_{osc})

$$\cos\theta_{osc} = \frac{\delta_0(2 + \delta_2) - \delta_1(2 + \delta_3)}{2(\delta_0 - \delta_1)}$$
$$A_{osc} = \frac{2V_{ref}}{\pi} \cdot \frac{\delta_0 - \delta_1}{\delta_2 - \delta_3} \tag{4.9}$$

Observe that, in this particular case, both oscillation parameters (frequency and amplitude) are related to all the δ_j coefficients, those one which control the modulator behaviour (δ_1, δ_0), and those one which are introduced for test purposes (δ_2, δ_3). The idea is simply to illustrate how feasible is to apply the OBT method to ΣΔ modulators. Therefore, for the sake of explanation, some specific values for δ_j have been selected. That is, the number of parameters has been reduced to two: δ_0 and δ_2. Despite it is not a practical situation, it allows us to extract some significant conclusions.

---

[1] Being V_{ref} the sum of the saturation levels of the comparator.

In this case, the oscillation parameters are reduced to[2]

$$\omega_{osc} = \frac{1}{T_S} \cdot \operatorname{acos}\left(\frac{2+\delta_2}{2}\right)$$

$$A_{osc} = \frac{2V_{ref}}{\pi} \cdot \frac{\delta_0}{\delta_2}$$

(4.10)

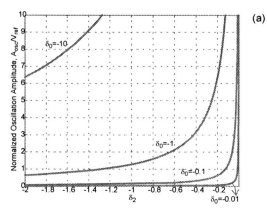

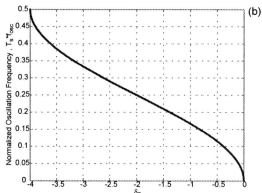

Figure 4.3: Normalized Oscillation Parameters

Both values, frequency and amplitude of oscillations can be selected by using the set of curves shown in Fig. 4.3. Values lying in the corner of Fig. 4.3-(a), and corresponding to a range of values of δ_0 between -0.1 and -0.01 can be a good choice for the normalized oscillation amplitude. Then, a plot similar to that in Fig. 4.3-(b) should be used to determine the expected (fault-free) oscillation frequency. Measuring both parameters have been proven efficient (Chapter 3) to detect faults in a discrete-time filter as the one in which the modulator has been converted to. Furthermore, a method similar to that reported in [29] can be employed to encoding the measured amplitude and frequency into digital bitstreams.

---

[2] Being Ts the system sampling period.

However, in the case of a modulator, the OBT method has tighter constraints than those in a filter. In essence, the difficulty relies in that higher-order harmonics can not be disregarded so simply, giving rise to nonlinear effects that must be taken into account. As

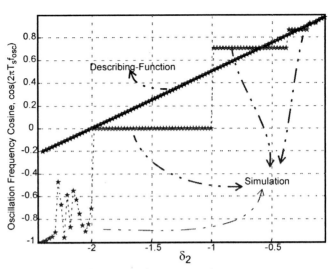

Figure 4.4: Simulation Example

discussed in previous chapters, a key issue for applying OBT is the ability to accurately predict the oscillation parameters as well as the impact of any change in a given component in terms of these parameters. Unfortunately, when the simple models introduced above are applied without any restriction, there can be significant disparities between what is predicted by the linearized model and what is actually observed. Fig. 4.4 depicts a more accurate (nonlinear) simulation of the oscillatory behaviour of a modulator for a particular feedback condition (in this case, $\delta_0 = -2$). As can be seen from this figure, there are discrete frequency variations instead of a continuous evolution as predicted by the describing function (see (4.10)). In addition, for values of $\delta_2 < -2$, chaotic behaviours can be observed that are not predicted by the linearized model.

This effect can be minimized by changing δ_0 (as illustrated in Fig. 4.5-(a) where the ideal line is split in more stretches), However, its existence is unavoidable since there is an intrinsic modelling inaccuracy due to a violation of the basic assumptions for the validity of this approximated method. This is due to an incomplete filtering action that does not eliminate higher-order tones.

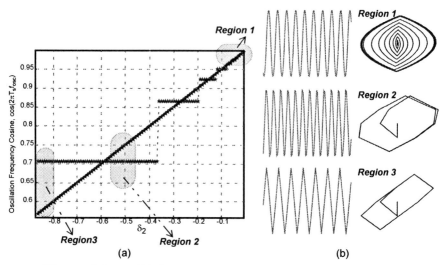

Figure 4.5: Oscillations and limit cycles in other example

We can also observe this phenomenon from the study of the involved limit cycles (see Fig. 4.5-(b)). The limit cycle obtained by simulation in *Region 1* is a clear example of an oscillatory behaviour given an almost pure sinusoidal signal. As can be seen, in this zone the oscillator practically satisfies the model obtained by the DF approach. However, in *Region 2* and *Region 3,* where we find out really some strong discrepancies between the DF approach and the simulations, the achieved limit cycles present a non-sinusoidal oscillatory behaviour.

In fact, assuming negligible all other harmonics than the fundamental is an important error source, since in conventional $\Sigma\Delta$ modulators the open-loop transfer function of its linear part does not filter out most harmonics. Instead, nonlinear oscillation modes can be easily excited as has been extensively proven in the literature [118]. It is true that these nonlinear modes are usually more robust than almost-linear ones, but their robustness lead to an insensitivity against some faults that may not be simple to investigate. The best way to cope with this problem is to resort to a different feedback strategy, which has to be used when the system parameters lead to this nonlinear mode.

4.1.2 Practical OBT scheme in low-pass 2nd-order ΣΔ modulators

As was explained in previous chapters an essential feature of the OBT technique must be to guarantee the spectral purity of the oscillator output (X_{in_comp} in Fig. 4.2) at least to an extent that the fundamental clearly dominates the output signal. Since we are dealing with a nonlinear system, the associated dynamics can be very complex unless we can ensure that high-order harmonics of the basic oscillator frequency are filtered out. Doing this, a secondary advantage is that simplified, quasi-linear analysis methods (like the DF approach) can be used to model the oscillator response.

The basic idea herein is to change the feedback path in order to provide the resulting linear transfer function with adequate properties to perform a filter action upon the undesired tones. Let us consider the oscillator shown in Fig. 4.6, where the solid lines give an equivalent representation of a conventional low-pass modulator and the dashed lines correspond to the extra feedback that is proposes to force oscillations without spurious harmonics. Observe that the feedback path has been split in two. The result is a modified linear system which allows to increase the filtering action and thus prevent nonlinear oscillation modes as much as possible. Notice that the nonlinear components required for the modulator normal operation have been re-used and only some linear extra elements have been added but no circuitry has been removed

We are taking advantage in Fig. 4.6 that, in the case of ΣΔ modulators, the nonlinear block is already available and connected in the SUT feedback loop, suggesting us to use the modulator structure without any extra nonlinear component [34]. In summary, the conclusions drawn in [41] were two-fold. First of all, in a low-pass ΣΔ modulator, the required nonlinear block is in place within the SUT, turning unnecessary to add this block. Second, the filtering action we require for forcing pure sinusoidal oscillations can not be, in general, provided by the linear block existing in the modulator, thus suggesting a modification of this linear part is needed to apply OBT. Using a similar view, we will analyse in next sections a similar strategy for bandpass modulators.

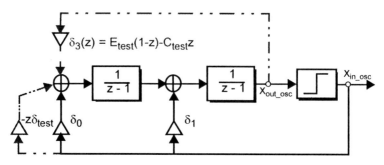

Figure 4.6: a) Low-pass second-order modulator (solid lines only) and b) the oscillator for OBT application (solid and dashed lines)

The transfer function coefficients of the linear part in Fig. 4.6 are related to those in (4.1) through the following linear transformations

$$k_2 = 0 \qquad k_1 = \delta_1 - \delta_{test} \qquad k_0 = \delta_0 - \delta_1$$
$$b_1 = C_{test} + E_{test} - 2 \qquad b_0 = 1 - E_{test} \tag{4.11}$$

There are several possibilities to select the new parameters to force oscillations; however, for the sake of simplicity and accuracy in the obtained oscillation, only the structure shown in Fig. 4.6 will be introduced herein with $\delta_{test} = \delta_0$. Observe that, in this case, the linear part of the OBT structure corresponds to a *BP01E transfer function* when $\delta_{test} = \delta_0$ (see Chapter 3 for more details). Then, the oscillation parameters and conditions are as given in Table 4.1, where the δ_1 and δ_0 parameters come from the modulator structure, while E_{test} and C_{test} form the extra circuitry added for test purposes.

Oscillation Parameters	Oscillation Conditions
$A_{osc} = \dfrac{2V_{ref}}{\pi}\left(\dfrac{\delta_1 - \delta_0}{E_{test}}\right)$	$\delta_1 - \delta_0 \neq 0$
	$\text{sign}(V_{ref}) = \text{sign}(\delta_1 - \delta_0)$
$f_{osc} = \dfrac{1}{2\pi T_s}\text{acos}\left[\dfrac{2 - C_{test}}{2}\right]$	$0 < C_{test} < 4$

Table 4.1: Simple Solution

The expressions in Table 4.1 have been derived assuming the extra loop test coefficient $\delta_{test} = \delta_0$. In the case where there is not an absolute matching between these elements, the expressions become slightly different in the oscillation frequency. This fact will be studied in detail in Sections 4.3.

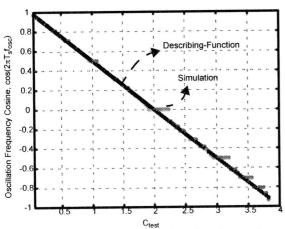

Normally, the resulting oscillator would also require a start-up strategy. But it can be provided by the modification suggested by the author's group elsewhere [86], [88] and developed in next chapters. Obviously, since the devised oscillator is similar to one of the oscillators proposed in Chapter 3, no additional consideration has to be made in this

Figure 4.7: Example using the alternative technique

sense. However, studying the oscillators obtained from the ΣΔ modulator structure, the phenomenon found in Fig. 4.5 shows up as well, a fact which stayed hidden in Chapter 3. Fig. 4.7 represents the oscillation frequency as a function of parameter C_{test} for values fulfilling the conditions in Table 4.1. We can again observe here that there exist certain disparities between the linearized DF approach theoretical results (Table 4.1) and the non-linear simulation results. However, it is important to remark that such disparities are now smaller compared to those obtained in Fig. 4.5.

4.1.3 Fault Analysis

As in filters, the effect of most faults translates into a change in either the frequency or the amplitude of the oscillator. Even in some cases both exhibit a significant variation. Additional work remains to be done yet to qualify the use of OBIST in the case of modulators, however it can be confirmed that a judicious election of the extra parameters (i.e., those not present in the original modulator structure) provides enough freedom to force oscillations which can be worthwhile for testing purposes.

To illustrate how faults are modifying the oscillation parameters, and how such a modification depends on the selected coefficients, let us consider the case of a variation in the integrator gains. If the δ_1 and δ_0 values are fixed (they are imposed by the modulator design) it has to be studied how to choose the value of the remaining parameters (E_{test} and C_{test}) in order to obtain a high fault coverage. It can be proved that a 10% change in the first or second integrator gain will affect both the oscillation amplitude and the oscilla-

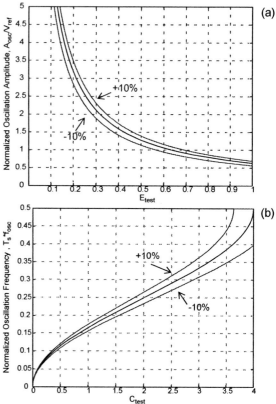

Figure 4.8: Changes in Oscillation Parameters for a 10% change in one integrator gain

tion frequency as well. Fig. 4.8-(a) depicts the movement of the normalized amplitude curves for this variation, and Fig. 4.8-(b) shows the corresponding curves for the normalized frequency. In both figures, the central curves correspond to the nominal case and the predicted changes depend on the value selected for E_{test} and C_{test}.

4.1.4 Fault Detection

Let us finally define an acceptability region. Such a region reports the zone in the oscillation parameter space where the fault-free circuits are located. To do this, let us consider small random deviations in the values of δ_0 and δ_1 and include small random deviations in the integrator gains as well. Then, depending on the allowed maximum deviations in δ_0, δ_1 and the gains of the integrators, different acceptability regions can be defined. For example:

Region#1: Normal Random Distribution with:

$$\sigma = \text{Nominal\_Value} \cdot 0.333 \cdot 10^{-3}$$

Region #2: Normal Random Distribution with:

$$\sigma = \text{Nominal\_Value} \cdot 0.500 \cdot 10^{-3}$$

Region #3: Normal Random Distribution with:

$$\sigma = \text{Nominal\_Value} \cdot 1.000 \cdot 10^{-3}$$

The nominal values of the involved parameters are shown in Table 4.2 as well as the deviations which determine the corresponding acceptability regions (Fig. 4.9). Notice that the values of E_{test} and C_{test} were chosen very small. As will be explained in next sections this is an important requirement to obtain not only a good accuracy in our predictions (given by the DF approach) but also an optimum value of the oscillation frequency from the test viewpoint (sufficiently small in relation to the sample frequency of the system).

Parameter	Nominal Value	Region #1	Region #2	Region #3
$f_{sampling}$	$1.82 \cdot 10^{-5}$ s	---	---	---
f_{osc}	1259.8 Hz	~0 %	>±0.10 %	>±0, 15 %
A_{osc}	5.098 V	>±0, 10 %	>±0, 18 %	>±0, 32 %
δ_0	−0.5		Random	
δ_1	−0.1			
Integrator Gain	1		Deviations	
Integrator Pole	1			
E_{test}	0.01			
C_{test}	0.02		Fixed	
δ_{test}	−2			

Table 4.2: Parameters for the example modulator

Observe from Fig. 4.9 and Table 4.2 that the oscillation frequencies deviates a little from its nominal values. However, the frequency expression

(Table 4.1) indicates that the oscillation frequency does not depend on δ_0 or δ_1. But this discrepancy, somewhat negligible ($>\pm 0.15\%$ even in the most unfavourable case, **Region #3**), could be owing to two reasons: either due to certain amount of inaccuracy linked to the DF approach or due to that the frequency expression in Table 4.1 was obtained considering $\delta_{test} = \delta_0$ and the gains of the integrator exactly equal to 1. These considerations allow us to simplify the involved expressions.

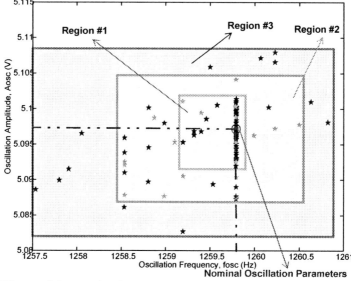

Figure 4.9: Involved Acceptability Regions

To finish this Section, let us also consider small random deviations in the elements incorporated for the test strategy, E_{test} and C_{test}. Then, the acceptability regions slightly enlarge. For example, the largest region (**Region #3**) extends in both dimensions (frequency up to $\pm 0.37\%$ and amplitude $\pm 0.58\%$) (see Fig. 4.10).

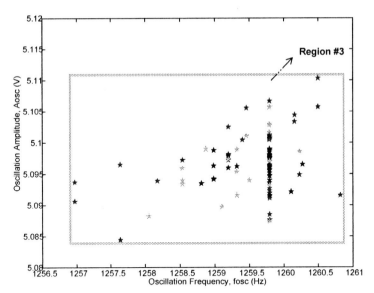

Figure 4.10: Acceptability Region incorporating random deviations in E_{test} and C_{test}

4.1.5 Extension to High-order Architectures

After paying attention to second-order structures, other architectures can be examined. As it has been pointed out by the authors in previous chapters, decomposing high-order filters (and hence, modulators) is the best way to guarantee a reasonable testability at system level. Following the same rationale, decomposing more complex modulator architectures seems to be a promising manner of coping with such architectures in what OBT is concerned. The contents of this section is exploratory, and only has to be considered as a way to illustrate the many avenues opened by the OBT concept when applied to modulators. In that sense, Figure 4.11 aims to show an example on how a 4th-order modulator can be split for testing as a combination of second-order cells. This requires to add a few extra components, but leads to obtain extra observability and thus enhanced testability. The linear part can be divided into second-order functions, and then, an extra feedback loop added to convert sequentially every second-order element into an oscillator similar to that in Fig. 4.6, as is illustrated in Figure 4.11-(b), which gives

an example of the starting configuration, while in the bottom parts of Figure
4.11 a sequence of the remaining configurations are illustrated[3].

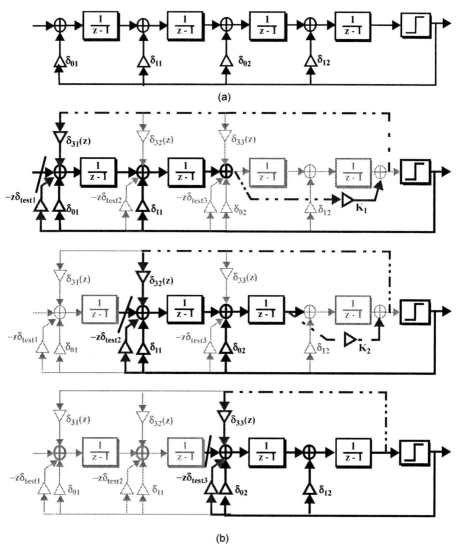

Figure 4.11: OBIST modification for a 4-th order single-loop
low-pass modulator

Furthermore, in Chapter 6 a general methodology has been described for
applying OBT/OBIST to filters of order higher than two and for reading the

---

[3.] In Fig. 4.11-(b) we represent the loops by $\delta_{3i}(z)$ which are $E_{testi}(1-z) - C_{testi}z$ just as in Fig. 4.6.

test output using a simple ΣΔ modulator. An extension of this technique can be also used in the generic high-order modulator in Fig. 4.11-(a). Then, part of the remaining blocks can be also used as a reduced-order modulator to generate a bit-stream version of the test output signal that can be interpreted as was suggested in [85], [87]-[90]. A smart re-structuring of the converter components can lead to obtain a meaningful information of the operational state of the modulator as well as to encode the test outcoming data. The key issue is to re-organize the overall converter in such a way that an oscillator followed by a simpler modulator can be always formed (during any test phase). This gives information about the functionality of any block as well as a digital encoding of that information. Of course, this will impose some constraints on the actual modulator design.

Fig. 4.12 illustrates how the 4th-order modulator in Fig. 4.11-(a) can be split into two lower-order modulators. The one at the upper part of Fig. 4.12 is re-organized as an oscillator and the one below is for encoding the test outcomes. It should be evident from this figure that an additional comparator needs to be introduced and that, depending on the modulator to test, the division into second-order cells can be more or less complex.

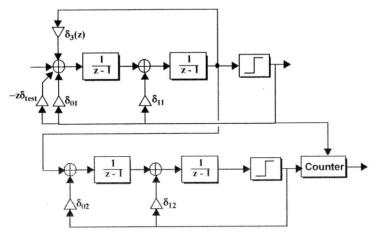

Figure 4.12: OBIST architecture for a 4-th order single-loop low-pass modulator with on-chip test interpretation

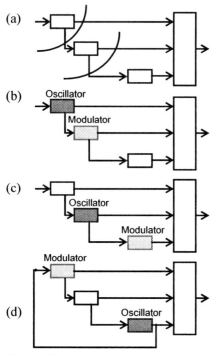

Figure 4.13: Examples of decomposition for a generic cascaded architecture

Extensions to other low-pass and band-pass architectures may also be targeted. Interpolative and cascaded modulator structures can be wisely reduced to a few second-order oscillators and their oscillation parameters can give a good idea about how the overall modulator is working. Furthermore, in complex structures, part of the system can be tested using totally or partially the remaining part of the modulators. Of course, the way to proceed depends on every particular structure what means that developing guidelines for dealing with any architecture is worthwhile. This opens many new testing possibilities, let us consider the 2-2-1 modulator in Fig. 4.13-(a). In this case, a sequence of configurations can be defined to test one-by-one the different cells forming the linear part of the converter. Examples of some of the test configurations for this converter can be seen in Fig. 4.13-(b), -(c) and -(d).

4.2 OBT CONCEPT IN BANDPASS DISCRETE-TIME ΣΔ MODULATORS

4.2.1 Background

Besides the many applications of low-pass modulators, ΣΔ bandpass modulators are deserving a great interest since they offer efficient signal processing for appealing applications like digital wireless devices. A primary motivation for the development of bandpass converters is their ability to deal with narrow-band signals. In particular, for communication systems, bandpass converters allow early conversion to digital, resulting in more robust devices and pushing the IF filters to the digital domain.

A bandpass ΣΔ modulator [106]-[107], [115] is basically formed by a resonator (i.e. a bandpass filter), a low-resolution quantizer (a symmetrical two-level comparator, for example),

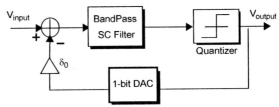

Figure 4.14: A bandpass Σ–Δ modulator

and a 1-bit DAC connected in a feedback loop with gain δ_0, as shown in Fig. 4.14. Feedback allows to shape noise away from an arbitrary passband; then, quantization noise can be filtered out and its contribution to the passband of interest can be made very small. The bandpass filter in Fig. 4.14 is built around one or several resonators, and the input signal can be shaped in different manners.

Most of the design process for bandpass modulators can be derived straightforward from the low-pass case. Approximate linear models can be borrowed from the latter, providing methods to select convenient Signal and Noise Transfer Functions (STF and NTF, respectively). However, there are some characteristic features coming up from the passband nature of the modulator. One of this is the central frequency for the noise notch filtering, f_o, which is usually selected to be an integer fraction of the sampling frequency, f_s. Typical numbers are 2 and 4, leading to simplified implementations [105]-[106]. For bandpass modulators, the oversampling ratio, OSR, is defined as

$$OSR = \frac{f_s}{2\Delta f},$$ where Δf represents the notch bandwidth.

The interest herein is in discrete-time versions of this modulator, specifically on how to apply to this kind of circuit the basic principles of OBT as an extension of the ideas presented in [34] for switched-capacitor low-pass modulators.

4.2.2 Basic OBT approach: forcing oscillations around the notch frequency

This section is intended to introduce the basic way to force oscillations in bandpass modulators, discussing the pros and cons of this alternative and leading to a more practical solution. We will start by considering the translation of the OBT concept to this bandpass case and then we will prove that some changes have to be devised in terms of proposing a manner to apply OBT to bandpass modulators.

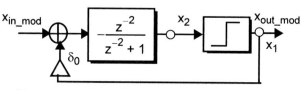

Let us consider the discrete-time second-order bandpass $\Sigma\Delta$ modulator shown in Fig. 4.15. This is the simplest structure we can devise. A second-

Figure 4.15: Discrete-time second-order bandpass $\Sigma\Delta$ modulator

order resonator is used as the loop filter, which has poles at $z = \pm j$, i.e., resonates at $\omega_0 = \pm\pi/2$. Then, the transfer function seen by the input signal has poles located at $\dfrac{f_s}{4}$, whereas the associated noise transfer function has zeros at the same locations, giving the desired notch at the same frequency.

Since δ_0 is selected by design to prevent the modulator to become unstable, we need to add at least an extra loop when OBT is applied. To investigate how this system can be forced to oscillate in an almost sinusoidal regime, two general feedback loops are added in Fig. 4.16-(a). The scheme depicted in Fig. 4.16-(b) is a simplified description of the system displayed in Fig. 4.16-(a), where $H(z)$ represents the involved transfer function of the circuit linear part. For the sake of generality we use this scheme, which includes as particular cases the two configurations with only one loop (feeding back to the input either x_1 or x_2 only).

As we discussed above, parameter δ_0 is usually chosen to optimize the normal modulator operation as well as to prevent instability; on the other hand, the additional elements, $G(z)$ and $F(z)$ are chosen to sustain oscillations when modulator is tested (dotted lines in Fig. 4.16-(a)). The problem is how $G(z)$ and $F(z)$ have to be selected; in particular, which are the simplest linear functions which can lead to a satisfactory implementation of OBT.

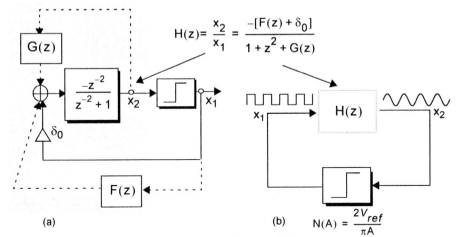

Figure 4.16: a) A bandpass Σ–Δ modulators OBT scheme.
b) Oscillator built around a bandpass Σ–Δ modulator

An exact analysis of the system presented in Fig. 4.16 would require a nonlinear study. However, as was explained a linearized analysis can be carried out using the DF method, where the comparator is replaced by an "equivalent" linear function, $N(A)$. This is a useful means to:

-derive the oscillation conditions to guarantee sustained and stable oscillations, as a function of δ_0, $G(z)$, and $F(z)$.

-estimate the main oscillation parameters (essentially, oscillation frequency and amplitude).

The characteristic function for the closed-loop feedback system in Fig. 4.16-(b) is $1 - N(A)H(z) = 0$. And this expression can be rewritten in terms of Fig. 4.16 as

$$z^2 + G(z) + 1 + [\delta_0 + F(z)]N(A) = 0 \qquad (4.12)$$

Since we are interested in simple solutions, only $G(z)$ and/or $F(z)$ formed by either constants or first-order delays are considered. Then, (4.12) can be replaced, in the most general of these cases, (identifying terms correctly) by

$$(z^2 - 2r\cos(\theta)z + r^2)(z + p_1) = 0 \qquad (4.13)$$

obtaining a pair of complex poles $z_{1,2} = r\cos(\theta) \pm jr\sin(\theta)$, (that obviously depend on $N(A)$ and δ_0), and a real pole, p_1. For oscillation ($r = 1$, poles have to be on the unit circle) this characteristic function can be solved for the oscillation phase, θ_{osc} and gain, $N(A_{osc})$. The resulting pole equations are summarized in Table 4.3.

Case	Local Feedback Loops	Pole Equations
A	$G(z) = \delta_2$ $F(z) = \delta_4$	p_1 does not exist Pair of complex poles given by: $z^2 = -(1 + \delta_2 + [\delta_0 + \delta_4]N(A))$
B	$G(z) = \dfrac{\delta_3}{z} + \delta_2$ $F(z) = \dfrac{\delta_5}{z} + \delta_4$	$\left.\begin{array}{l} p_1 - 2r\cos(\theta) = 0 \\[4pt] r^2 - 2r\cos(\theta)p_1 = \delta_2 + 1 + (\delta_0 + \delta_4)N(A) \\[4pt] r^2 p_1 = \delta_3 + \delta_5 N(A) \end{array}\right\}$

Table 4.3: Two different choices for the local feedback loops

Now, all possible modes of oscillation for both cases can be computed. The resulting values for the oscillation parameters are summarized in Table 4.4 (being f_{osc} the oscillation frequency and f_s the sampling frequency of this closed-loop discrete system).

Observe that the predicted oscillation frequency for case *A* results to be either $f_s/2$ or $f_s/4$ (depending on the modulator design parameters, see Fig. 4.17), since these are the values selected to place the modulator notch.

Case	Oscillation Mode	$\|p_1\|$	θ_{osc}	$f_{osc} = \dfrac{\theta_{osc}}{2\pi} f_s$
A	$\|z\| = 1$ Poles in $\pm j$ $N(A_{osc}) = \dfrac{-\delta_2}{\delta_0 + \delta_4}$	--------	$\dfrac{\pi}{2}$	$\dfrac{f_s}{4}$
	Poles in ± 1 $N(A_{osc}) = \dfrac{-2-\delta_2}{\delta_0 + \delta_4}$		π	$\dfrac{f_s}{2}$
B	$r = 1$ $p_1 - 2\cos(\theta_{osc}) = 0$ $-2\cos(\theta_{osc})p_1 = \delta_2 + (\delta_0 + \delta_4)N(A_{osc})$ $p_1 = \delta_3 + \delta_5 N(A_{osc})$	$\|p_1\| < 1$	$acos\left(\dfrac{p_1}{2}\right)$	$\dfrac{f_s}{2\pi} acos\left(\dfrac{p_1}{2}\right)$

Table 4.4: Oscillation Mode Solutions

Experiments performed by simulation prove a reasonable agreement with predictions from the linearized model. However, there are a few problems related to this structure. First of all, the oscillation frequency is insensitive to the passive components in the feedback path; only the amplitude exhibits a significant deviation with these components. Additionally, the high values achieved for the oscillation frequency are not so convenient for OBT for different reasons:

a) Since the oscillation frequency is so near to the Nyquist limit, the number of points available for analysing the test outcome is too small.

b) Equivalently, dealing with high-frequency signals (near the maximum signal frequency for which the modulator was designed) is not so convenient for test.

c) For the same reason, the resonator bandpass action does not guarantee the validity of the describing function method.

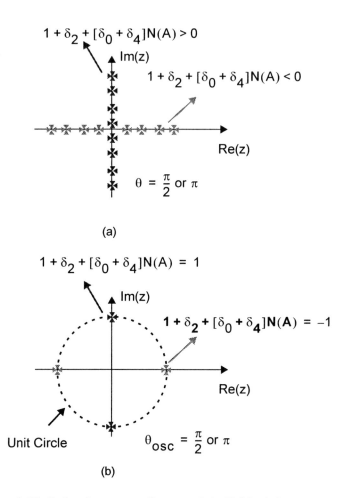

(a)

(b)

Figure 4.17: Pole placements for case A in Table 4.4

Concerning the case B in the lower row of Table 4.4, the oscillation frequency value can (apparently) be controlled by means of the real pole p_1. However, this is not valid either. Decreasing the oscillation frequency requires increasing the absolute value of p_1, but p_1 is related to θ_{osc} in such a way that $\cos(\theta_{osc}) = p_1/2$ (Table 4.4). It is simple to prove that trying to force oscillations fulfilling $f_s \gg f_{osc}$ leads to $\mathrm{acos}(p_1/2) \rightarrow 0$, or equivalently, $|p_1|/2 \rightarrow 1$. Then, the real pole moves out of the unit circle ($p_1 > 1$, approaching $r = 2$) and, the closed-loop system becomes unstable.

Notice from all these oscillation results that the idea of employing the type of global feedback loops proposed in Fig. 4.16-(a), must be disregarded. The SUT oscillates, but there are no efficient ways to relate variations in the expected oscillation signal with modulator parameters. Of course, this strategy (Fig. 4.16) may be explored using more complex functions $G(z)$ and $F(z)$, but this would not be a satisfactory manner to apply OBT. In conclusion, mechanisms to achieve sustained oscillations must be devised by considering the detailed resonator structure in order to introduce partial feedback loops.

4.2.3 Practical OBT scheme: downsizing the oscillation frequency

The previous section has shown that a global feedback does not allow simple OBT solutions. Therefore, we have to set up a new means to convert the bandpass ΣΔ modulator in an effective oscillator. The idea is to turn to the actual implementation of the second-order resonator in Fig. 4.16-(a) and then, consider how we can inject other local feedback signals to convert the SUT into an oscillator. For example, we can consider the particular second-order resonator structure shown in Fig. 4.18-(a), where the resonator is built as shown in Fig. 4.18-(b).

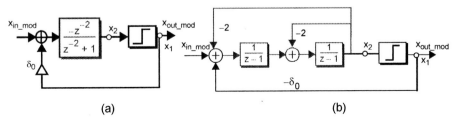

(a)   (b)

Figure 4.18: a) A bandpass ΣΔ modulators OBT scheme. b) The same scheme for a given resonator structure

Forcing this closed-loop system to oscillate can be carried out by using various local extra loops (displayed in dotted red lines in Fig. 4.19).

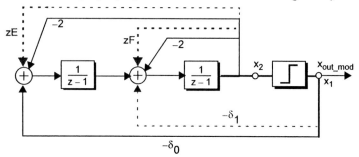

Figure 4.19: A bandpass $\Sigma\Delta$ modulators OBT scheme

Establishing again the closed-loop characteristic equation, we obtain the following result

$$z^2 + \left(\frac{F - E + N(A)\delta_1}{1 - F}\right)z + \frac{1}{1 - F}(1 + N(A)[\delta_0 - \delta_1]) = 0 \qquad (4.14)$$

Notice that, in this case, the z-term is present. It makes us suspicious that oscillations valid for OBT are possible. If that is really the case, we can achieve an oscillation mode where OBT is feasible. This mode is given by the subsequent system of equations

$$-2\cos(\theta_{osc}) = \frac{F(2\delta_1 - \delta_0) - E(\delta_1 - \delta_0)}{(1 - F)(\delta_1 - \delta_0)}$$

$$N(A_{osc}) = \frac{F}{\delta_1 - \delta_0} \qquad (4.15)$$

However, there are a set of problems related to this proposed oscillation strategy if we want to apply the OBT technique. Observe from (4.15) that at least the loop called F is required (on the contrary, the oscillation amplitude would be zero). But, if we study in detail the difficulties associated with this structure in itself, we find that merely to implement the local loops Ez and Fz is not possible unless you change the system structure. On the other hand, the second-order resonator circuit in Fig. 4.18-(b) is only a very particular case which does not summarize all the problems that may appear when one is trying to develop an OBT methodology suitable to a wide range of bandpass $\Sigma\Delta$ modulators.

According to this, after searching among the many different implementation structures available in the literature, we found out that those based on a cascade-of- LDI-phased resonators [106]-[107] are very appealing. The simplest modulator from this class is the one shown in Fig. 4.20-(a), corresponding to a second-order bandpass. Our interest relies in the fact that the core of the modulator in Fig. 4.20-(a) is rather similar to that used in previous sections for the lowpass case [34]. Forcing oscillations in this modulator can be done by adding the dotted feedback loops in Fig. 4.20-(b). There are two different actions involved. First, the feedback paths existing in the SUT are modified by adding some branches (parameters E_{test}, $-zR_{test}$, and $z\delta_{test}$). Second, the regular input is disconnected and x_1 is also injected through a_0 and a_1. Although it may look complex in the block diagram of Fig. 4.20, the implementation in a switched-capacitor modulator is simply performed by adding a few capacitors and switches. The only difficulty is the connection and disconnection of the input, which can be done by the method we will discuss in the next Section.

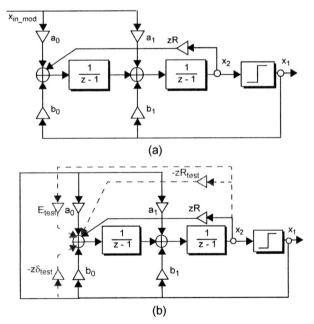

Figure 4.20: a) Cascade-of-resonator $\Sigma\Delta$ modulator b) Oscillator for the OBT method proposed in this book

Let us now analyse the basic quadratic structure in Fig. 4.20, since as was discussed above is the basic oscillation element after decomposing any modulator filter. Handling the characteristic equation and replacing the expression of the comparator describing-function $N(A)$, it can be solved for the oscillation conditions

$$\delta_{test} = a_0 + b_0$$
$$a_1 - a_0 + b_1 - b_0 \neq 0$$
$$sign(V_{ref}) = sign(a_1 - a_0 + b_1 - b_0) \qquad (4.16)$$
$$-1 < \frac{2 + R + E_{test} - R_{test}}{2} < 1$$

and for the oscillation parameters

$$A_{osc} = \frac{2V_{ref}}{\pi} \cdot \left(\frac{a_1 - a_0 + b_1 - b_0}{E_{test}} \right)$$

$$f_{osc} = \frac{f_s}{2\pi} \cdot acos\left[\frac{2 + R + E_{test} - R_{test}}{2} \right] \qquad (4.17)$$

where V_{ref} corresponds to the comparator reference voltage (if the comparator has two saturation levels given by $\pm V$, then $V_{ref} = 2V$).

Since a_0, a_1, b_0, and b_1 are fixed by design, the oscillation amplitude can be controlled by the additional parameter E_{test}. Similarly, R is also fixed and R_{test} gives enough freedom to control the oscillation frequency. Although it is hidden in this approximate linear analysis, both f_{osc} and A_{osc} depend on the resonator and comparator parameters as well. An important result is that (4.17) indicates that the oscillation frequency can be moved to lower values than f_s by playing with the parameter R_{test}. This means we can move downwards this frequency avoiding most of the inconveniences discussed in the previous Section.

4.2.4 Structural Test and Fault Analysis

Two kind of experiments have to be done in terms of validating our results. First of all, the accuracy of the describing-function model has to be proven. Second, the quality of the OBT technique has to be demonstrated through fault simulation experiments.

In any practical situation all parameters are fixed by the modulator design [106]-[107] except, of course, those used for test purposes, E_{test} and R_{test}. These parameters can fix the most suitable oscillation condition (that is, an acceptable oscillation range for test purposes). To prevent amplitude values too high, small values of E_{test} are demanded. On the other hand, we are mainly interested in frequency values much smaller than the clock frequency because the larger the oscillation frequency, the less accurate the linearized model and the more difficult to achieve measurements to confirm the expected value.

A detailed simulation study (performed by SIM-ULINK and using FFT to determine the oscillation frequency) confirms that the validity of our linearized model based on the describing-function method [29], [34] also demands small values of the E_{test} parameter. To illustrate the model, simulations have been carried out for a second-order modulator using E_{test} fixed to 0.01 (Fig. 4.21) or 0.1 (Fig. 4.22), while R_{test} is sweeping all possible values fulfilling the oscillation conditions (4.16). Under these assumptions, the different achieved oscillations must have the same amplitude

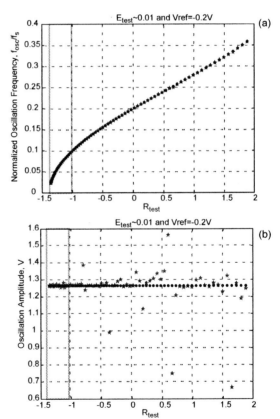

Figure 4.21: Oscillation Parameters sweeping R_{test} ($E_{test}\sim0.01$)

although cover all the possible frequency values (see (4.17)).

In Fig. 4.21 and Fig. 4.22 the theoretical data (given by (4.17)) are represented by dots while the simulation results are drawn with stars. It should be clear from Fig. 4.21-(a) and Fig. 4.22-(a) that a good choice for E_{test} can be found around 0.01 (avoiding a degradation in the oscillation frequency), while R_{test} can be taken not higher than -1 (avoiding a degradation in the oscillation amplitude) (see Fig. 4.21-(b) and Fig. 4.22-(b)). In fact, the main effect of increasing E_{test} is the reduction on the region where the linearized model is valid. Increasing R_{test} above -1 leads to undesired nonlinear modes and decreasing below -1.5 causes the oscillation to disappear.

On the other hand, to illustrate how faults are modifying the oscillation parameters and how such a modification depends on the selected coefficients (those ones coming from the modulator structure and the remaining ones used exclusively for testing), let us consider the case of a variation in the integrator gains. Fig. 4.23 depicts how the nominal oscillation parameters are altered by the changes in the first integrator gain. Looking at them, it can be seen that whether a ±10% change in the first integrator gain takes place, it will affect both the oscillation amplitude and the oscillation frequency as well. The deviation range is approximately

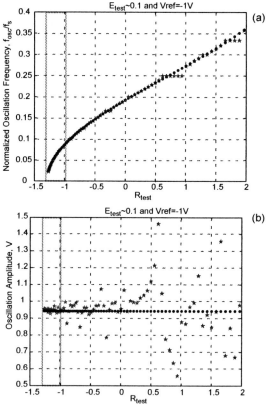

Figure 4.22: Oscillation Parameters sweeping R_{test} (E_{test}~0.1)

a ±5 % change in the frequency and more than a ±30 % change in the amplitude. Regarding this, and depending on the discrimination rank for the experimental test measurements, a good fault coverage can be achieved. Another issue to be borne in mind is that due to the right choice of the remaining coefficients E_{test} and R_{test}, added for test purposes, the theoretical curves (dots) match up practically with the simulations (stars).

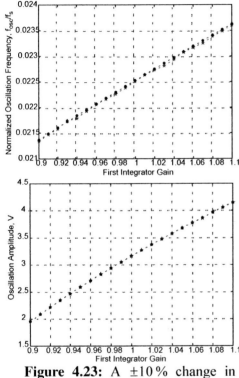

Figure 4.23: A ±10 % change in one integrator gain

Depending on the parameter, either amplitude or frequency can be preferential for test. For example, if one of the modulator coefficients, a_i or b_i is varied producing a ±10 % change in its nominal value, a variation is noticed in the amplitude but never in the frequency. Examples where only the frequency is changed can be given; both situations are just the limit cases for the OBT technique.

4.2.5 Fault Detection

In order to assess on the validity of this approach for detecting faults, two experiments were finally carried out for a modulator with the parameters listed in Table 4.5. Firstly, small deviations were randomly injected for all the feedback coefficients relating to the modulator. Let us consider slight random deviations in the values of δ_0 and δ_1 and include slight random deviations in the integrator gains as well. Then, depending on the allowed maximum deviation in δ_0, δ_1 and the gains of the integrators, different acceptability regions can be defined. Examples of such acceptability regions are described as follow:

Region#1: Normal Random Distribution with:

$$\sigma = \text{Nominal\_Value} \cdot 0.333 \cdot 10^{-3}$$

Region #2: Normal Random Distribution with:

$$\sigma = \text{Nominal\_Value} \cdot 0.500 \cdot 10^{-3}$$

Region #3: Normal Random Distribution with:

$$\sigma = \text{Nominal\_Value} \cdot 1.000 \cdot 10^{-3}$$

The nominal values of all the involved parameters are shown in Table 4.5 as well as the deviations which determine the corresponding acceptability regions (Fig. 4.24). Observe that the values of E_{test} and R_{test} were chosen in such a way that we may guarantee that it is obtained not only a good accuracy in our predictions (given by the DF approach) but also an optimum value of the oscillation frequency (sufficiently small in relation to the sample frequency of the system, see (4.17)).

Parameter	Nominal Value	Region #1	Region #2	Region #3
f_s	55 kHz	---	---	---
f_{osc}	1.259 kHz	>±2.25 %	>±4.90 %	>±5.60 %
A_{osc}	3.164 V	>±1.30 %	>±2.20 %	>±2.80 %
a_0	-0.1701			
b_0	-0.1576			
a_1	-0.2388		Random Deviations	
b_1	-0.0149			
R	-1.3940			
E_{test}	0.0149			
δ_{test}	0.3277			
R_{test}	-1.3591		Fixed	
Integrator Gain	1			
Integrator Pole	1			

Table 4.5: Parameters for the example modulator

Again, any of the above acceptability regions illustrates how to proceed for determining the actual acceptability region as a function of the required accuracy of the modulator specifications and its performance. It is not straightforward to relate the modulator gains and other coefficients at system level with the technological parameters and their tolerances in the specific technology. In fact, we would have to select an implementation at transistor level to define an acceptability region which contemplates the transistor mismatches.

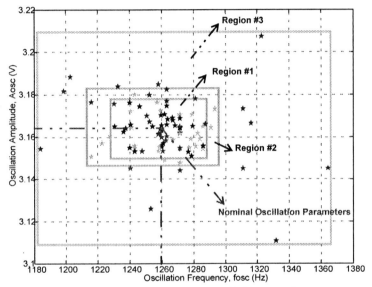

Figure 4.24: Involved Acceptability Regions

Also, some larger changes were injected for all the feedback coefficients as well as for the integrators gain and the integrator pole positions (Table 4.6).

Parameter	Typical Value	Variation (Small)	Variation (Large)
f_s	55 kHz	---	---
f_{osc}	1.25 kHz	150-250 Hz	100-800 Hz
A_{osc}	3.2 V	< 0.8 V	> 0.8 V
a_0	-0.1701	1-5%	>20%
b_0	-0.1576	1-5%	>20%

Table 4.6: Parameters for the example modulator

Parameter	Typical Value	Variation (Small)	Variation (Large)
a_1	-0.2388	1-5%	>20%
b_1	-0.0149	1-5%	>20%
R	-1.3940	0.5-3%	8-10%
E_{test}	0.0149	<10%	>20%
δ_{test}	0.3277	<10%	>20%
R_{test}	-1.3591	<10%	>20%
Integrator Gain	1	1-5%	>10%
Integrator Pole	1	0.1-0.5%	>5%

Table 4.6: Parameters for the example modulator

The results are shown in Fig. 4.25 where two possible faulty regions (as well as the called **Region #3**) have been plotted. One of these regions is the rectangle formed by intersecting the lower and higher limits for both the amplitude and the frequency. The discrimination is not difficult since these regions ranges from 1.1 to 1.4 kHz (frequency measurements) and from 2.5 to 4 Volts (amplitude measurements). There is some dissymmetry in Fig. 4.25 because the fault-free modulator does not lie midway in both coordinates.

The width and height of these regions will depend on the desired test accuracy and the intended yield, and are illustrative in this example. In this case, for instance, points inside **Region #3** correspond to "good" modulators, points in blue define a tolerance window corresponding to "acceptable" modulators, and points outside this window define the faulty region corresponding to modulators that should be rejected. The distribution of the points is of no significance since its shape is due to the manner we have performed the experiment. Good circuits are clusterized around the nominal and large changes move the circuits far away from the acceptability region.

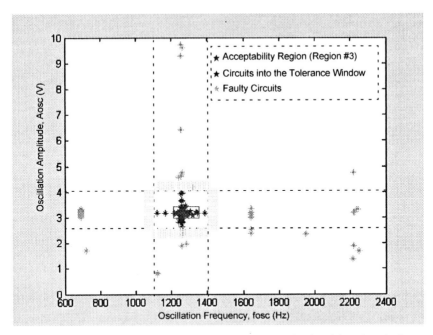

Figure 4.25: Example of two Faulty Regions obtained by simulation

4.2.6 Extension to Higher order structures

To illustrate the feasibility of this approach for higher-order modulators, we will consider in what follows the fourth-order structure whose detailed implementation is given in Fig. 4.26. As was discussed in [25]-[28], [30]-[37], it is difficult to determine the oscillation conditions as well as to predict the oscillation parameters if a high-order structure has to be handled. For this reason the OBT concept is applied by splitting up the overall fourth-order structure into smaller blocks (second-order subsystems). Oscillations can be separately forced in every second-order block by adding the dotted feedback loops in Fig. 4.27. Then, using a multiplexed sequential testing or/and swopamps [82]-[83] to bypass signals from a point to another point of the system [37], both subsystems should be tested.

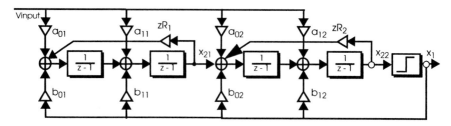

Figure 4.26: A fourth-order $\Sigma\Delta$ bandpass modulator structure

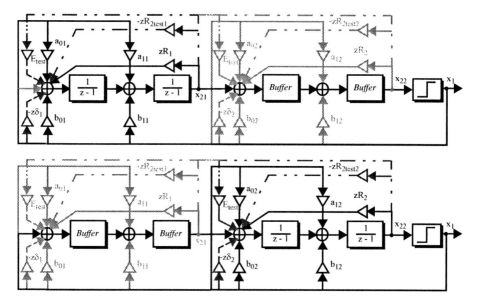

Figure 4.27: OBT method applied to a fourth-order structure

4.3 PRACTICAL OBT SCHEME FOR ANY TYPE OF MODULATORS

First of all, notice that practical schemes for both kinds of $\Sigma\Delta$ modulators (low-pass and band-pass) considered in previous sections can be represented by the generic oscillator shown in Fig. 4.28. If the CUT is a low-pass modulator, then $R = 0$, whereas if the CUT is a band-pass modulator, $R \neq 0$.

Observe again that this proposed OBT structure has been inspired by Chapter 3 and its linear part is a BP01E transfer function. The oscillation parameters and conditions obtained by the DF approach are given in Table 4.7, where the δ_1, δ_0 and R parameters come from the modulator structure, whereas E_{test} and C_{test} from the extra circuitry added for test purposes. For the sake of clarity, let us suppose the specific case when $\delta_{test} = \delta_0$.

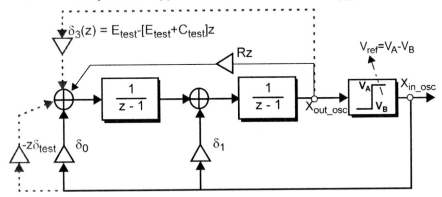

Figure 4.28: Generic Oscillator for ΣΔ modulators

Oscillation Parameters	Oscillation Conditions
$A_{osc} = \dfrac{2V_{ref}}{\pi} \cdot \left(\dfrac{\delta_1 - \delta_0}{E_{test}}\right)$	$\delta_1 - \delta_0 \neq 0$
	$\text{sign}(V_{ref}) = \text{sign}(\delta_1 - \delta_0)$
$f_{osc} = \left(\dfrac{1}{2\pi T_s} \cdot a\cos\left[\dfrac{2 + R - C_{test}}{2}\right]\right)$	$-1 < \dfrac{2 + R - C_{test}}{2} < 1$

Table 4.7: Oscillator features

Practical guidelines for the OBT implementation[4] (Fig. 4.28) were given in Chapter 3. These guidelines include: a) feedback sign condition, b) start-up requirements, c) choosing the adequate ranges of the values of the extra parameters, E_{test} and C_{test} and d) fault coverage considerations.

In what follows, these two last issues (c and d) are further developed.

---

[4] Observe that Fig. 4.28 is similar to Fig. 4.20-(b), but now considering $R_{test} = C_{test} + E_{test}$.

4.3.1 Theoretical Normalized Oscillation Parameters

Let us define the oscillation amplitude normalized with respect to both the gain of the comparator and the coefficients of the modulator:

$$A_{osc}\big|_{norm} = \frac{A_{osc}}{|V_{ref}||\delta_1 - \delta_0|} = \frac{2}{\pi} \cdot \frac{1}{E_{test}} \tag{4.18}$$

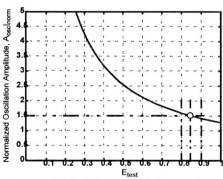

Figure 4.29: Theoretical Normalized Oscillation Amplitude

The purpose is to obtain an expression only related to the extra element, E_{test}. In principle, for realizability reasons, the E_{test} parameter has to be positive and its value has no influence in the oscillation conditions (see Table 4.7).

Graphically, (4.18) is plotted in Fig. 4.29. Observe, for example, that an oscillation amplitude of around $1.5V$ with $|V_{ref}||\delta_1 - \delta_0| = 1$ can be obtained by choosing $E_{test} \in [0.8, 0.9]$.

On the other hand, the oscillation frequency is defined with respect to the sampling frequency, f_s

$$f_{osc}\big|_{norm} = \frac{f_{osc}}{f_s} = \frac{1}{2\pi}\text{acos}\left[\frac{2 + R - C_{test}}{2}\right] \tag{4.19}$$

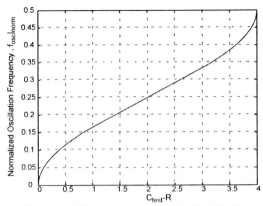

Figure 4.30: Normalized Oscillation Frequency

Notice that, in this case, the normalized oscillation frequency depends only on the R and C_{test} parameters. We can see that, in order to fulfil the oscillation conditions, we have to impose that $C_{test} - R$ varies in the interval [0,4]. Observe, however, that our interest herein is that $f_{osc} \ll f_s$ not only to achieve a good discrimination between both frequencies but also to simply use f_s in the test interpretation. Since this last expression reports the relationship between the oscillation frequency and the sampling frequency, it will be object of study in the next lines.

Considering simultaneously the above-mentioned four guidelines, we have to select the solution (i.e a set of frequency-amplitude) which seems more appropriate.

Observe from the theoretical expression of the normalized oscillation amplitude (4.18) that its value is determinate by only the E_{test} parameter. Therefore, when the DF linearized model is considered valid, if E_{test} remains a fixed value, the normalized amplitude has to remain a fixed value. On the other hand, likewise, the normalized oscillation frequency evolution, regardless of the fixed value of E_{test}, must vary according to the values of C_{test} and R.

We can exhaustively evaluate the ranges of E_{test} and $C_{test} - R$ where the linearized model is valid for specific values of δ_0 and δ_1. Supposing, for example, $|\delta_0| = 1$ and $|\delta_1| = 2$, we will study the behaviour of the system by a more accurate nonlinear simulation and we will compare such results with the theoretical results predicted by the describing-function approach.

Firstly, we will consider four cases where the E_{test} parameter is fixed to various values while the $C_{test} - R$ value is changed sweeping the proposed significant range [0,0.3] (see Table 4.8). These cases provide sufficient evidence (see Table 4.8 and Fig. 4.31) to state that we have to select the E_{test} parameter in the interval [0,0.01] and then, the $C_{test} - R$ parameter must be in the interval [0,0.04].

<--C_{test}-R-->	E_{test}	Frequency Validity	Amplitude Validity
Case #1	0.001		Good agreement mainly in the interval [0,0.05]
Case #2	0.01	Good agreement	The simulation results follow theoretical predictions only in the interval [0,~0.04]
Case #3	0.05	The simulation results follow theoretical predictions only in discrete intervals	No very good agreement
Case #4	0.1		Very bad agreement

Table 4.8: Oscillation parameters validity (I)

But returning to Fig. 4.29, we can observe that when the E_{test} parameter is in the interval [0, 0.01], the normalized oscillation amplitude is very high. However, fortunately, we can control its value by the V_{ref} parameter. Notice, on the other hand, that we are interested in a small value of the normalized oscillation frequency. Then, if the $C_{test} - R$ value is in the interval [0, 0.04], the required result of the normalized oscillation frequency is carried out.

Next Fig. 4.31 illustrates the facts shown in Table 4.8. When E_{test} is chosen too high, the simulation oscillation frequency results follow the theoretical predictions only in discrete intervals, independently on the value of $C_{test} - R$; and the theoretical oscillation amplitude is not valid any more. But as E_{test} goes smaller, then a good agreement in the oscillation frequency is found and when $C_{test} - R$ is small enough, then a good agreement in the oscillation amplitude is found as well.

A parallel study can be made by considering four cases where the $C_{test} - R$ coefficient is fixed to various values while the E_{test} parameter is changed sweeping the proposed significant range $[0,0.3]$ (see Table 4.9). We can seen from Fig. 4.31 that for the proposed values of these parameters the oscillation amplitude is the expected by the linearized model whereas the oscillation frequency deviates as the E_{test} increases. Again these cases provide sufficient evidence (see Table 4.9 and Fig. 4.31) to state that the $C_{test} - R$ coefficient has to be selected not higher than 0.01 while the E_{test} parameter must be also small.

<--E_{test}-->	C_{test}-R	Theoretical $f_{osc}\vert_{norm}$	Frequency Validity	Amplitude Validity
Case #1	0.01	0.0160	The simulation results follow theoretical predictions only in the interval [0,0.025]	
Case #2	0.05	0.0357	The simulation results follow theoretical predictions only in the interval [0,0.006]	Good agreement
Case #3	0.10	0.2341	The simulation results follow theoretical predictions only in the interval [0,0.005]	
Case #4	0.20	0.2180	The simulation results follow theoretical predictions only in the interval [0,0.002]	

Table 4.9: Oscillation parameters validity (II)

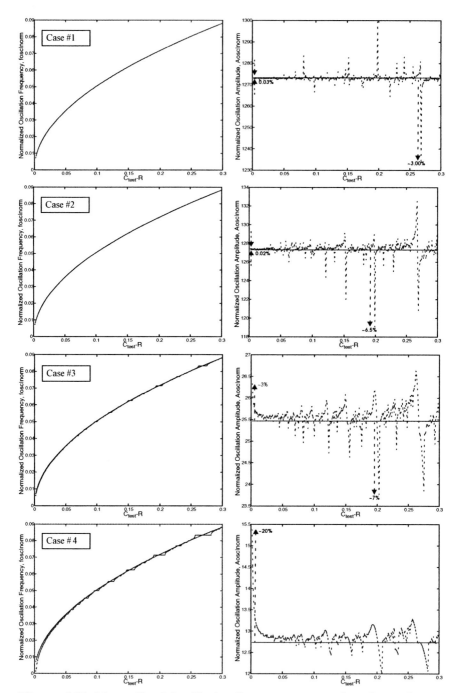

Figure 4.31: Normalized Oscillation Parameters sweeping $C_{test} - R$

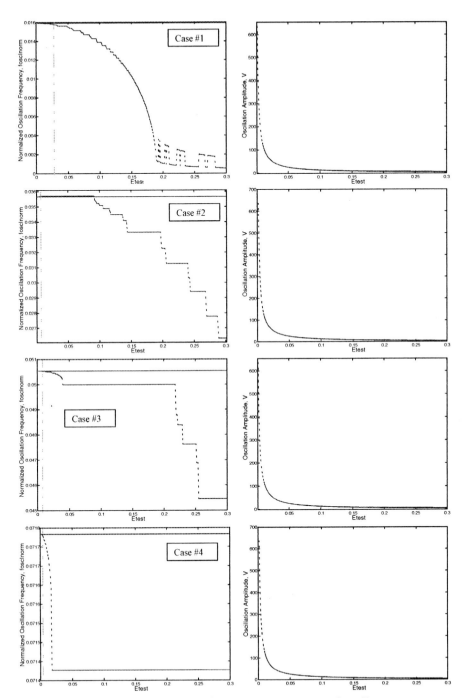

Figure 4.32: Normalized Oscillation Parameters sweeping E_{test}.

4.3.2 Fault Coverage considerations

The above expressions (Table 4.7) were derived assuming the extra loop coefficient δ_{test}, to be exactly equal to δ_0. However, if there is not an absolute matching between these elements, (4.19) (the case of the oscillation frequency) becomes slightly different. Specifically, if δ_{test} is $\delta_0 + \Delta\delta_{test}$, we can prove:

$$\cos\left(2\pi\frac{f_{osc}}{f_{sampling}}\right)\bigg|_{actual} = \cos\left(2\pi\frac{f_{osc}}{f_{sampling}}\right)\bigg|_{ideal} + \Delta\delta_{test}\frac{E_{test}}{\delta_0 - \delta_1} \qquad (4.20)$$

This error term gives us an opportunity to tune the sensitivity of the expected frequency variations when testing. Therefore, for a given value of the sum $\delta_0 - \delta_1$, the larger the value of E_{test}, the bigger the influence of the additional element, δ_{test}. In fact, this point is very significant because shows that we must not underestimate the role of the δ_{test} parameter in the definitive value of the oscillation frequency.

On the other hand, suppose a deviation in δ_0 or δ_1 (that is, $\Delta\delta_0$ or $\Delta\delta_1$). Then we obtain a new oscillation amplitude given by:

$$A_{osc}\big|_{actual} = A_{osc}\big|_{ideal} + \frac{2}{\pi}\cdot\frac{|V_{ref}||\Delta\delta_i|}{E_{test}} \qquad i = 0, 1 \qquad (4.21)$$

Observe, from this last expression, that the value of the E_{test} parameter controls the importance of a δ_i deviation in the oscillation amplitude. However, this importance also depends on the exact value of the $A_{osc}\big|_{ideal}$ which is governed, not only by the parameters δ_i and V_{ref}, but also by the E_{test}. Therefore, a good estimation of the influence of a δ_i deviation is better given by the calculation of the corresponding sensitivities.

Likewise, we can study the influence of a deviation of R in the oscillation frequency. In this case we have

$$\cos\left(2\pi\frac{f_{osc}}{f_{sampling}}\right)\bigg|_{actual} = \cos\left(2\pi\frac{f_{osc}}{f_{sampling}}\right)\bigg|_{ideal} + \frac{\Delta R}{2} \qquad (4.22)$$

Therefore, Table 4.10 gives an exhaustive information about the behavior of the oscillation parameters versus the oscillator coefficients. To consider all the possibilities we can calculate the amplitude sensitivity as well as the frequency sensitivity (see Table 4.10).

$S_{\delta_1}^{A_{osc}} = \dfrac{100}{\|\delta_1 - \delta_0\|}$	$S_{\delta_1}^{\cos(\Theta_{osc})} = 0$
$S_{\delta_0}^{A_{osc}} = \dfrac{-100}{\|\delta_1 - \delta_0\|}$	$S_{\delta_0}^{\cos(\Theta_{osc})} = 0$
$S_{E_{test}}^{A_{osc}} = \dfrac{-100}{E_{test}}$	$S_{E_{test}}^{\cos(\Theta_{osc})} = 0$
$S_{R}^{A_{osc}} = 0$	$S_{R}^{\cos(\Theta_{osc})} = \dfrac{-100}{-2 - R + C_{test}}$
$S_{C_{test}}^{A_{osc}} = 0$	$S_{C_{test}}^{\cos(\Theta_{osc})} = \dfrac{100}{-2 - R + C_{test}}$

Table 4.10: Sensitivities in frequency and amplitude

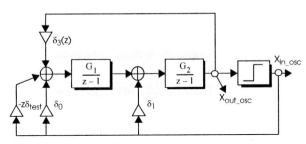

Figure 4.33: Oscillator Scheme

A last consideration refers to the importance of the oscillation frequency as a test parameter. Observe from Table 4.10 that the oscillation frequency is not a function of the modulator gains, at least in the case of a low-pass modulator ($R = 0$) where it exclusively depends on the extra parameter C_{test} added to built the test oscillator (see Table 4.7). In fact, from expressions in Table 4.7, the oscillation frequency may seem a secondary test parameter. However, the expressions in Table 4.7 were deduced without considering some issues. They were derived regarding the integrators of the modulators as ideal. Let us now suppose, for example, that the gains of the integrators are taken into account (G_1 and G_2 in Fig. 4.33). Then, the actual expressions of the oscillation parameters are

$$A_{osc} = \frac{2V_{ref}}{\pi} \cdot \left(\frac{\delta_1 - G_1\delta_0}{G_1 E_{test}} \right)$$

$$f_{osc} = \frac{1}{2\pi T_s} \cdot \text{acos}\left[\frac{2 + G_1 G_2 (R - C_{test})}{2} \right]$$

(4.23)

We can observe how G_2 only influences the oscillation frequency. Therefore, this parameter will be necessary to detect deviations (or faults) in the second integrator gain. In fact, it was indirectly studied in the sections concerning to the fault analysis where some graphics showing this situation were displayed (Fig. 4.8 and Fig. 4.23).

4.4 SUMMARY

In this Chapter an extension of the OBT/OBIST concept has been presented to be applied to Oversampled $\Sigma\Delta$ modulators, exploiting previous experience coined through the implementation of OBT/OBIST in SC integrated filters. This approach is based on the analogy between a filter and the linear part of the analog modulator core. After presenting some problems related to the OBT strategy, when global feedback loops are used to design the required oscillator, a basic method and its associated equations have been discussed, and the outcome seems to be appealing for future development. Analytical and simulation results demonstrate that it is always feasible to find out an OBT scheme for a typical discrete-time second-order modulator structure without adding any substantial extra circuitry, but only resorting to local feedback loops. A feedback strategy can be chosen providing enough freedom to force oscillations, which can be worthwhile for testing purposes. The conversion of a second-order modulator into an oscillator allows to use the existing knowledge on OBIST as was coined for filters.

The technique has been presented and illustrated through some particular second-order structures at block level. Initial fault analysis shows that an interesting difference in this case is the availability of extra freedom through the selection of the added components in order to increase the sensitivity of (oscillator) frequency and amplitude to faults. More detailed experiments at a circuit level must be carried out to fully validate this technique. Much work has to be done before qualifying this technique both in terms of its practical fault coverage and of its validity for any kind of oversampling modulator, but the initial results as well as the simplicity of the method make it promising.

Since this test methodology is not strongly correlated to the converter's performance parameters we should complement in the time being the above analysis taking into account the following points: more realistic test threshold for the CUT measurements, potential test escapes and possible decrease in yield. Furthermore, like most defect-oriented approaches, future work has to be devoted to connect this defect-driven test method with the conventional specification-driven test measurements required to characterize analog and mixed-signal circuits.

The basic concept is applicable to modulators of higher order and with more complex structure. The main result for second-order can thus be combined with extensions on partitioning and re-use, such as it was introduced by the authors in previous papers (see [34], for instance), in order to apply the concept reported herein to test any modulator.

Finally, in the last part of the Chapter, critical points of the proposed OBT solution have been considered in order to establish some guidelines useful to define a systematic way to implement this test approach for any kind of ΣΔ modulators.

Chapter 5

OBT Implementation in Discrete-Time Filters
Circuits and Examples

IN THIS CHAPTER different low-order filters forming the filter banks of a Dual-Tone Multifrequency (DTMF) receiver will be used as examples to theoretically validate the basis of the OBT methodology itself and the test proposals developed in previous chapters. Likewise, this particular system[1] will be employed in next chapters to demonstrate experimentally all the theoretical results obtained in this Chapter.

5.1 A SPECIFIC CIRCUIT

Let us study in detail a particular example of an OBT application. Until now, in this book, only the FL-topology has been considered at component level but never at capacitor level (see Chapter 3). Fig. 5.1 displays the generic Switched-Capacitor (SC) biquad that will be employed in the following sections to implement the second-order functions which will be tested by the OBT approach. The purpose is to link the OBT analysis discussed in Chapter 3 with the involved capacitors of the proposed biquad structure. The biquad shown in Fig. 5.1 can be customized to operate as required for any filter stage by adequately sizing (or even removing) the existent capacitors. This circuit is an integrator-based second-order biquadratic section with two available outputs (one per integrator). It is possible to implement a given function employing V_{o1} or V_{o2} as the circuit output. Then, taking advantage of this feature, both the normal output, V_{o2} as well as the secondary output, V_{o1} can be used to implement the required oscillators to exploit the OBT approach. This last mentioned issue is very valuable owing to the fact that forcing oscillations in some biquad configurations is not feasible whereas in other biquad configurations is relatively straightforward.

---

[1] The DTMF receiver will be describe in detail in Chapter 6 and Chapter 7.

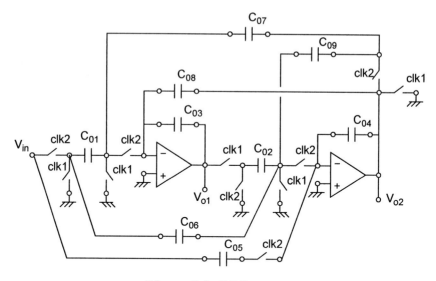

Figure 5.1: SC Structure

For the sake of clarity, let us simplify this structure with a z-domain equivalent circuit shown in Fig. 5.2 where, now, the drawn capacitors are normalized with respect to C_{03} or C_{04}, respectively, in such a way that

$$C_1 = \frac{C_{01}}{C_{03}} \qquad C_2 = \frac{C_{02}}{C_{04}} \qquad C_5 = \frac{C_{05}}{C_{04}}$$

$$C_{56} = \frac{C_{56}}{C_{04}} \qquad C_7 = \frac{C_{07}}{C_{03}} \qquad C_8 = \frac{C_{08}}{C_{03}} \qquad C_9 = \frac{C_{09}}{C_{04}}$$

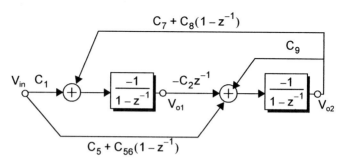

Figure 5.2: Biquad z-domain equivalent circuit

If we compare this z-domain equivalent circuit shown in Fig. 5.2 with the z-domain equivalent circuit of the FL implementation used in Chapter 3

(Fig. 5.3), we observe that the first one is simply a modified version of the second one. We just have to impose the following constraints[2]

$$H = 0 \qquad G = C_1 \qquad A = C_2$$
$$I = C_5 + C_{56} \qquad J = C_{56} \qquad C = C_7$$
$$E = C_8 \qquad B = D = 1 \qquad F = C_9$$

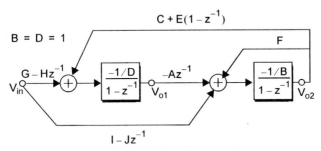

Figure 5.3: FL implementation

The generic transfer functions at each output are shown in Table 5.1. The coefficients in both numerator and denominator are related to the normalized capacitors.

$H_{01}(z) = \dfrac{V_{o1}}{V_{in}} = \dfrac{k_{21}z^2 + k_{11}z + k_{01}}{z^2 + b_1 z + b_0}$	$H_{02}(z) = \dfrac{V_{o2}}{V_{in}} = \dfrac{k_{22}z^2 + k_{12}z + k_{02}}{z^2 + b_1 z + b_0}$
$k_{21} = -C_1 + \dfrac{(C_5 + C_{56})(C_7 + C_8)}{1 + C_9}$	$k_{22} = \dfrac{-(C_5 + C_{56})}{1 + C_9}$
$k_{11} = \dfrac{C_1 - C_8(C_5 + 2C_{56}) - C_{56}C_7}{1 + C_9}$	$k_{12} = \dfrac{C_5 + 2C_{56} - C_1 C_2}{1 + C_9}$
$k_{01} = \dfrac{C_{56}C_8}{1 + C_9}$	$k_{02} = \dfrac{-C_{56}}{1 + C_9}$
$b_1 = \dfrac{-2 - C_9 + C_2(C_7 + C_8)}{1 + C_9}$	$b_0 = \dfrac{1 - C_2 C_8}{1 + C_9}$

Table 5.1: Biquad coefficients in relation to the involved capacitors

---

[2] Observe that, in this case, we do not consider $A = 1$ as in Chapter 3.

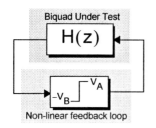

Figure 5.4: OBT Scheme

Let us consider again the proposed scheme of OBT shown in Fig. 5.4. That is, closing a non-linear feedback loop around the biquad under test by using a comparator with two level of saturation, V_A and $-V_B$. In this case, the oscillation parameters in function to the biquad coefficients extracted by the DF approach are given by (see Chapter 3)

$$A_{osc} = \frac{2V_{ref}}{\pi} \cdot \frac{k_2 - k_0}{1 - b_0}$$

$$f_{osc} = \frac{1}{2\pi T_s} \cdot acos\left[\frac{1}{2} \cdot \frac{b_1(k_2 - k_0) + k_1(b_0 - 1)}{k_0 - b_0 k_2}\right]$$

(5.1)

being T_s the system sampling period and $V_{ref} = V_A + V_B$.

On the other hand, the oscillation conditions are (see Chapter 3)

$$k_2 \neq k_0$$

$$sign(V_{ref}) = sign(k_2 - k_0) = sign(k_2 b_0 - k_0)$$

(5.2)

Let us study the general case where all capacitors are present. Table 5.2 shows the dependency of the oscillation parameters with respect to all the involved capacitors. An important fact can be derived from it: depending on the selected output for establishing the feedback loop, the oscillation parameters give more or less test information and the oscillation conditions are more or less restrictive. For instance, the parameter $A_{osc}[H_{o2}]$ is not a function of the capacitors C_1 and C_7. It would reduce, in a general case, the fault coverage. Notice, however, this is not the same for $A_{osc}[H_{o1}]$ and $f_{osc}[H_{o1}]$ because both parameters are function of all normalized capacitors involved in the structure. It can be very positive in order to efficiently apply the OBT strategy. On the other hand, notice that for H_{o2}, the requirements imposed by the oscillation conditions are stronger because they involve less coefficients. Moreover, if C_5 is not present, then, the structure can not oscillate.

From (5.1) and (5.2), it must be clear that examining the general case is not straightforward. Moreover, the general case does not allow us to deduce practical conclusions. Therefore, for the sake of convenience, let us consider some particular examples which will be employed in next chapters as vehicles to experimentally demonstrate the feasibility of the OBT approach.

	A_{osc}	f_{osc}	Oscillation Conditions
H_{o1}	$f(C_1, C_2, C_5, C_7, C_8, C_9, C_{56})$		$-C_1 + \dfrac{C_5 \cdot (C_7 + C_8) + C_{56}C_8}{1 + C_9} \neq 0$ $sign(V_{ref}) = f(C_1, C_2, C_5, C_7, C_8, C_9, C_{56})$
H_{o2}	$f(C_2, C_5, C_8, C_9, C_{56})$	$f(C_1, C_2, C_5, C_7, C_8, C_9, C_{56})$	$\dfrac{-C_5}{1 + C_9} \neq 0$ $sign(V_{ref}) = f(C_2, C_5, C_8, C_9, C_{56})$

Table 5.2: Dependences of the oscillation parameters with the involved capacitors

5.2 SOME PRACTICAL EXAMPLES

Let us now consider the two filter banks shown in Fig. 5.5. They consist of a cascade of several second-order functions or biquads[3] (called in Fig. 5.5 *Not#1, Not#2, LG#1, LG#2, LG#3, HG#1, HG#2* and *HG#3*). Each filter bank is a high-order filter whose properties make this kind of filters very suitable not only to be used as a benchmark for the OBT technique but also to extract conclusions that can be extended to other applications with the same characteristics.

As can be seen in Fig. 5.5, three groups of filters are involved: the Dialing Filter, the Low-Band Filter and the High-Band Filter. In our actual example, all the biquads are implemented using the generic topology shown in Fig. 5.1.

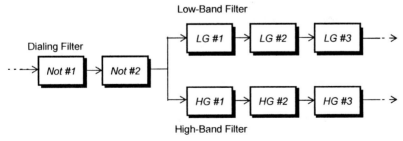

Figure 5.5: Analog Filter Banks of a DTMF

[3.] As will be detailed in next chapters, the analog core of a Dual-Tone Multi-Frequency (DTMF) receiver is essentially composed of this dual structure.

Let us show in Table 5.3 the involved capacitors for each biquad group.

Group	Biquad	C_{01}	C_{02}	C_{05}	C_{07}	C_{08}	C_{09}	C_{56}
Dialing Filter	Not #1			▓		▓		
	Not #2			▓			▓	
Low-Band Filter	LG #1			▓				▓
	LG #2						▓	
	LG #3	▓					▓	
High-Band Filter	HG #1			▓			▓	▓
	HG #2			▓		▓	▓	
	HG #3	▓		▓			▓	

Table 5.3: Capacitors present in every type of biquad. Shadow squares mean that the capacitor is deleted from the actual filter (Fig. 5.1)

Additionally, let us derive the specific transfer function coefficients in relation to the capacitor values (see Table 5.4 and Table 5.5).

Biquad	k_2	k_1	k_0	b_1	b_0
LG #1, HG #1	$-C_1$	C_1	0	$-2 + C_2\langle C_7 + C_8\rangle$	$1 - C_2 C_8$
Not #1	$-C_1 + \dfrac{C_{56}C_7}{1+C_9}$	$\dfrac{C_1 - C_{56}C_7}{1+C_9}$	0	$\dfrac{-2 - C_9 + C_2 C_7}{1+C_9}$	$\dfrac{1}{1+C_9}$
LG #2, HG #2, Not #2	$-C_1 + C_{56}(C_7 + C_8)$	$C_1 - C_{56}(C_7 + 2C_8)$	$C_{56}C_8$	$-2 + C_2\langle C_7 + C_8\rangle$	$1 - C_2 C_8$
HG #3	$C_{56}(C_7 + C_8)$	$-C_{56}\langle C_7 + 2C_8\rangle$	$C_{56}C_8$	$-2 + C_2\langle C_7 + C_8\rangle$	$1 - C_2 C_8$
LG #3	$C_5(C_7 + C_8)$	$-C_5 C_8$	0	$-2 + C_2\langle C_7 + C_8\rangle$	$1 - C_2 C_8$

Table 5.4: Biquad coefficients for Ho1

Biquad	k_2	k_1	k_0	b_1	b_0
LG #1, HG #1	0	$-C_1 C_2$	0	$-2 + C_2 \langle C_7 + C_8 \rangle$	$1 - C_2 C_8$
Not #1	$\dfrac{-C_{56}}{1+C_9}$	$\dfrac{2C_{56} - C_1 C_2}{1+C_9}$	$\dfrac{-C_{56}}{1+C_9}$	$\dfrac{-2 - C_9 + C_2 C_7}{1+C_9}$	$\dfrac{1}{1+C_9}$
LG #2, HG #2, Not #2	$-C_{56}$	$2C_{56} - C_1 C_2$	$-C_{56}$	$-2 + C_2 \langle C_7 + C_8 \rangle$	$1 - C_2 C_8$
HG #3	$-C_{56}$	$2C_{56}$	$-C_{56}$	$-2 + C_2 \langle C_7 + C_8 \rangle$	$1 - C_2 C_8$
LG #3	$-C_5$	C_5	0	$-2 + C_2 \langle C_7 + C_8 \rangle$	$1 - C_2 C_8$

Table 5.5: Biquad coefficients for Ho2

We can recognize several types of biquads existing in the DTMF core as shown in Table 5.6.

Group	Biquad	H_{o1}	H_{o2}
Dialing Filter	Not #1	GENERAL	HP-Notch
	Not #2	GENERAL	HP-Notch
Low-Band Filter	LG # 1	BP00	LP01
	LG #2	GENERAL	LP-Notch
	LG #3	GENERAL	BP00
High-Band Filter	HG #1	BP00	LP01
	HG #2	GENERAL	HP-Notch
	HG #3	GENERAL	HP

Table 5.6: Types of transfer functions present in the filter banks

Analysing every specific second-order block into the filter banks, we can establish which output can be used to implement every oscillator and under which conditions oscillations can be sustained. Table 5.7-Table 5.9 show the results. Notice we have studied for every biquad (every row), the two possible cases: the H_{o1} transfer function and the H_{o2} transfer function.

Biquad	H_{o1}	Oscillation Conditions	H_{o2}	Oscillation Conditions
Not #1	GENERAL	IFF $sign(V_{ref}) = sign(-C_1(1+C_9) + C_{56}C_7)$ $-4 < \dfrac{C_1 C_9^2 - C_2 C_7 C_1(1+C_9) + C_2 C_7^2 C_{56}}{C_1(1+C_9) - C_{56}C_7} < 0$	HP-Notch	NO OSCILLATIONS
Not #2	GENERAL	IFF $sign(V_{ref}) = sign(-C_1 + C_{56}C_7)$ $-4 < \dfrac{(C_2 C_7)(C_1 - C_{56}C_7)}{(C_1 - C_{56}C_7)(C_2 C_8 - 1) - C_2 C_8^2 C_{56}} < 0$	HP-Notch	

Table 5.7: Oscillations for the Dialing Filter Group

Biquad	H_{o1}	Oscillation Conditions	H_{o2}	Oscillation Conditions
LG # 1	BP00	$sign(V_{ref}) < 0$ IFF $\;-4 < \dfrac{-C_2 C_7}{1 - C_2 C_8} < 0$	LP01	NO OSCILLATIONS
LG #2	GENERAL	IFF $sign(V_{ref}) = sign(-C_1 + C_{56}C_7)$ $-4 < \dfrac{(C_2 C_7)(C_1 - C_{56}C_7)}{(C_1 - C_{56}C_7)(C_2 C_8 - 1) - C_2 C_8^2 C_{56}} < 0$	LP-Notch	
LG #3	GENERAL	IFF $sign(V_{ref}) > 0$ $-4 < \dfrac{C_2 C_7^2}{(C_7 + C_8)(C_2 C_8 - 1)} < 0$	BP00	IFF $sign(V_{ref}) < 0$ $-4 < \dfrac{-C_2 C_7}{1 - C_2 C_8} < 0$

Table 5.8: Oscillations for the Low-Band Filter Group

Biquad	H_{o1}	Oscillation Conditions	H_{o2}	Oscillation Conditions
HG #1	BP00	$\text{sign}(V_{ref}) < 0$ IFF $-4 < \dfrac{-C_2 C_7}{1 - C_2 C_8} < 0$	LP01	
HG #2	GENERAL	IFF $\text{sign}(V_{ref}) = \text{sign}(-C_1 + C_{56} C_7)$ $-4 < \dfrac{(C_2 C_7)(C_1 - C_{56} C_7)}{(C_1 - C_{56} C_7)(C_2 C_8 - 1) - C_2 C_8^2 C_{56}} < 0$	HP-Notch	NO OSCILLATIONS
HG #3	GENERAL	$\text{sign}(V_{ref}) > 0$ IFF $-4 < \dfrac{C_2 C_7^2}{(C_7 + C_8)(C_2 C_8 - 1)} < 0$	HP	

Table 5.9: Oscillations for the High-Band Filter Group

From Table 5.7-Table 5.9, it can be observed that V_{o2} is not normally useful for implementing the closed-loop system required in the OBT technique excepting the biquad named LG #3. It is due to the fact that the capacitor C_5 is not present in the structure of all biquads excluding LG #3 (see Table 5.4). It is the determining factor which causes that all the biquads present an inadequate transfer function in the second output. This issue could be considered as a DfT rule. For example, the required transfer functions could be designed taking from the constellation of possible solutions, those ones involving the use of C_5.

In summary, Fig. 5.6 shows in thick line the biquad outputs that can be employed to build the oscillators.

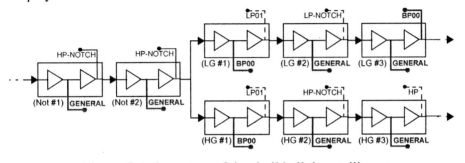

Figure 5.6: Outputs useful to build all the oscillators

5.3 FAULT COVERAGE CONSIDERATIONS

Table 5.10 and Table 5.11 show for every oscillator of the DTMF core how a ±10% deviation in each individual capacitor influences each oscillation parameter (these tables outline all the results extracted from a graphical analysis made in Appendix 5.A).

Deviation		Not #1	Not #2	LG #1	LG #2
±10% C_1	Freq.	~0%	3.4%, 1.8%	~0%	0%, 0.8%
	Amp.	~±10%	18%, 15%	~±10%	22%, 28%
±10% C_2	Freq.	6%, 6.5%	6.5%, 7.4%	~±4.5%	4.6%, 5.1%
	Amp.	~0%	9%, 10%	~±10%	9%, 10%
±10% C_5	Freq.				
	Amp.				
±10% C_7	Freq.	6%, 4%	1.7%, 0%	~±4.5%	4.0%, 5.1%
	Amp.	16%, 20%	22%, 28%	~0%	18%, 15%
±10% C_8	Freq.		3.4%, 3.6%	~0%	0.8%, 0%
	Amp.		9%, 10%	~±10%	9%, 10%
±10% C_9	Freq.	~0%			
	Amp.	~±10%			
±10% C_{56}	Freq.	~0%	1.8%, 3.4%		0.8%, 0%
	Amp.	16%, 20%	22%, 28%		18%, 15%

Table 5.10: Impact of the capacitor deviations in the oscillation parameters (I)

Deviation		LG #3	HG #1	HG #2	HG #3
$\pm 10\%\ C_1$	Freq.		~0%	~0%	
	Amp.		~$\pm$10%	14.5%, 16.9%	
$\pm 10\%\ C_2$	Freq.	4%, 6.5%	~$\pm$4.5%	5.2%, 4.7%	5%, 5.5%
	Amp.	9%, 10%	~$\pm$10%	9%, 10%	9%, 10%
$\pm 10\%\ C_5$	Freq.	~0%			
	Amp.	9%, 10%			
$\pm 10\%\ C_7$	Freq.	4%, 6.5%	~$\pm$4.5%	~$\pm$4.5%	~0%
	Amp.	~0%	~0%	21%, 27%	9%, 10%
$\pm 10\%\ C_8$	Freq.	~0%	~0%	~0%	4.5%, 5%
	Amp.	9%, 10%	~$\pm$10%	9%, 10%	9%, 10%
$\pm 10\%\ C_9$	Freq.				
	Amp.				
$\pm 10\%\ C_{56}$	Freq.			~0%	16%, 20%
	Amp.			21%, 27%	9%, 10%

Table 5.11: Impact of the capacitor deviations in the oscillation parameters (II)

For the sake of clarity, Table 5.12, a summary of Table 5.10 and Table 5.11, is included as well.

$\pm 10\ \%$ in a capacitor		Frequency Deviation		
(Total: 36 cases)		<2%	2% to 9%	"" $\geq$ 10%
Amplitude Deviation	Cases Number /Total Cases	17/36	18/36	1/36
		Cases Number/Amplitude Deviation Cases		
<2%	4/36	0/4	4/4	0/4
2% to 9%	0/36	0/0	0/0	0/0
"" $\geq$ 10%	32/36	17/32	14/32	1/32

Table 5.12: Summary of the impact in the oscillation parameters

From Table 5.10, Table 5.11 and Table 5.12, it is simple to see that if only one of the oscillation parameters is measured we can not detect all $\pm 10\%$ capacitor deviations for all the biquads. When only the oscillation frequency is measured, 47.22% of the cases (17 of 36 cases) do not show up, while for amplitude-only measurements the percentage of undetected cases is 11.11%[4]. Fortunately, there is no overlapping between the cases undetected by frequency-only and amplitude-only measurements[5] giving a high fault coverage (almost 90% for a measurement accuracy of 10% and 100% for an accuracy of 2-9%).

On the other hand, let us also outline in Table 5.13 the results extracted from the analysis of the Bode Diagrams obtained for every biquad. From such graphics we can compare the resulting oscillation frequencies with the peak frequencies of the biquads (see Apendix 5.A). From Table 5.13 it should be clear that in most cases the oscillation frequency is close to the involved peak frequency. But if we study Table 5.13 thoroughly, we can observe that Not #1 (which may seem equal to Not #2, LG #2 and HG #2) presents, however, an oscillation frequency very different from its peak frequencies. It is, perhaps, due to the effect of the capacitor C_9, which is present in Not #1 but not in Not #2, LG #2 and HG #2 (see Table 5.7-Table 5.9).

BIQUAD	TYPE	FEEDBACK OUTPUT	$f_{peak}\big|_{V_{o1}}$ (Hz)	$f_{peak}\big|_{V_{o2}}$ (Hz)	f_{osc} (Hz)	Distance (%)	
LG #1	BP00-LP01		846.42	821.98	816.00	3.60	0.73
HG #1			1350.34	1320.99	1363.00	0.93	3.08
Not #1		V_{o1}	352.21	313.21	504.00	30.12	37.86
Not #2	GENERAL-		567.61	430.62	666.00	14.82	35.38
LG #2	HPNOTCH		978.53	968.75	975.68	0.36	0.64
HG #2			1169.33	929.61	1188.00	1.57	21.75
LG #3	GENERAL-BP00	V_{o2}	611.62	596.95	616.00	0.71	3.09
					218.00	64.36	63.48
HG #3	GENERAL-HP	V_{o1}	1683.06	1702.63	1728.00	2.60	1.47

Table 5.13: Involved frequencies in the different obtained oscillators

---

[4] Let us remark that, in this case, the amplitude covers more capacitor deviations than the frequency.

[5] Observing Table 5.12 we have that the four capacitor deviations where the amplitude deviation is zero can be detected by using the frequency if it can be measured with a 2-9% of precision.

5.4 OSCILLATOR MODELLING ACCURACY

A first step is validating the accuracy of the oscillation parameters as predicted by the DF approach. Observing all the Figures obtained in Appendix 5.B we can make a first determination about how accurate is the linearized model given by the DF approach with regard to a more meticulous simulation approach (see the corresponding Table 5.14).

Group	Biquad	Frequency (Hz) (DF approach)	Frequency (Hz) (Matlab)	Frequency Deviation	Amplitude (V) (DF approach)	Amplitude (V) (Matlab)	Amplitude Deviation
Dialing Filter	Not #1	503.66	503.66	~0.00%	6.523	6.545	0.34%
	Not #2	666.34	629.62	5.51%	1.378	1.532	10.03%
Low-Band Filter	LG # 1	816.19	796.51	2.41%	1.272	1.276	0.31%
	LG #2	975.68	967.51	0.84%	1.927	1.926	0.06%
	LG #3	616.89	603.90	2.11%	5.064	5.072	0.17%
High-Band Filter	HG #1	1363.12	1345.49	1.29%	1.306	1.310	0.31%
	HG #2	1188.59	1181.89	0.56%	1.526	1.533	0.45%
	HG #3	1728.93	1710.38	1.07%	7.130	7.228	1.36%

Table 5.14: Comparison between the oscillation parameters given by Matlab/Simulink and those obtained by the DF approach

The DF approach usually yields an accurate approximation to any oscillation's analytical description with the evident exception of the biquad called Not #2. In this case, simulation results given by Simulink [124] differ appreciably from the theoretical predictions using such a DF method. However, for other similar cases as LG #2 and HG #2, the theoretical oscillation parameters practically match up with the simulation results. This is a clear example of a case where the results given by the DF approach are contradictory for the same type of biquad. These biquads present the same type of transfer function, but however, the order the magnitude of their involved capacitor values are different. When the DF approach was revisited in Chapter 2, it was found that depending on the specific values of the transfer function coefficients, the error bounds of this method can vary a lot for a same type transfer function. That means that although apparently we have the same situation with Not #2, LG #2 and HG #2, the accuracy of the DF method is not enough for Not #2. As a curiosity, if the three involved Bode Diagrams are observed (see Fig. 5.7), for the Not #2, around the oscillation frequency value the system does not behave as a bandpass while it happens in the other two cases.

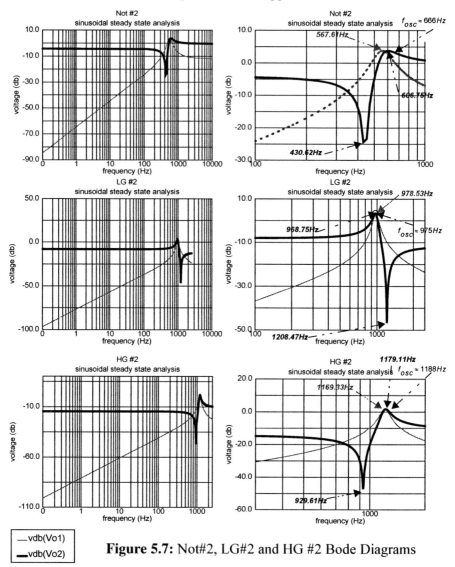

Figure 5.7: Not#2, LG#2 and HG #2 Bode Diagrams

In short, a first-order DF analysis predicts that the studied closed-loop systems oscillate at a frequency very close to the filter pole frequency. All the studied biquads of the DTMF core (with the exception of the biquad called Not #2) are examples where the predicted oscillations have a good agreement with the simulation results, and the DF approach is valid. However, we could confirm that, in cases such as Not #2 (see Chapter 2), we would need a readjustment mechanism. Then, the strategy would be to use the DF result as a first seed for further simulations. But problems could appear when even the simulations were not useful (see Chapter 4 where we find some oscillator examples in which, under some conditions, their oscillation results predicted by simulation deviate substantially from the oscillation results predicted by the DF method). Then, a higher order DF analysis or a more accurate nonlinear theory would be required to get close to the exact oscillation signal.

5.5 DTMF BIQUAD VALIDATION

The next step would be the validation of the DF model when the DTMF biquadratic cells are used to build oscillators but from a more practical viewpoint. For this purpose, the setup block diagram of Fig. 5.8 has been employed. Notice it differs slightly from that considered in previous sections. However, this new scheme is more generic and realistic because it takes into

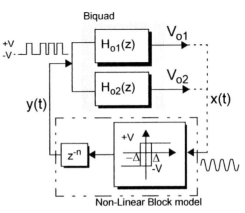

Figure 5.8: Realistic oscillator model

account important factors that have an impact on the idealized model proposed in former sections. It considers the existence of a certain hysteresis in the comparators used in the non-linear feedback element as well as possible delays in the feedback loop. In fact, Fig. 5.8 takes into account the existing zero crossing detectors in the DTMF core. As will be seen in next chapters, these zero crossing detectors will be re-used to build the OBT oscillators.

When a comparator with hysteresis is considered the describing-function is then: $N(A) = (4V/A\pi)e^{-j\theta}$, where $\theta = \mathrm{asin}(\Delta/A)$, representing Δ the value of the hysteresis window. In this case, analytical equations involving amplitude and frequency become more complex.

With the exception of the case where no delay is present in the structure (case in which practical analytical oscillation equations are possible), for the other cases, the resulting system of equations makes very difficult to cope with symbolic yet generic calculations to obtain the oscillation parameters. For this reason, calculations for every biquad have been made numerically using the Matlab software tool [126].

The validation has been performed in two steps. First of all, the resulting oscillator has been computed using the linearized model of the non-linear block but considering different extra delays (n=0, 1 and 2). Although it may (to some extent) inaccurate, it can give us an approximated idea about the behaviour of the corresponding oscillator. The results in Table 5.15 correspond to the use of Vo1 as the feedback signal and in Table 5.16 to the Vo2 case.

Biquad	Feedback Sign	extra delays (z^{-n})					
		n=0		n=1		n=2	
		A	f_{osc}	A	f_{osc}	A	f_{osc}
Not #1	+	0.67	837	0.72	790	0.77	751
	-						
Not #2	+	1.57	623	1.66	612	1.73	604
	-			0.33	27966	0.32	13838
LG #1	+					0.04	9477
	-	1.27	853	1.27	840	1.26	828
LG #2	+			0.03	7271	0.04	4681
	-	1.88	974	1.84	968	1.78	963
LG #3	+	1.42	267				
	-					0.03	8212
HG #1	+					0.05	9528
	-	1.27	1360	1.27	1333	1.24	1307
HG #2	+	1.49	1187	1.54	1179	1.57	1172
	-					0.04	12994
HG #3	+	7.13	1727	7.37	1697	7.34	1669
	-					0.38	12290
		V	Hz	V	Hz	V	Hz

Table 5.15: Amplitude and frequency parameters using linear analysis when Vo1 is used as the feedback signal. An empty box indicates no oscillation. The amplitude (A) is relative to the output levels (V) of the non-linear function in Fig. 5.8

Biquad	Feedback Sign	extra delays (z^{-n})					
		n=0		n=1		n=2	
		A	fosc	A	fosc	A	fosc
Not #1	+			1.30	27967	1.30	14143
	-			1.28	2418	1.30	27967
Not #2	+			1.24	27967	1.24	14020
	-			1.33	1276	1.41	981
LG #1	+			0.12	1760	0.24	13732
	-						
LG #2	+			0.33	27967	0.33	14006
	-						
LG #3	+					0.08	9425
	-	4.16	614	4.16	608	4.14	602
HG #1	+			0.19	2217	0.37	1819
	-						
HG #2	+			0.38	27967	0.38	14010
	-			0.58	1517	0.76	1351
HG #3	+			1.30	27967	1.31	14058
	-			2.63	2325	3.86	2016
		V	Hz	V	Hz	V	Hz

Table 5.16: Amplitude and frequency parameters using linear analysis when Vo2 is used as the feedback signal. An empty box indicates no oscillation. The amplitude (A) is relative to the output levels (V) of the non-linear function in Fig. 5.8

It is important to re-use the original circuitry in order to avoid extra area and power overhead. That is, it would be convenient that all oscillators use the same circuitry in the feedback path. Moreover, many of the possible oscillators require extra delays in the feedback path to work properly (see Table 5.15 and Table 5.16). However, it is not a problem because the SC comparators, that are normally used to implement the zero crossing detectors, introduce it (this is what really happens in the DTMF demonstrator). In addition, it was shown in other sections that the OBT was successful if the oscillation frequencies are close to the biquad pole frequencies. On the other hand, the final configurations should facilitate their implementation in the whole system. Consequently, those configurations that not fulfil the above points have been discarded from a second validation step consisting on an accurate computation of the oscillation parameters. Therefore, non-linear analysis has been performed using Matlab and the complete model of Fig. 5.8. The hysteresis levels of the comparator have been also used to study the robustness of the oscillators showing that, under realistic variations (deduced from Monte Carlo analysis of a typical SC comparator), the oscillation parameters are very stable (they are not significantly altered).

The results obtained for the selected configurations are in Table 5.17. In this, the amplitude (A_{osc}) corresponds to the peak-voltage of the output. Concerning the frequency, it has been measured in two ways. The first one has been computed using the FFT (Fast-Fourier-Transform) of the output, while the second one corresponds to measurements over several periods of the squared-wave output once it is settled (for this reason a maximum, mean and minimum values appear). Both types of measurement make sense depending on how the test is evaluated. Of course, if an external evaluation is performed, both types of measurements can be performed. However, the second type seems, although with some loss of information, more appropriate if internal evaluation is performed (OBIST), because of the simplicity of the measurements. Again, from a more realistic oscillator scheme (Fig. 5.8), it can be inferred from the results using the linear and non-linear models, that a linear model of the oscillator is not usually enough to calculate the oscillation parameters.

Biquad	Output / Feedback Sign	extra delays									
		n=1					n=2				
		A_{osc} (Vp)	f_{osc} (FFT)	f_{osc} ($1/T_{square-wave}$)			A_{osc} (Vp)	f_{osc} (FFT)	f_{osc} ($1/T_{square-wave}$)		
				max	mean	min			max	mean	min
Not #1	Vo1 / +	0.85	710	717	717	717	0.85	710	699	699	699
Not #2	Vo1 / +	1.83	601	589	589	589	1.83	574	583	583	583
LG #1	Vo1 / -	1.27	819	835	829	823	1.27	819	823	823	823
LG #2	Vo1 / -	1.83	956	981	966	964	1.81	956	964	964	964
LG #3	Vo2 / -	4.17	601	608	605	601	4.15	601	601	595	599
HG #1	Vo1 / -	1.27	1338	1332	1332	1332	1.25	1311	1300	1300	1300
HG #2	Vo1 / +	1.55	1174	1190	1165	1179	1.57	1174	1190	1172	1165
HG #3	Vo2 / -	3.63	2048	2071	1998	2046	4.00	1939	1928	1928	1928
		V	Hz				V	Hz			

Table 5.17: Amplitude and frequency parameters using non-linear analysis. The amplitude (A) is relative to the output levels (V) of the non-linear function in Fig. 5.8

5.1 Fault coverage considerations

Another issue to consider is the magnitude to measure during the test phase. It has been discussed in Chapter 1 and extensively proven not only in several references [25]-[41], but also in previous sections, that frequency-only measurements may lead to insufficient fault coverage, thus rendering the OBT technique of little use. As was said, the alternative is the combined measurement of both frequency and amplitude. In our case, we have validated how OBT can be applied to the DTMF cell by extensive fault simulation [36]. For the most common faults affecting any of its biquads we can summarize the results obtained, when frequency and amplitude of the first harmonic of the oscillations have been measured. Results are displayed in Table 5.18, where the percentage of faults detected is compared when only the frequency or amplitude are measured (first row in Table 5.18) and when both are simultaneously measured (second row). In either case, a test accuracy of 5% was assumed. From this table, it should be clear that considering two test measurements is advantageous. Then, we can clearly postulate again that evaluating both, the frequency and amplitude of oscillations, is required to obtain high fault coverage.

	Detected Faults (%)							
Biquad	Not #1	Not #2	LG #1	LG #2	LG #3	HG #1	HG #2	HG #3
Only frequency or amplitude measured	86.7	80.0	71.7	76.7	81.6	83.4	88.3	78.3
Both frequency and amplitude measured	98.3	100	98.3	100	100	100	100	98.3

Table 5.18: Fault coverage for every biquad in the DTMF core

5.2 Test Quality

On the other hand, an estimation of the overall test quality of the OBT technique is required to support its use. In terms of fault coverage, there is a lack of a widely accepted criterion. Alternatively, we try to give an assessment by injecting meaningful faults into every biquad and proving that all of these faults can be detected. This assessment is complemented by showing that a fault-free biquad is not graded as faulty by using this approach.

As a first step, an acceptability region has to be defined as the place where all the good circuits must lie. During the design process of an analog IC, designers use the worst-case processing parameters given by the silicon foundry, to find a robust solution which optimizes production yield. If a filter is well-designed and correctly centered in the technology, the influence of worst-case electrical parameters should be minimized. On the other hand, capacitor mismatchings may have an important influence, too. In practice, centering a design is equivalent to defining a region in the design space which must be within the test acceptability region for the application specifications. For OBT we need to map such an inner zone into our test space, where the coordinates are deviations in the oscillation parameters with respect to the nominal ones. To determine where the fault-free circuits are within this test space, a Monte Carlo analysis was performed for each biquad considering process parameter variations and a 0.2% mismatching in capacitor ratios. The black box in the lower-left corner of the plot in Figure 5.9 shows a typical situation corresponding to Not #1. Although Monte Carlo's predictions are always pessimistic, all fault-free circuits exhibit a rather small deviation in the oscillation parameters. The results for all of the biquads are shown in Table 5.19. It can be seen from this Table that the frequency is very stable in all the cases, while the amplitude varies slightly more. These values will be used as the minimum tolerance margin allowed to decide whether a fault may or may not be detected, and to define an inner region within the acceptability region of the SUT.

	NOT #1	NOT #2	LG #1	LG #2	LG #3	HG #1	HG #2	HG #3
ΔA_{osc} (%)	4.7	4.3	1.7	3.9	1.9	2.2	3.4	4.4
Δf_{osc} (%)	2.5	0.7	0.8	0.8	0.9	1.1	0.9	1.2

Table 5.19: Maximum parameter deviations for a 0.2% of mismatch in capacitor ratios and process parameter variations

For all of the biquads in the above Table, we undertook an exhaustive validation of the OBT technique. We have considered small parametric changes which can potentially take the circuit out of specs and cause great changes or even change the circuit topologies. Fault simulations have been carried out with SWITTEST [59] for every biquad. Shorts and opens for switches and capacitors have been considered, as well as the influence of a permanent stuck-at ON of switches. Additionally, deviations from 5% up to 50% in capacitor values have also been injected. Therefore, both hard and soft faults have been considered in order to cover a wide range of possibilities[6]. From the simulator results, both the frequency and amplitude of the first harmonic of the oscillations have been measured. All these results are shown in Fig. 5.10 to Fig. 5.12 (and in Fig. 5.9 for the particular case, Not #1), where only those faults causing a deviation below 50% have been displayed.

A more complex issue is dealing with the test yield. The "gray" zone between the acceptability region and the space where the faulty circuits lie corresponds to circuits slightly out of specs. The ability to adjust test measurements will determine a higher or lower yield. But this is common to any test procedure and must be handled similarly for both a conventional functional test and for OBT. The tighter the measurements are, the higher is the risk, of rejecting good circuits, and vice versa.

Going back to Figure 5.9, a tolerance window has been drawn in gray illustrating where we place the acceptability region. The selected window allows us to discriminate all faulty circuits in this example. Similar results are shown for the remaining biquads in Fig. 5.10 to Fig. 5.13.

From all these Figures it should be clear that only one single frequency measurement is not always enough for an exhaustive testing. It can be seen from the figures that there are faults very close to the horizontal axis. They correspond to faults causing an oscillation frequency nearly identical to that of the fault-free biquad. In fact, only with an extremely precise measurement can some faults be detected. In reality, many of the above-mentioned faults could coincide with the deviations allowed by the Monte Carlo analysis. Therefore, if only the oscillation frequency is estimated, we could not note the differences between those faults and a tolerated deviation. Something similar happens when only amplitudes are measured. However, when both frequency and amplitude are considered, most of the injected faults are easily detected. Even in the most pessimistic situation, an accuracy of around 5% in frequency and 10% in amplitude is required for separating good from bad circuits. When this is the case, the percentage of detected faults (from those injected by simulation) ranges from 98.3 to 100% (depending on the biquad).

---

[6] The fault models are described in Appendix 5.C.

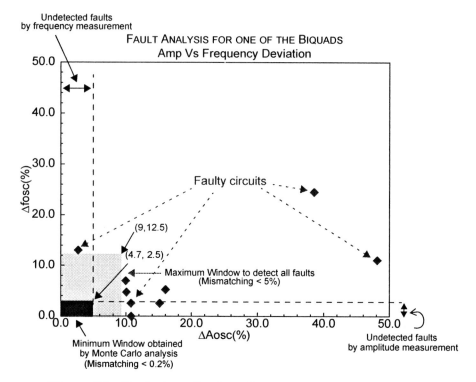

Figure 5.9: Typical amplitude vs frequency errors (%) in a biquad.
For clarity, only faults causing deviations below 50% are displayed

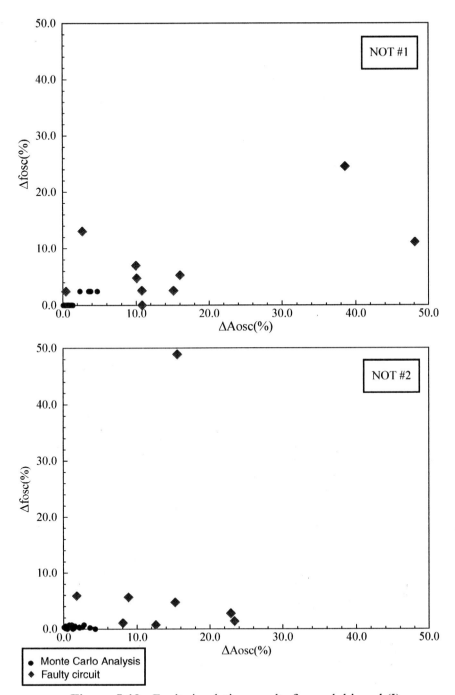

Figure 5.10: Fault simulation results for each biquad (I)

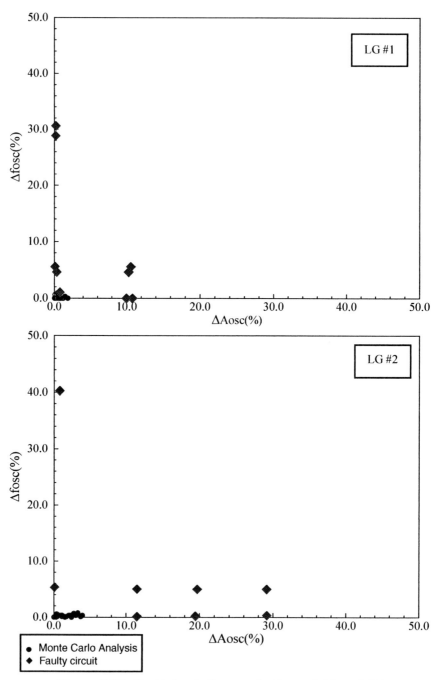

Figure 5.11: Fault simulation results for each biquad (II)

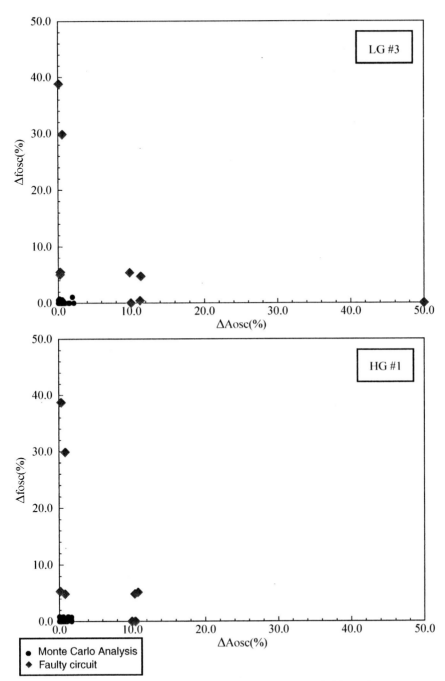

Figure 5.12: Fault simulation results for each biquad (III)

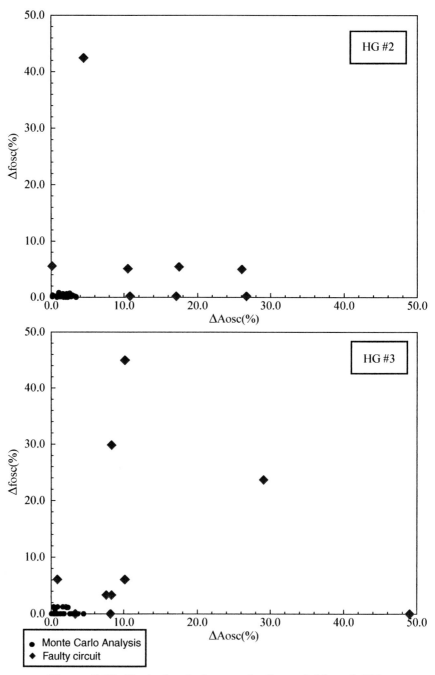

Figure 5.13: Fault simulation results for each biquad (IV)

5.6 SUMMARY

A generic biquadratic filter has been studied using both symbolic expressions and specific numerical data as well. The goal was to comprehensively evaluate the practical implementation of the OBT scheme proposed in Chapter 3 for a particular filter structure. Many conclusions about the OBT implementation in discrete-time filters, such as establishment of the test parameters, validation of the oscillator model, fault coverage, test quality, etc. have been extracted.

Summing up, the main conclusions are:

-the accuracy of the DF approach in this specific case is high. Only a few examples deviate slightly from the predicted theoretical results. In these cases, a more accurate solution can be extracted by using the DF result as first seed in a re-iteration simulation process.

-the defect coverage is high using both, amplitude and frequency tests, although is strongly dependent on measurement accuracy.

-the possibility of using other oscillation parameter (i.e. harmonic distortion) as a third parameter to detect faults may be explored to improve the OBT approach.

Chapter 6

Practical regards for OBT-OBIST implementation
Simulation Prototypes

THE OBT CONCEPT is an analog and mixed-signal testing technique which can be used in conventional off-line testing or as the core of the so-called Oscillation-Based Built-In-Self-Test (OBIST). In high-complexity analog and mixed-signal systems, where the accessibility to the system internal nodes is critically restricted and where there is a extremely limited number of test pins, the best option is to implement OBT by means of a BIST approach. Therefore, an OBIST solution would allow, not only to minimize the number of external test pins, but also to enhance the observability of the faults and the controllability of the test procedure.

The main goal of this Chapter is to describe an example of the integration of the OBT-OBIST technique into the frame of analog-core-based design of complex mixed-signal ICs. A Dual-Tone Multi-Frequency (DTMF) embedded macrocell will be used to illustrate the potentiality of applying OBIST in a complex industrial circuit. As a result, a complete and viable OBIST strategy will be reported.

In particular, we pursue to find an OBIST test solution satisfying four main objectives:

-an on-chip stimulus generation.

-an on-chip control of the test strategy (test circuitry, test configurations, test facilities, etc.)

-a digitally encoded test outcome.

-an on-chip test interpretation.

Many BIST structures need specific test stimulus generators. But, the implementation of an on-chip stimulus generator normally requires a noteworthy investment in hardware. This point can become very critical in some cases, not only because of the involved additional area which can increase significantly the production cost, but also because of the risk of the circuit performance degradation. Consequently, achieving a BIST technique which is able to operate without a test stimulus generator is a very promising BIST approach.

Because of the own philosophy of the OBT concept, no test stimulus is required. Since the SUT is generating its particular test stimuli, the only difference between OBT and OBIST lies in whether test interpretation is carried out on or off-chip. In fact, OBT can handle BIST without the penalty of dedicated, additional on-chip signal generation hardware.

The OBIST concept seems very appealing since it allows to make the test of the SUT relatively independent of external tester. Test signals are internally generated and test interpretation can be pre-processed using, for example, the digital circuitry available on-chip. In fact, if we pursue a low-cost, yet efficient OBIST strategy, the main requirement is to reuse the own circuitry of the SUT and avoid adding extra test elements. But, obviously, it would require an additional effort of design. Let us show herein that this effort is minimal and perfectly redeemable.

Our purpose is to devise a structural test approach for a specific demonstrator macrocell. Because of manipulative limitations or high simulation times, handling a circuit structure of a complexity higher than second-order can seem cumbersome. However, it will see herein that the OBIST concept can be easily applied to certain kind of complex systems.

A practical way to carry out an effective implementation of the OBT-OBIST will be discussed as well as guidelines for its application will be given. The practical problems behind the use of such an approximation will be discussed at length, and principles will be argued in terms of *Design Decisions* which must be taken during the design process. After presenting a practical OBIST solution, its compatibility with a functional test approach is studied as well as a comparison is undertaken considering the common design practice used for analog and mixed-signal circuits.

Although a particular demonstrator macrocell has been selected as the test vehicle, the experience acquired with this system can be useful for many other applications. Therefore, this Chapter must not be understood as a special case or a specific example. Most of the inferred *Design Decisions* can be used in a more general situation.

6.1 DEMONSTRATOR MACROCELL

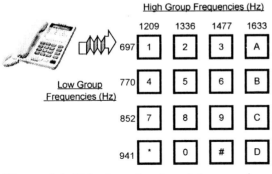

Figure 6.1: Telephone keyboard frequencies

This chapter will deal with a complex macrocell required in many communication SOC's. This system is a Dual-Tone Multi-Frequency (DTMF) receiver intended for decoding the dialling information in telephony and aiming a broad use for dedicated terminals.

This cell is gaining more and more importance in the world of telecommunications since it is required in paging systems, repeaters, mobile radio, credit card systems, remote control, personal computers, telephone answering machines, etc. This circuit can be found either as a stand-alone, mass-produced chip or as a core to be embedded in a complex SOC. Then, the test support around this cell must be flexible enough to allow the user to select a test strategy as a function of his needs, instead of forcing the use of a fixed test methodology.

DTMF receivers convert 16 different types of DTMF signals into 4-bit binary serial data. Its input is an audio signal composed by the superimposition of two tones which are selected by line-and-column addressing of a keyboard (Fig. 6.1). The output of the receiver is a digital code carrying information on the present signals at the input.

The DTMF signal is first processed by a bandsplit filter which separates the high and low frequencies of the received pair as it is shown in Fig. 6.2. Each frequency is then square-shaped and decoded separately. The decoder task is to establish whether the present frequencies are recognized as a

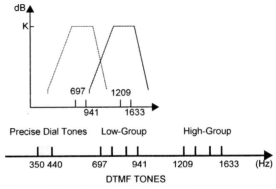

Figure 6.2: Band splitting in a DTMF receiver

DTMF tone (it verifies both the frequency and duration of the received tones before passing the resulting 4-bit code to the output bus). The criteria to be followed by the decoder should be neither too relaxed (to avoid the recognition of no-DTMF signals as DTMF ones) nor very strong (to cope with received tone imperfections and noise). So the detection algorithm becomes a trade-off between:

- Tolerance for the accepted tones, and
- Immunity to tones simulated by speech (talk-off).

In general the DTMF receiver consists of an analog processing part (a two-channel filter bank), followed by a digital decoding mechanism. Fig. 6.3 illustrates the entire DTMF receiver block diagram which has been used as the circuit demonstrator.

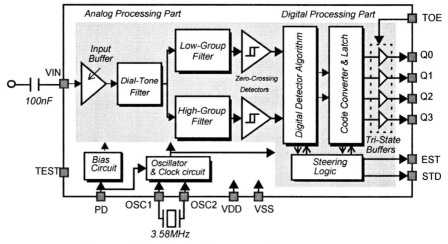

Figure 6.3: Block diagram and I/O pins of the originally conceived DTMF receiver

In the analog part, the square-wave versions of the dialing tones are obtained and, in the digital part, such waveforms are decoded. For the sake of completeness, the two off-chip components, a quartz crystal resonator and a decoupling capacitor are also shown in Fig. 6.3. The I/O pins and their names are displayed as well[1]. As can be observed, the full DTMF receiver integrates both the bandsplit filters and the decoder function into a single 14-pin package. However, the system only requires 13 pins. But since we are forced to

---

[1.] The role of the I/O pins will be explained in next sections

use a 14-pins package, the goal is to employ the free-pin (called TEST in Fig. 6.3) for the application of the OBIST technique and/or other test facilities.

Finally, a more detailed block diagram corresponding exclusively to the analog processing part is depicted in Fig. 6.4 where two filter paths can be distinguished, the so-called low-group and high-group to discriminate the low and the high band frequencies, respectively. Notice, moreover, that both filter paths are preceded by a built-in dial tone rejection circuit (Dialing Filter) which is provided to eliminate the need for pre-filtering. Each filter bank is formed by a cascade of several second-order functions where many typical transfer function types are employed (LP, BP, HP, LP-notch and HP-notch to be precise).

Two zero-crossing detectors make the interface to the digital part, which detects the presence of correct dialling tones and validates a tone accomplishing with the time requirements. This digital part performs a time evaluation of the upcoming signals and uses digital counting techniques to detect and decode all the 16 DTMF tone pairs into a 4-bit code. The structure of the decoder makes this very adequate for implementing a digital BIST as well as for interpreting the test outcome from the analog subcircuit.

In what follows, only the analog subsystem will be considered, since the digital part is designed and tested using more conventional techniques. The implementation of every block for this analog subsystem is briefly described to understand the consequent modifications that must be include for the implementation of the OBIST technique.

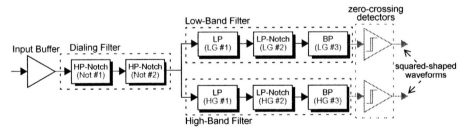

Figure 6.4: Block diagram of the band splitting filter of the DTMF detector

a) Input Buffer:

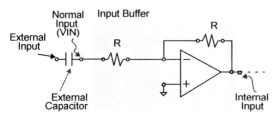

Figure 6.5: Input buffer schematic

The first block in the system of Fig. 6.4 is an input buffer. This circuit is employed for decoupling the DC levels of the external input from the internal levels. Fig. 6.5 shows its structure. It is no more than a conventional inverting amplifier that, together with the external decoupling capacitor, forms a HP first-order section.

b) Band Split Filters:

The bandsplit filters are composed by two paths of cascaded second order sections (biquads that we will call, from this point forward, Not #1, Not #2, LG #1, LG #2, LG #3, HG #1, HG #2 and HG #3, see Fig. 6.4). Both paths share the dialing notch filter (4th-order system) while the band-splitting is performed by two 6th-order BP filters with passbands adjusted accordingly to the low and high group of frequencies. Each 6th-order BP filter consists of three different biquads. Therefore, although each filter path is a 10th-order system, eight internal subsystems (biquads) can be perfectly distinguished.

Obviously, the analog filters are one of the key point in the DTMF design. Consequently, several design considerations need to be discussed in order to achieve a good DTMF receiver performance. But here, however, the points that arouse the interest are not those regarding the DTMF normal design. On the contrary, we are exclusively interested in studying those aspects regarding just the OBIST implementation. Therefore, only those issues related to the OBIST realization will be considered in this Chapter.

Every second-order section has been carried out using the generic switched-capacitor (SC) biquad shown in Fig. 6.6. This structure, which has been largely considered in previous chapters, was chosen to cope with the accuracy, low power, small area, etc., requirements of the DTMF core. Accordingly, every DTMF filter stage has been customized to operate as required by adequately sizing the capacitors from the main biquadratic structure.

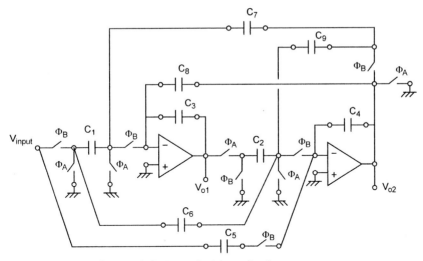

Figure 6.6: Generic biquadratic structure

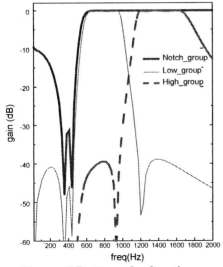

Figure 6.7: Transfer functions

Clock phases Φ_A and Φ_B in Fig. 6.6 are interchanged for sharing-load considerations between preceding and succeeding stages. It introduces additional delays between stages which, although are not important in the normal operation mode of the DTMF, can be a decisive factor in the OBIST mode. The impact of the possible existence of additional delays was previously studied in Chapter 5.

Fig. 6.7 shows the transfer functions obtained with SWITCAP [74] simulations at the output of each filter path. The ripple in the passband is 0.3dB for the low-group and 0.5dB for the high-group, and the minimum stop-band attenuation is approximately 39dB.

On the other hand, Table 6.1 summarizes the main data for each filter section.

Filter Group	Notch		Low_Group			High_Group		
Section	NOT #1	NOT #2	LG #1	LG #2	LG #3	HG #1	HG #2	HG #3
Type	HP-NOT	HP-NOT	LP	LP-NOT	BP	LP	HP-NOT	HP
C_{TOTAL}	80.5	68.5	59.8	41.1	38.2	44.7	61.7	19.4
C_{SPREAD}	25.5	28.0	33.3	22.9	15.5	25.5	41.0	5.9
Nº CAP	7	7	6	7	6	6	7	6
C_{TOTAL}	413.9							
C_{SPREAD}	41.0							
Nº CAP	52							

Table 6.1: Filter design summary. Values are referred to unit capacitors

c) Zero-Crossing Detectors:

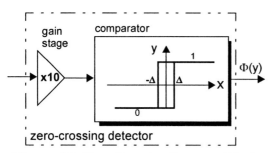

Figure 6.8: Zero-Crossing Detector Block

The zero crossing detector must reject signals below a certain threshold (including the offset of the prior stages) and compensates its own offset. The threshold must be selected accordingly to the desired sensitivity (lower the threshold, larger the sensitivity). But a trade-off is required to avoid interferences of unwanted signals. For these reasons, it has been designed as a gain amplifier followed by a latched comparator with hysteresis. Fig. 6.8 shows the block diagram of such a zero-crossing detector implementation.

6.2 APPLYING THE OBT-OBIST METHODOLOGY TO THE DTMF MACROCELL

Let us focus first on how to interpret the OBT-OBIST principles for being effective in the described DTMF cell. In fact, as was discussed above, in order to increase the efficiency of the OBT method and to implement a practical BIST solution for this mixed-signal system, the testability must be early considered during the design process and never completely isolated from it. That is the reason why different design decisions will be taken in parallel with the development of the new chip design, presenting, thus, the modifications needed for incorporating the different parts of OBIST structure to the original DTMF design.

Our OBT proposal relies on converting the DTMF (or part of it) in an oscillator by adding a non-linear feedback path and modifying the circuit (adding or removing some passive components) [36]. Then, the system splits operation into two modes:

-operational, in which the system connects to its regular input, and

-test, in which a closed non-linear feedback loop encircles the DTMF and the regular input is disconnected.

Removing components can render them untestable, leading to the following rule:

Design Decision #1

DD#1- Apply OBT without removing components from the normal signal path

---

A first approach to test the DTMF cell could be converting the complete SUT into an oscillator. In this case the SUT results in a 10-th order transfer function, because this is the order of each filter path. However, as was in detail explained in preceding chapters, the analytical design of a high-order oscillator is quite complex and must be dismissed. For this reason, breaking up the whole filter into its component biquads and applying the OBT-OBIST concept to every second-order structure (biquad) seems more reasonable from the viewpoint of both, analytical calculations and simulations;

Design Decision #2

DD#2- Applying OBT-OBIST requires splitting the filter into individual component biquads

---

Keeping in mind **DD#1** and **DD#2**, two complementary problems must be considered. One is related to the biquad level, it is, how to make *any* biquad to oscillate independently of its transfer function and (if possible) using a feedback element which can be shared by other biquads. This point must be solved making use of the results postulated in previous chapters but now regarding the specific demonstrator features. A second problem concerns the filter level, specifically the difficulty of combining the biquad-level test in order to verify the whole filter.

In order to develop an economic and efficient OBT-OBIST technique, applicable to a complex system under test, it must be pursued to re-use the original existing hardware, not only to avoid extra area and power overhead, but also because this would be a means to re-check more than once the correctness of the circuitry. Then, re-employing the present circuitry and avoiding to add extra components must be one of the main objectives. As will be seen, it is (relatively) simple in cases like the DTMF demonstrator or $\Sigma\Delta$ modulators, where the main required additional circuitry for applying the OBIST procedure is available within the system.

6.2.1 Biquad-Level Test

As was seen through the analytical results obtained in Chapter 3, band-pass transfer functions are easily converted into oscillators just adding some non-linear feedback mechanism. But when this is not the case, it is still feasible to force oscillations without altering the filter transfer function if the biquad has (at least) two opamps. In particular, for the structure shown in Fig. 6.6 it is possible to implement a given function taking V_{o1} or V_{o2} as the corresponding output. Then, either the normal output or the secondary output (taken at an "internal" node) can be used to force oscillations (see [36]).

a) Non-Linear Block:

One of the essential requirements to convert a biquad into a robust oscillator is to choose the most suitable non-linear feedback loop. In the case of the biquads associated with the DTMF demonstrator, it was seen that an adequate nonlinear block can be formally described by a 1-bit ADC followed by a 1-bit DAC and realized by an analog comparator. Such as was described in Chapter 3, an important fact derived from this oscillator strategy is that the amplitude of oscillations can be controlled by the reference voltage of the 1-bit DAC. Therefore, opamps can be maintained in the linear range avoiding saturation. A transfer function gain larger than unity is required not only to adjust the oscillation amplitude, but also, to accomplish the demanded gain in the feedback loop to successfully carry the oscillation strategy out. Therefore, if a 1-bit DAC is used, the amplitude value can be chosen to achieve the best testing conditions in the demonstrator filter. Apart from that, the DF approach for this non-linear block allows to give in many cases a good approximation for the analytical description (see previous chapters).

Therefore, a third design rule for applying the OBT to the demonstrator can be proposed:

Design Decision #3

DD#3- *Use to force oscillations a non-linear feedback block formed by cascading a 1-bit ADC and 1-bit DAC*

However, the main pursued goal is to fully exploit all the resources available on a first version of the DTMF demonstrator without including the OBT technique. Consequently, observing the existent circuitry into the previous DTMF design it can be re-used the zero-crossing detector as the 1-bit ADC and only the 1-bit DAC must be incorporated. So, the proposed non-linear feedback contains two stages as shown in Fig. 6.9. One is the zero-crossing detector (a gain stage followed by a voltage comparator with hysteresis regulated by the voltage reference Δ) already present in the DTMF and the other one is an extra block named voltage limiter acting as an 1-bit DAC and restricting the highest and the lowest values of the square wave to $\pm|V_{ref}|$. This value may be adequately fixed for every biquad and plays the same role than the saturation level V_{ref}, of the non-linearity considered in Chapter 3.

Now, a non-linearity with hysteresis is being considered and, therefore, the resulting equations for the oscillation conditions and the oscillation parameters must be modified. But, as was shown in Chapter 5, in this case the obtained results slightly differ from those oscillation results predicted in Chapter 3.

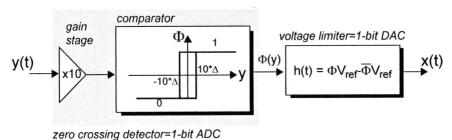

Figure 6.9: Modified Non-linear block

b) Voltage Limiter:

The levels of the square-wave signals must be regulated to appropriate values not only to provide the satisfactory OBT feedback values but also to maintain the amplitude of the oscillations into the linear operating ranges of the opamps. It has been performed with the circuit shown in Fig. 6.10. This circuit is a modified voltage amplifier with offset and gain compensation. Φ_1 and Φ_2 represent the two non-overlapped phases of the filter operation. The positive input of the opamp is switched to V_{ref} or to ground during Φ_1 (depending on the signal $\Phi(y)$ in Fig. 6.10)[2] in order to generate $+V_{ref}$ or $-V_{ref}$. In this way, the generated squared signal (called *Vsq* in Fig. 6.10) is limited to $\pm V_{ref}$ as needed[3]. Notice that this is made by using only one voltage reference, V_{ref}.

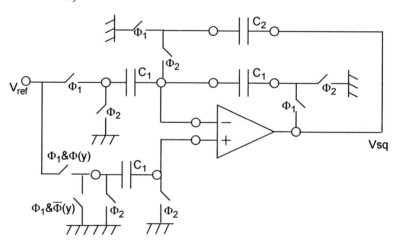

Figure 6.10: Voltage Limiter schematic

6.2.2 System-Level Test

After finding a unified way to force any particular biquad to oscillate, it has to be examined how to give an efficient support to the OBT-OBIST

[2.] Signal $\Phi(y)$ is the output of the zero crossing detector which is valid during Φ_1 and maintained until the end of Φ_2.

[3.] Depending on the required feedback sign (positive or negative) in the OBT strategy, the control switches of the voltage limiter are commanded by $\Phi(y)$ or $\overline{\Phi}(y)$.

technique in a filter formed by a cascade of biquadratic filters. Obviously, a direct way to use OBT on such a cascaded filter, involves "disconnecting", during the test mode, each biquadratic section from the filter signal path and "connecting" it to the non-linear feedback loop.

Two possibilities can be planned for accomplishing these modifications. One is to simultaneously convert all the biquadratic filters to oscillators as is shown in Fig. 6.11-(a) (a parallel test). This strategy is based on closing a feedback loop around each biquad after isolating that biquad from the rest of the system. Therefore, a non-linear block per biquad would be required, simplifying the routing of comparison signals but increasing the active area. A simultaneous evaluation can be performed, in this case, reducing the test time to the maximum settling time of the oscillators plus the evaluation time required. In spite of the fact that the number of I/O pins are normally very reduced, we can overcome this problem by adding all the outputs to obtain a multitone signal (Vmt as is illustrated in Fig. 6.11-(a)).

On the other hand, a second option consists on a multiplexed sequential testing. That is, converting all the biquadratic filters sequentially, one after the other (a sequential test), to an oscillator as displayed in Fig. 6.11-b. In this approach, only one non-linear block is required. Besides the time needed for connecting and disconnecting the feedback network, the total test time is higher than the sum of the settling time of all oscillators plus the minimum time to evaluate each of the oscillation values. Hence, this second method requires less extra hardware, but spends a longer test time.

From these Figures (Fig. 6.11-(a,b)), it should be clear that there is a remaining difficulty due to the fact that now several switches are needed to connect and disconnect all local feedback loops. An effective design for such switches is a critical point because their insertion could degrade the overall circuit performance. Therefore, minimizing this problem must be a key objective. Later on, we will discuss in detail how to overcome this difficulty by using the so-called Switchable Operational Amplifier (swopamp) concept.

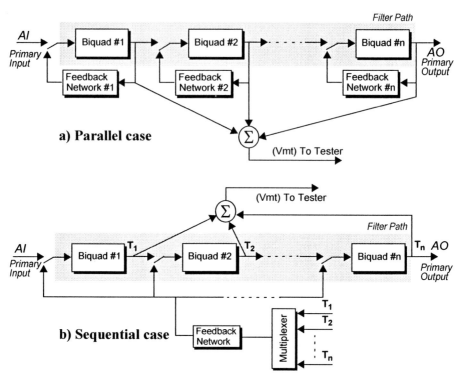

Figure 6.11: Two possibilities to convert the SUT in an oscillator

To come up with a new Design Decision, it must be considered the potential performance degradation as well as the area/power penalty of each approach. In the case of the DTMF demonstrator, there is the possibility of re-using the existing zero-crossing detectors for testing and employing a global feedback loop as will be proved later. Then:

Design Decision #4

DD#4- *Select a sequential test structure in order to reduce the number of additional test components to a minimum*

The next step is to find a way to implement the sequential test concept with minimal cost (in terms of area, power, performance, etc). Fortunately, it can be applied an idea reported a few years ago and successfully brought into play for DfT [37], [82]-[83]. Fig. 6.12 shows what has been called a swopamp (from Switchable Opamp) which is an operational amplifier with a configurable internal architecture.

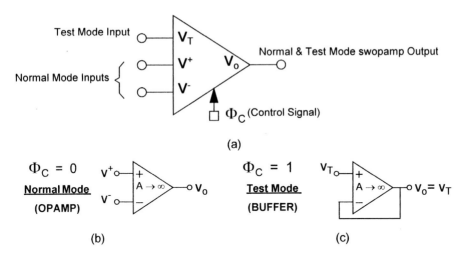

(a)

(b)   (c)

Figure 6.12: SWOPAMP Concept

This device can operate under two different modes, controlled by a logic signal named Φ_C in Fig. 6.12. In its regular mode (when $\Phi_C = 0$), the swopamp performs as a conventional opamp, amplifying the difference between its regular inputs (Fig. 6.12-b). In the test mode, a third signal, V_T, is enabled by the control signal, Φ_C, to be, in fact, passed to the swopamp output (Fig. 6.12-(c)). Therefore, in the test mode, the swopamp operation is similar to that a unity-gain buffer.

Exploiting the swopamp functionality, we can devise an improved system-level test scheme which allows us to implement more efficiently the OBT-OBIST strategy. The new approach is to substitute one or more regular opamps in every subsystem by swopamps. Two options are possible:

-the first alternative consists in a multiplexed sequential testing where every stage is sequentially transformed to an oscillator by a proper control of swopamps and multiplexing switches. The main idea is using swopamps to consecutively close the oscillation feedback loops (Fig. 6.13) but employing switches to select every time the specific SUT.

-in the second choice, swopamps are used to bypass signals from their inputs (see Fig. 6.14). There is just a single feedback loop, but only one stage is configured as an oscillator at once. Sequential evaluation is then needed.

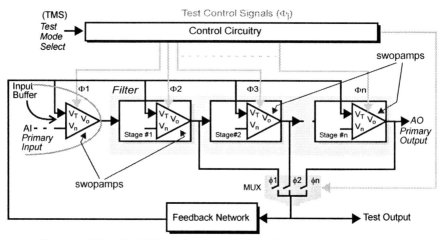

Figure 6.13: Multiplexed sequential testing using switches

Therefore, the second opamp of every biquad (the opamp providing the output to the next stage) can be substituted by one of these swopamp structures just to provide the stage with a mechanism to select either its regular input or a signal generated somewhere in the circuit. The specific biquad under test can be isolated and independently reconfigured as a closed-loop system regardless of the remainder biquads. The input of every swopamp which does not form part of the biquad under test, can be connected to its previous stage. The goal is that each test oscillation signal covers all the filter paths in such a way that this not be affected by the circuitry which does not belong to its associated biquad under test (see Fig. 6.14 for more details). A similar structure was reported by some of the authors in [37]. With this last alternative the number of extra switches is reduced.

> ### Design Decision #5
>
> *DD#5- Employ swopamps to selectively close the feedback loop*

Figure 6.14 shows a filter structure after carrying out these changes. This filter now has a single feedback loop, but only one stage (the j-th stage) can act as an oscillator every time. The other stages play the role of buffers either to bypass the feedback signal to the j-th stage input or finally to bypass the oscillator output to the filter primary output. Notice that these changes also simplifies the sequential structure in Fig. 6.14 because the filter no longer requires multiplexing at the feedback block's input.

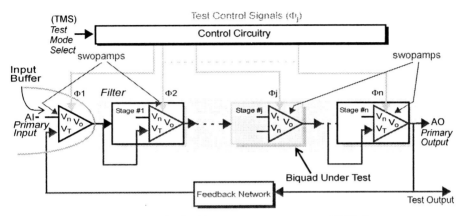

Figure 6.14: Block diagram to convert a SUT in an oscillator using swopamps

The control is very simple, when the j-th stage is under test, all the swopamps, excluding the swopamp belonging to this particular stage, must have their control signals high. Therefore, the test control signals are always a bitstream of 1s and exclusively 0 in the stage under test position, j. The method is as follows: just a logic 1 is shifted along a shift-register at a pace depending on the oscillation frequencies (depending on the specific biquad under test).

An important advantage of this procedure displayed in Fig. 6.14 is that all the added components are inherently tested during this process. When the j-th biquadratic filter is under test, the test loop contains all added switches within an opamp, so the test checks them too. Hence, along the entire test process, the test checks all the elements. Therefore, our next design rule recommends this single-loop configuration:

Design Decision #6

DD#6- *Transform the OBT sequential structure into a single-loop, complex oscillator*

Observing the structure in Fig. 6.14, one immediately wonders which is the impact of the swopamp chain: the inserted offset. But, this problem will not be significant because, as will be seen, the swopamp design may include a mechanism for offset compensation. Moreover, the hysteresis of the zero crossing detectors will help to eliminate the potential offset in the filter paths of the DTMF core.

6.2.3 A modified System Architecture

After stating all of the preceding Design Decisions (*DD*s), we are in a position to incorporate them into the core demonstrator. Architectural support is needed for connecting/disconnecting the feedback paths, connecting/disconnecting the remaining circuit (complete filter) and even testing the extra components. Since there are two paths in this system, the different *DD*s can be performed through a double oscillation loop as shown in Fig. 6.15. As can be seen from this Figure, our first proposal is using the zero-crossing detectors available in each filter path connected to a single 1-bit D/A (a voltage limiter as shown in Fig. 6.9). Thus, testing a specific biquad is achieved by closing the adequate loop. That is, testing the biquads in the upper (alternatively, in the lower) path is carried out by closing the upper (respectively, the lower) loop. On the other hand, testing any of two biquads shared by both paths can be achieved by closing any of the two loops. It is worth noting that this implementation only requires the addition of a DAC in the common part formed by the feedback loop, plus a simple extra control circuitry. Therefore, the silicon area overhead is quite small. In terms of the design of those elements present in both the "conventional" and the OBT design, they are required to have a similar performance, which means that the design effort is not increased too much. Therefore:

Design Decision #7

DD#7- *Reduce the OBT cost by wisely reusing every element of the system structure in the oscillation feedback path*

To understand the proposed OBT-DTMF structure, Fig. 6.15-(a) illustrates how the system must be reconfigured to test, for instance, the low-pass notch placed in the lower path. The corresponding loop is closed using the associated comparator and the biquad under test (BUT) is left unaltered. All the remaining stages (in the loop) emulate a buffer through their swopamps, in accordance with Fig. 6.14. Therefore, the overall closed-loop system (emphasized by thicker lines in the figure) corresponds to the oscillator formed around this biquad. Blocks located in the upper path are depicted by a dashed line because in this case they are not part of the oscillation loop. However, as will be explained next, these upper blocks could be used to read the test signal from the BUT while its test is being performed.

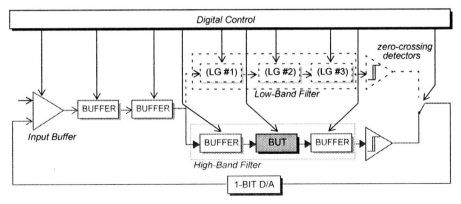

(a) Testing the lower stages

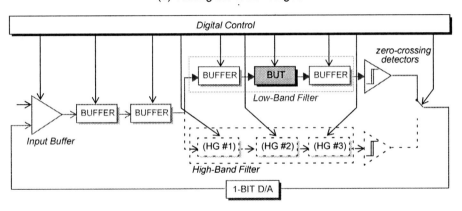

(b) Testing the upper stages

Figure 6.15: Modified DTMF receiver showing the test of a biquad (shaded box)

Similarly, Fig. 6.15-b shows how an upper biquad (to be precise, the low-pass notch placed in the upper path) can be transformed in an oscillator by closing the corresponding upper loop. The same procedure is carried out: the remaining upper stages act as buffers while such a biquad is being tested.

There are practical limitations to this approach. First of all, every buffer in the chain (during test modes) introduces a delay at the test frequency. Nevertheless, for frequencies much lower than the unity-gain frequency, this delay is quite small and the accumulated loop delay can be neglected. If this is not the case, it can be estimated and taken into account for analytical calculations.

A second problem comes from the accuracy of the DF method and the signal purity at any point of the global feedback loop. This problem leads to a distorted oscillator signal, but is not very significant in most cases. As far as the DTMF demonstrator is concerned, there is enough filtering, but it does not actually matter since we can model (with a reasonable approximation) the distorted signal at any biquad output.

Another issue is that we have to guarantee that the amplitude of the oscillations must not saturate none of the amplifiers in the filter chain. It is controlled by the value of the voltage level in the DAC (the value of V_{ref} in the voltage limiter).

There is another important feature to consider. We must guarantee that every component, either in the filter or in the feedback network, is tested. It should be clear from Fig. 6.15 and the test operation described above that every biquad is tested once along the test procedure. Additionally, all the swopamps are tested in both modes (normal and test). In fact, the test path is tested n-1 times (being n the number of biquads), which can help in diagnosis, reducing the impact of the extra components on system testability. Only the input terminal and its connection to the signal path must be additionally checked out.

On the other hand, to finish this Section, another point to be contemplated is how to give support to frequency and amplitude measurements. Reading an analog signal is feasible, although the measurement accuracy is limited by noise and even it may impose extra design requirements which are not desirable if we wish to avoid an additional design effort. Unfortunately, the frequency information is easily coded into digital information, but this is not the case for the oscillation amplitude. However, the only way to obtain a good fault coverage is to be able to measure with precision both parameters (frequency and amplitude), which means translating all the signal information into digital data.

A convenient way to achieve such a coding is to use an oversampling ADC [29], [85], [87]-[90]. For our purposes, a medium-resolution converter should be enough. Using blocks similar to those necessary for the nonlinear feedback previously discussed, we can synthesize a $\Sigma-\delta$ modulator capable of providing a 1-bit digital version of the test output. Looking at the complete circuit in Fig. 6.4, we have two comparators available. Although this is not explicitly shown in Fig. 6.4, an extra gain stage is connected between every filter channel and the corresponding comparator. Thus, we can again use a swop-amp instead of a regular buffer. In one of its operational modes such a swopamp will act as a buffer, and in the other mode it is used to implement a discrete-time integrator. A local feedback loop shall be closed in this latter mode to provide a simple $\Sigma\Delta$ modulator which generates a digital bitstream, which can be either read by an external tester or fed into a digital interpretation circuit. These integrators can be switched at a higher frequency to comply with the oversampling requirements, although in [87]-[90] it was shown that in most practical situations using the same sampling rate as in the filter can be enough. In particular, Fig. 6.16 illustrates the case when one of the biquads in the lower filter bank is under test (specifically, the same biquad as in Fig. 6.15-(a)). We then use one of the comparators to close the oscillation feedback loop and the other one to implement the testing ADC. An equivalent connection is used when the biquads in the upper filter bank are tested. In other words, we arrive to a new design rule.

Design Decision #8

DD#8- *Around every comparator we can build up a low-accuracy $\Sigma\Delta$ modulator for reading-out the test oscillatory outputs*

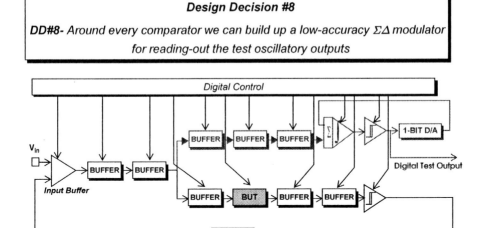

Figure 6.16: Example: Generating a bit-stream test output

6.2.4 An alternative implementation

In the previous Section, we have proposed a systematic way (a general idea) to devise an OBIST-DTMF system-level scheme. But, this particular solution described in Fig. 6.15 does not take fully advantage of the peculiarities existing in the DTMF structure. Therefore, let us design a more detailed system-level scheme for the DTMF system. Three aims are mainly pursued:

-to adapt the biquad-level OBT requirements with the possible OBIST system-level approaches.

-to make reasonably simple the programming of the test mode and the extra routing.

-to reduce the test time as much as possible.

As was described, a peculiarity of the FL-biquad is that it has two available outputs (see Fig. 6.6) to apply the OBT strategy. In the case of the DTMF filter, choosing the adequate output in the biquads allows to achieve sustained oscillations. Therefore, the selection shown in Table 6.2 has been taken because it gives the simplest and more straightforward OBIST configuration at system-level (see Chapter 5 for details).

	NOT #1	NOT #2	LG #1	LG #2	LG #3	HG #1	HG #2	HG #3
V_{o1}	OK	OK	OK	OK		OK	OK	
V_{o2}					OK			OK

Table 6.2: Selected outputs for each biquad for the system level configuration in Fig. 6.17

An alternative to the generic structure in Fig. 6.15 is the system-level architecture shown in Fig. 6.17 where, for the sake of simplicity, only the test inputs of the swopamps are displayed. In this scheme, the opamps of each biquad as well as the opamps of the input buffer have been substituted by their corresponding swopamps. Therefore, the test inputs of those swopamps which provide the oscillator output (V_{o1} except for LG#3 and HG#3) are used to propagate the oscillator outputs to the inputs of the comparators. The remainder swopamps have their test inputs connected to the output of the comparators and are used to close the feedback paths. Notice as well that the zero crossing detectors (comparators) existing in the normal design are being used to implement the non-linear function, thus, avoiding extra hardware. The way in which the whole filter is tested is described as follows.

As can be deduced from Fig. 6.17, the OBIST-DTMF system-level scheme has been divided in two parts. The first one is composed by Not #1 and all the Low-Band biquads. The second part is thus composed by the rest of the components, that is, Not #2 and all the High-Band biquads. Because two comparators are available, it is possible to test two biquads simultaneously[4].

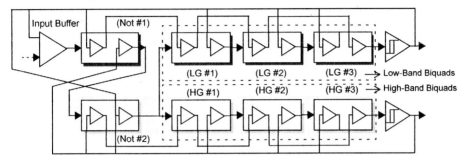

Figure 6.17: System level configuration to implement the OBT strategy in the DTMF filter

Fig. 6.18(a-d) depicts the four possible test configurations of the system (four test groups with two OBT signals each one), where the emphasized lines and shadowed blocks designate the activated circuitry. Every test configuration can be selected by properly programming the digital signals that control the swopamps. The oscillator corresponding to each biquad under test is built by driving the comparator by the correspondent output of the biquad under oscillation (propagating its output through the swopamps in the consecutive sections) and feeding back, through the second swopamp in the previous biquad. Moreover, notice that the start-up strategy introduced in previous chapters is also implemented; when all the swopamps are set in the buffer mode at the same time (activation of the start-up), all the biquads under test are initialized by the actual output of the correspondent comparator.

In short, we can observe from Fig. 6.18 and Fig. 6.19 that the test operation sequence is as follows (one biquad of the upper loop is simultaneously tested together one biquad of the lower loop):

Group #1.- Testing Not #2 and LG #3.
Group #2.- Testing LG #2 and HG #1.
Group #3.- Testing LG #1 and HG #2.
Group #4.- Testing Not #1 and HG #3.

[4.] Now, each comparator in Fig. 6.17 incorporates a voltage limiter (an 1-bit DAC) as shown in Fig. 6.9.

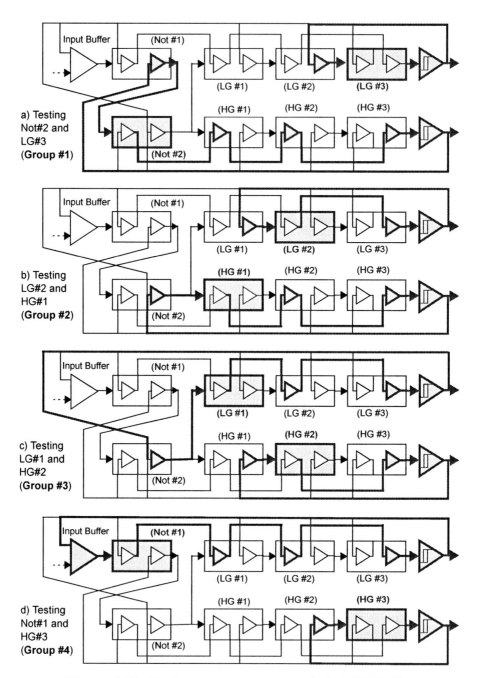

Figure 6.18: Configurations to test the whole DTMF Filter using the OBT technique

We can also observe from Fig. 6.18(a-d) that the number of swopamps used for each test group is always five except for the group #3 whose structure allows to save one swopamp. But, on the other hand, the signal paths involved in each testing configuration are very diverse because they depend on the location of the specific biquads under test inside the filter banks. However, this testing organization was carefully designed in such a way that optimizes the test time and the required extra hardware for programming all the testing groups.

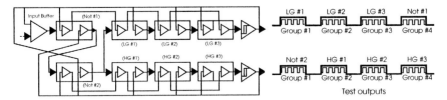

Figure 6.19: Output sequence in the test mode

6.2.5 Cells adaptation for OBIST implementation

According to the test-planning shown in the previous figure, it is necessary to redesign some circuitry in order to adapt the design to the OBIST facility. Such cells needing a modification are the Input Buffer and the biquad cells to implement the multiplexing and the start-up strategies by the use of the swopamp. In addition, the zero-crossing detector outputs can not be used directly as the feedback signal, and it is necessary to adjust the voltage levels of the square-wave signal in order to avoid the saturation of the filters.

-New Input Buffer:

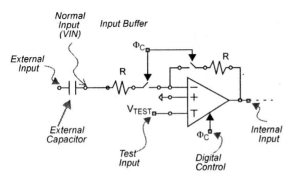

Figure 6.20: Modified Input Buffer

The modified version of the Input Buffer is shown in Fig. 6.20. As can be seen, the main modifications are the substitution of the normal opamp by a swopamp and the introduction of a couple of switches in the normal signal path. The extra input (V_{TEST}) provides the required feedback path while the switches have been introduced to isolate the output of the buffer from the input (V_{IN}). Notice that both switches are placed in small signal paths (virtual ground), thus avoiding the alteration in the behaviour during normal operation because of the non-linear characteristics of the MOS transistors.

-New Biquad cells:

Fig. 6.21 shows the modified biquad schematic accordingly to the requirements of the OBT application. The change is that the normal opamps have been substituted by swopamps. Moreover, as will be seen in next sections, a couple of switches, controlled by the same digital signals than the swopamps, have to be added to the biquad structure for the star-up process of the oscillators.

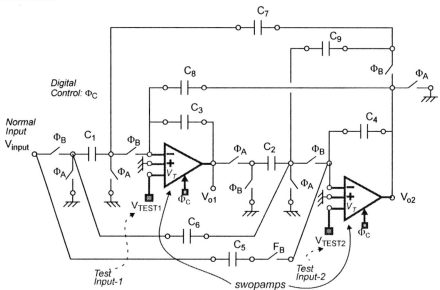

Figure 6.21: Modified biquad structure schematic

-Opamp and Swopamp Designs:

The design of the opamps / swopamps is an important question to be considered into the system design and, for this reason, this point must be carefully handled. First of all, a low voltage supply- as imposed by the DTMF original design- requires the use of simple structures that must avoid cascoded transistors. On the other hand, they are responsible of major current consumption in the chip, so a careful design is needed to cope with the low power requirement of the application. Because the number of opamps is also high, the area occupied by them is of vital importance. However, the low-offset require-ments force the use of not too-small transistor sizes. So, the design of the opamps must be a trade-off between the factors above mentioned.

Depending on the use of the opamps the requirements also change. For example, it must be taken into account if they are loaded with capacitors or/ and resistors. In this sense, two types of opamp have been designed, one for the SC part (capacitance loads) and other for the input buffer and the bias circuitry (resistance and capacitance loads).

A simple two-stage opamp with RC compensation has been used. Fig. 6.22-(a) shows its schematic and Fig. 6.23-(a) its layout. For opamps with resistive loads, a buffering output stage was added. This same opamp has been used in all the SC part except in the gain x10 amplifier, where the output stage was adjusted to cope with the large gain bandwidth required. Table in Fig. 6.22 gives the main sizes of the design parameters for a $0.6\mu m$ CMOS technology and $V_{DD} = -V_{SS} = 2.5V$.

A specific design alternative of swopamp has been chosen (other types of swopamp implementations can be found in literature [82]-[83]). It is based on the partial duplication of the input stage (the input differential pair) and the incorporation of some internal NMOS switches to disconnect such a duplicate block from the active load when it is necessary. Its schematic is in Fig. 6.22-b (and its layout in Fig. 6.23-b). The size of transistors implementing the additional switches (M_s) is 10μm/0.8μm. The operation can be described as follows. When the swopamp control signal, Φ_C, is 0, the switches connect a normal differential block to the rest of the circuit, and the system works in opamp mode. On other hand, if the control signal, Φ_C, is 1, the duplicate differential block is then connected to the device while the normal block is isolated from the rest of circuitry. In this case, the swopamp operates in its buffer mode.

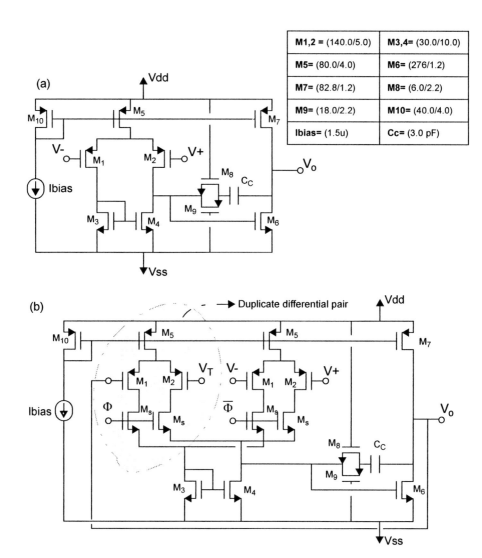

M1,2 = (140.0/5.0)	M3,4= (30.0/10.0)
M5= (80.0/4.0)	M6= (276/1.2)
M7= (82.8/1.2)	M8= (6.0/2.2)
M9= (18.0/2.2)	M10= (40.0/4.0)
Ibias= (1.5u)	Cc= (3.0 pF)

Figure 6.22: Opamp (a) and swopamp(b) schematics

Fig. 6.23 shows both layouts, the opamp design is on the right and the swopamp circuit on the left. This Figure is useful to compute the additional area percentage. Both layouts have the same width but the swopamp is a little longer. So, the swopamp area is a 26% larger than the opamp area. In fact, this means in the DTMF layout an increase of 26% in the global area appointed in the opamp array. This extra area comes from the auxiliary switches and from the necessary duplicate differential stage to design the required swopamps.

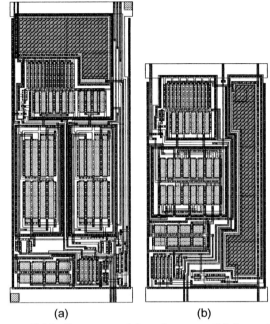

(a) (b)

Figure 6.23: Swopamp (a) and opamp (b) layouts

On the other hand, Fig. 6.24-(a) shows the frequency responses obtained with SPECTRE simulations for the opamp and the swopamp. And Fig. 6.24-b the frequency response of the extracted swopamp. Responses for the opamp mode with a load capacitor $C_L = 25pF$ and a load resistor $R_L = 1M\Omega$. The opamp mode operation for the swopamp is very close to the reference in both cases. As can be observed, phase deviations occur at very high frequencies but are negligible.

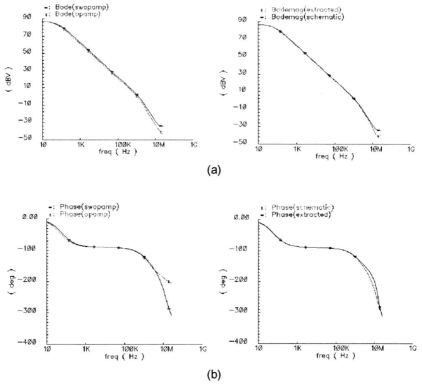

(a)

(b)

Figure 6.24: Magnitude and phase responses from a) both schematics in Fig. 6.22 (opamp and swopamp) and b) schematic and post-layout simulations (swopamp)

The main parameters extracted from these curves are summarized in Table 6.3.

	OPAMP SCHEMATIC	SWOPAMP SCHEMATIC	POST-LAYOUT SWOPAMP
Low-Frequency Gain (dB)	86.72	86.66	86.7
Unity- Gain Bandwidth (MHz)	1.201	1.441	1.375
Phase Margin (Deg)	56.65	54.9	58.39

Table 6.3: Comparison between opamp and swopamp main features

Other parameters should be considered to measure the impact of the modifications on the original circuit. Among them, offset, settling time (T_s), slew rate (SR), output swing (OS), and common-mode input range (CMR) are the most important. The prototypes have been simulated. In general, degradation is not important and even imperceptible in the values of all parameters. We can conclude that the extra circuitry has not appreciable impact in the device performance because it has not special sensitivity to the switches.

In summary, the main characteristics of the designed opamp-swopamp extracted from electrical simulations at nominal (27 °C) and extreme (-25°C, 85°C) temperatures as well as worst case models and voltage supplies (2.7V to 5.0V) are in Table 6.4. Although omitted in Fig. 6.3 for clarity, the bias circuitry (that will be explained later) was included in the simulations. In such conditions, it is clear that the opamp is well centered in the technology and fulfills all the requirements for the application.

		OPAMP	SWOPAMP	
Low-Frequency Gain		86.7		dB
Unity- Gain Bandwidth		1.1		MHz
Phase Margin		56		Deg
Offset Voltage		3.8		μV
Current		42		μA
Slew Rate	+	0.62	0.61	$V/\mu s$
	-	0.69	0.70	
Settling Time	+	456	482	ns
	-	545	551	
Area		118 x 219	118 x 289	μm^2

Table 6.4: Worst case simulation results of the opamp and swopamp using SPECTRE. CL= 25pF. Vsupply=2.7V

-Switches, resistors and capacitors:

Concerning the switches for the SC part, they have been built using complementary transistors of equal size (2.0μm/0.6μm) to cancel the first-order effect of clock feedthrough [121].

The technology used provides high-linear double-poly capacitors with $0.86fF/\mu m^2$ approximately. The unit capacitor for the SC part is ~0.45pF, which is also composed by three pieces of ~0.15pF each.

Resistors are made by using polysilicon with 24Ω per square. Both, capacitors and poly-resistors are placed over n-well to reduce the noise due to the substrate coupling.

-Bias Circuitry:

The bias circuitry provides the bias current for the opamps (or swopamps in our case), the reference voltage for the zero crossing detectors and the

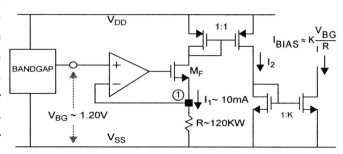

Figure 6.25: Schematic for the generation of bias current

analog ground voltage. The bias current of the opamps is of crucial importance if, as desired in the specifications, the circuit must operate properly in the range from 2.7V to 5.0V of supply voltages but maintaining a low power consumption. Consequently, it is necessary to consider that the extreme simplicity of the opamps used in the chip helps to cope with these requirements. The main properties and performance of the opamps can be preserved if a constant current bias is maintained in all the range of supply voltages. Thus, the current bias should be as independent as possible from the supply voltages. A circuit coping with this requirement is shown in Fig. 6.25. It uses a temperature compensated bandgap circuit that provides a constant voltage (V_{BG}~ 1.2V) with respect to V_{SS} in the range of interest of supply voltages (2.7V to 5V). Transistor M_F is used to feedback the output of the opamp to its inverting output. Because of the virtual ground property of the opamp, the voltage at node ① tends to be V_{BG} as well. Thus, current I_1 is fixed by the relation $I_1 = V_{BG}/R$. Using current mirrors as illustrated in the figure, a scaled current bias of value $I_{BIAS} = K(V_{BG}/R)$ is generated. As said before, the actual circuitry shown in Fig. 6.25 was included in the simulations of opamps.

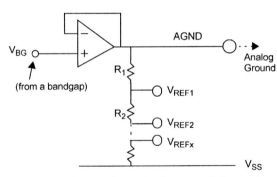

A low-impedance analog ground and the voltage references for the zero-crossing detectors and the voltage limiters have been obtained from the bandgap output voltage as shown in Fig. 6.26.

Figure 6.26: Voltage Reference Schematic

6.2.6 Start-up problem

To study the behaviour of a non-linear oscillator circuit, the start-up issue has to be consider. It is particularly critical in an oscillator built for testing purposes because of initialization requirements. And it can be a significant problem if it is not given special attention to.

The maximum time the oscillator requires to reach stable operation within specifications has to be taken into account. This time includes the transition time until the system start (*start-up time*) as well as the time until the system settles at the frequency and amplitude which satisfies the specifications (*settling time*). Both times depend on the initial conditions of the system.

It is required to reduce both the start-up and the settling times[5] as much as possible for test purposes since this extra time influences on the overall test time. Obviously, the OBT strategy has to be applied once the oscillation is perfectly established. In fact, one of the most important factors that make a test approach feasible is the time duration. Therefore, it is necessary to devise an start-up technique in order to guarantee a fast and reliable start-up (to safe time not only at practical level but also at simulation level). In practice, it will be shown below that using swopamp facilitates forcing a start-up condition and shortening the oscillation buildup time under any situation.

As was explained above, not all biquadratic sections of the DTMF core reconfigured as oscillators fulfil the mentioned initial conditions [86], [88]. Although, in principle, it is expected that this type of oscillators will be

---

[5] From this point forward, we will refer both parameters (start-up and settling times) as start-up time.

activated with the only action of the system turn-on, it has been observed by simulation that under certain conditions of charge feedthrough, dc offset, initial voltages, or power supply turn-on sequencing, oscillations may not start and/or the transient time to settle them is too large. This has been verified via analysis of the transient response of these oscillators. For this reason, it is required to devise a strategy that guarantees, under any circumstance, the start-up of the oscillations in a time comparatively short with respect the complete test time.

Obviously, the start-up problem (safe and short transient-time) can be solved if the oscillator operation is forced to start from a proper initial state. The start-up objective is to place the oscillatory system state in the vicinity of the steady-state. In fact, the closer the initial state to a steady-state point, the shorter the transient time. In fact, we have to store previously the suitable initial values in any passive-memory element inside the oscillator.

Accordingly, we have exploited the functionality of the swopamp to implement a start-up approach in our demonstrator. Fig. 6.27 illustrates how the start-up strategy is performed using the swopamp concept. As can be seen, every DTMF biquad is modified in such a way that an additional switch (st-up

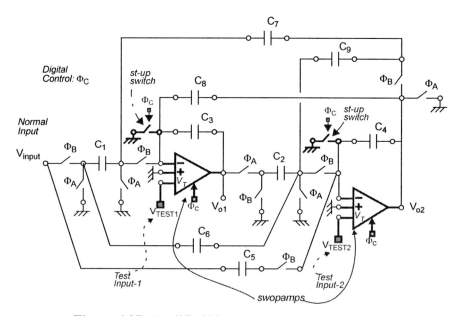

Figure 6.27: Modified biquad structure schematic

switches) and a swopamp are used instead of the normal opamp. Both, the switch and the swopamp are controlled by the same digital signal (Φ_C) which must be synchronized with the biquad master clock ($\Phi_{A/B}$).

The start-up approach works as shown in Fig. 6.28. In the test mode before the biquad is reconfigured as an oscillator, the logic signal Φ_C is settled to HIGH during a couple of periods. The swopamp is then working in its buffer mode and the bottom plate of the capacitor is grounded (Fig. 6.28-b). At the same time, a voltage V_{init} is applied to the buffer through the additional input V_T. It means that the output V_{out} is then set to V_{init} and, consequently, capacitor C is charged to that value. This configuration forces that V_{out} follows approximately the equation:

$$V_{out}(t) = V_{init}(t)\left(1 - e^{-\frac{t}{GB}}\right) + V_{out}(0)e^{-\frac{t}{GB}} \tag{6.1}$$

where GB is the amplifier gain bandwidth and $V_{out}(0)$ represents the value of V_{out} just when the start-up process begins. Since the GB of the opamps is normally much larger than the circuit maximum operating frequency, the speed of the charging process is very high. In fact, it can be considered around five times the time constant ($\sim 5/GB$).

Once the capacitor has been charged, the oscillator mode is settled while the swopamp is returned to its normal mode ($\Phi_C = LOW$) (see Fig. 6.28-(c)). In this configuration, the st-up switches are open. Therefore, considering the charge conservation law, and taking into account that the bottom plate of capacitor C switches between ground and virtual ground, the integrator starts to operate with an output voltage $V_{out} = V_{init}$. That means that the oscillator starts then with an initial state given by V_{init}. Thus, selecting properly V_{init}, the oscillator can run correctly.

Although the swopamp offset voltage and other factors can cause an error in the stored charge, this error is usually very small and normally can be discarded.

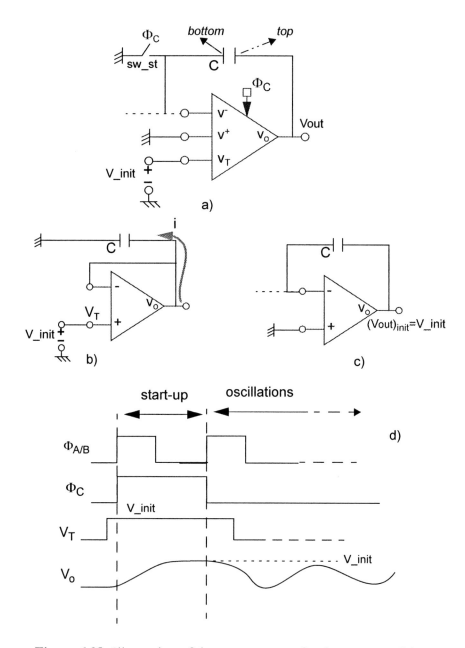

Figure 6.28: Illustration of the swopamp use for the start-up of the oscillators. a) New configuration of the integrator. b) Equivalent situation during charging mode. c) Equivalent situation when the oscillator is turned on. d) Involved waveforms

Let us finally analyse by simulation the influence of the initial conditions on the oscillatory result as well as how the efficiency of our presented start-up mechanism. In Fig. 6.29 the oscillations obtained with the proposed start-up strategy (wave-form in red line) and the oscillations

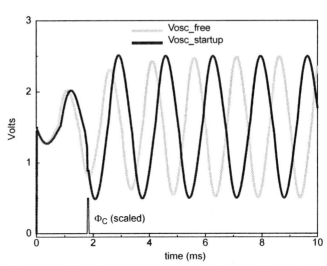

Figure 6.29: Generated waves without (Vosc_free) and with (Vosc_startup) the proposed start-up strategy

achieved freely (waveform in gray line) are displayed for one of the biquad. Observe that the free oscillations need a response time before they can be finally considered well-established.

Fig. 6.29 demonstrates the usefulness of the proposed start-up approach. Signal Vosc_free is obtained by forcing the biquad to oscillate just closing the feedback loop. Notice that the steady-state is not reached until the fourth period. Obviously, the value of this transient time depends on the particular biquad taken into account. But, it is vital to save this transient time because, otherwise, these four periods must be added to the overall test time. On the other hand, signal Vosc_startup is obtained when the start-up strategy is activated (with a delay after closing the feedback loop) as indicated by the high value of the signal control Φ_C. In this case, the steady-state is accomplished immediately.

6.2.7 The DTMF integrated prototype

After going into details describing all the modifications needed in the DTMF analog part to apply satisfactorily OBT-OBIST approach, let us summarize all those required changes in Table 6.5. This table allows us to get an idea of how simple and feasible is to transform the DTMF core in order to successfully implement our proposed OBIST strategy.

CHANGES	REGULAR DTMF DEVICE	OBIST DTMF DEVICE
Re-designed Blocks	Opamps	Swopamps
	Input Buffer	Modified Input Buffer (with swopamp and some extra switches)
	Bias Circuitry	Modified Bias Circuitry (voltage references)
	Digital Circuitry	Modified Digital Circuitry
Added Blocks	------------------	Control OBIST Circuitry
		Voltage Limiter
		Some Extra Switches

Table 6.5: Main involved changes

Observe that no essential changes in the DTMF structure are demanded, a lot of the original system circuitry is reused for the test facilities, nothing is removed from the primary prototype and the added new blocks are relatively simple and their incorporation does not mean a substantial increase of area.

Two IC prototypes have been designed (see Fig. 6.30): one of them implementing the DTMF and the second one incorporating the OBIST technique. Both circuits were built within a research project.[6]

Observe from Fig. 6.30 that the analog part of the normal prototype occupies an area which is only a ~7% smaller than the prototype incorporating the OBIST test strategy. Additional area is mainly that one shown within the dotted regions.

Notice as well that the so-called digital part of the normal device is slightly smaller than the digital part corresponding to device with the OBIST scheme owing to the fact that the OBIST strategy demands some extra control digital circuitry.

[6.] *ASTERIS: Advanced Solutions in Test Engineering Research for next generation Integrated Systems.* ESPRIT-CEE, N° 26354-1293/AA. Centro Nacional Microelectrónica (IMSE), Lancaster University, Austria Mikro System (AMS), Dolphin Integration (Francia). 1998-2000.

Moreover, both systems have the same number of pins (a single 14-pin package). Pins called PD, TEST and TOE in the first device must not be confused with the pins PD, TEST and TOE in the second device since these three pins have less functions for the first case. Consequently, despite of the fact that apparently both systems seem similar, it is not completely exact because the second device contains, however, more operation modes (it will be studied in detail in the next section). In the last circuit (Fig. 6.30-b) we have managed to integrate the OBIST test strategy without increasing the number of pins by means of an appropriate modification of the peripheral circuitry as well as the digital part, and optimizing, likewise, the use of the pins called PD, TEST and TOE (read next sections for details).

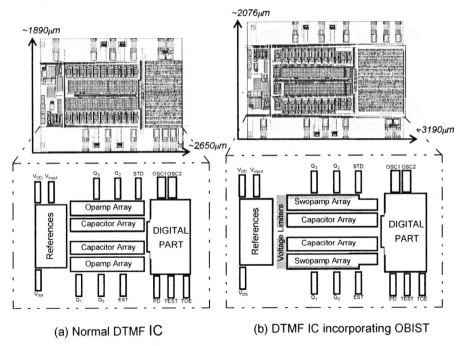

(a) Normal DTMF IC   (b) DTMF IC incorporating OBIST

Figure 6.30: DTMF Prototypes

6.3 ON-CHIP EVALUATION OF THE OBT OUTPUT SIGNALS

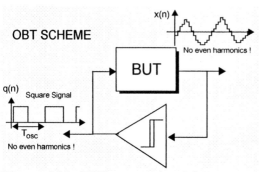

Figure 6.31: OBT output signal features

As was seen, the test output waveform coming from OBT (see Fig. 6.31) is an oscillatory signal (more or less pure) which is characterized by its frequency, amplitude, DC level, distortion, etc., parameters that are directly related to the system performance. Therefore, it is an essential issue to devise any kind of mechanism to extract the value of the oscillation parameters.

The most simple and straightforward method to estimate the frequency of an OBT output signal is to convert the set of uniformly spaced points (obtained either by simulation or experimentally) from the time domain to the frequency domain. This is, analysing the achieved output oscillation data by employing the Fast Fourier Transform (or FFT). However, implementing an on-chip circuit to evaluate the FFT involves an important additional area.

The final goal of this section will be to propose a general on-chip evaluation scheme. However, regardless of this optimum test interpretation mechanism we have studied by simulation other possible alternatives to measure and diagnose the oscillation parameters. Among them, let us present those two which best exemplify our work: to determine the oscillation frequency we have used a mere counter and to measure the oscillation amplitude we have used a peak detector.

6.3.1 Using a Frequency Measurement Counter

Let us study an easy way to determine the oscillation frequency. The employed method is well-known and relatively straightforward. Notice from Fig. 6.32 that after applying OBT technique we have available the square version of the oscillations (output of the zero-crossing detector, called $q(n)$).

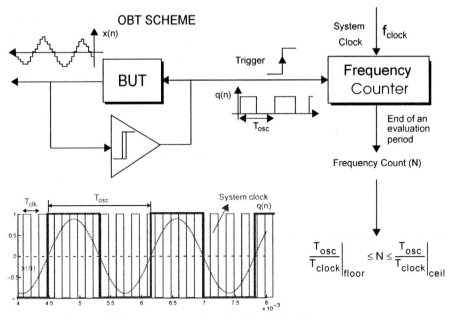

Figure 6.32: Frequency Evaluation Algorithm

Obviously, this square version of the oscillation signal has the same frequency than the sine signal. Thus, it is used to trigger a counter that will calculate the number of system clock periods within a signal period. Since the clock frequency is known, the signal frequency will be immediately deduced. In fact, the counter output will fluctuate between the two integers limiting the exact value of the oversampling ratio. This means, the state of the counter at the end of each period of $q(n)$ can be formally expressed as

$$\left\lceil \frac{T_{osc}}{T_{clock}} - 1 \right\rceil \leq count\text{-}freq = N \leq \left\lfloor \frac{T_{osc}}{T_{clock}} + 1 \right\rfloor \tag{6.2}$$

being T_{osc} the oscillation period and T_{clock} the sampling period.

Notice that the smaller the clock period compared with the square-wave period, the higher the precision of the frequency count. That is, the accuracy is determined by the oversampling ratio (T_{osc}/T_{clock}). For example, for an oversampling ratio as low as 30, the accuracy of the frequency measurement is around 3%, approximately. This precision is large enough for many applications, like the DTMF example where a typical accuracy can be 10% [120]-[121].

6.3.2 Using a Peak Detector to determine the amplitude

Basically, we employ the circuit shown in Fig. 6.33 whose result is displayed in Fig. 6.34. This device detects and stores the minimum signal value of its input. To carry it out successfully the detection of the oscillation amplitude must be made when we are sure that the oscillations are well-established. An important detail that has to be taken into account in the circuit of Fig. 6.33 is that the reset must be performed during the phase called here '*ph*1 '.

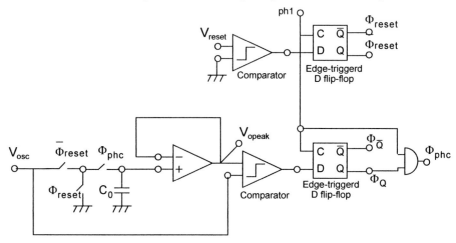

Figure 6.33: Peak Detector

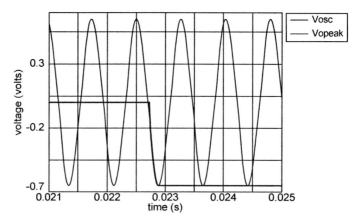

Figure 6.34: Example of Peak Detection Strategy

6.3.3 Using a low-accuracy $\Sigma\Delta$ modulator

Let us underline some features about the OBT output signals. Fig. 6.35 shows again how the BUT is reconfigured to oscillate by means of the OBT strategy. As was explained, this is obtained by closing a non-linear feedback loop (e.i. a comparator). The effect of this non-linear feedback loop is to convert its input wave-form into an amplitude limited square-wave which is then filtered by the BUT. Let us suppose then that a periodic π-symmetric solution is obtained. That is, the output signals contain only odd harmonics (6.3). Moreover, only the fundamental harmonic will be important and the rest of them will remain small (being even negligible). This fact will ease further developments

$$x(n) = B + \sum_{k_{odd}} A_k \sin(k\omega_{osc}nT_s + \theta_k) \approx B + A_{osc}\sin(\omega_{osc}nT_s + \theta) \quad (6.3)$$

On the other hand, the result of introducing this discrete oscillation, $x(n)$, in a $\Sigma\Delta$ modulator will be a digital version of it, called in Fig. 6.35 $d(n)$, which consists in a modulated pulse train

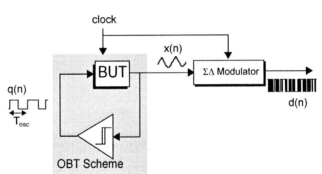

Figure 6.35: Using a $\Sigma\Delta$ modulator

containing all the waveform information. Therefore, $d(n)$ is a digitally-coded version of $x(n)$. Depending on the modulator features (its order, the sampling ratio, the quatization levels, etc) such a digital version will have a higher or lower accuracy.

Let us describe how we can take advantage of the scheme in Fig. 6.35. For the amplitude evaluation, the idea is basically employing a synchronous integration: the rising edge of the square wave, $q(n)$, starts the integration (acting as the integrator trigger). On the other hand, the falling edge inverts the polarity of the integration, which is equivalent to integrate a rectified wave-form (see Fig. 6.36). At the end of the evaluation period, the output of the integrator is proportional to the amplitude and the oversampling ratio. Again, as in the case of the frequency counter described above, the precision of the measurements depends on the oversampling ratio. As we can see in Fig. 6.36 and taking into account (6.3) at the end of an evaluation period we would have

$$\text{Amplitude Count}= \sum_{1}^{|V_1|}x(n) - \sum_{N_1+1}^{|V}x(n) \approx \frac{2}{\pi}AN \qquad (6.4)$$

being $N = \dfrac{f_{clock}}{f_{osc}}$, the oversampling ratio.

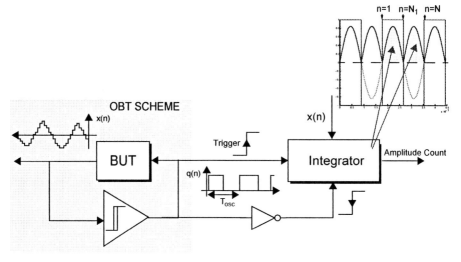

Figure 6.36: Amplitude Evaluation Algorithm

Likewise, for the evaluation of the DC-level we could again use synchronous integration (see Fig. 6.37). Again, the square wave is used to trigger the integrator, and at the end of an evaluation period, the integrator output will be the product between the DC-level and the oversampling ratio

$$\text{DC Count}= \sum_{1}^{|V_1|}x(n) \approx BN \qquad (6.5)$$

being again $N = \dfrac{f_{clock}}{f_{osc}}$, the oversampling ratio.

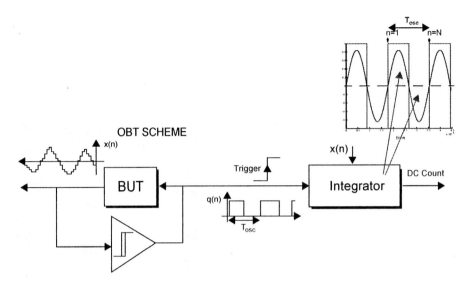

Figure 6.37: DC Evaluation Algorithm

But we have to disregard an analog integrator because we are looking for a digital technique. The proposal would be to employ the OBT scheme followed by a first-order sigma delta modulator (just as in Fig. 6.35) and then to directly integrate the output of such a modulator.

Fig. 6.38 shows the simple model of a first order $\Sigma\Delta$ modulator. The basic behavioural equation can be deduced. The input waveform has been normalized to the full-scale of the feedback 1-bit DAC of the $\Sigma\Delta$ modulator. The term $e(n)$ models the nonlinear error introduced by the comparator and the DAC (the quantization error) as a noise source. Therefore, the modulator output would be a bit-stream of positive and negative ones given by the following expression

$$d(n) = x(n) + e(n) - e(n-1) \tag{6.6}$$

We can calculate an approximation to the integral by summing up N samples of the modulator output. This is done by counting the number of 1's and -1's

$$\sum_{0}^{N} d(n) = \sum_{0}^{N} x(n) + e(N) - e(0) \tag{6.7}$$

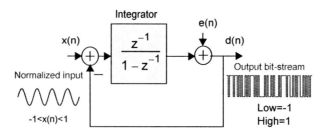

Figure 6.38: First-order $\Sigma\Delta$ Model

Notice that if there is not overflow in the integrator, the quantization error will remain between 1 and -1 ($e(N) \in [-1, 1]$).

Observe, then, that the amplitude evaluation scheme in Fig. 6.36 is conceptually equivalent to the scheme of Fig. 6.39, where the integrator fed by the OBT signal is replaced by a counter fed by the $\Sigma\Delta$ modulator bit-stream ($d(n)$). Notice, moreover, that the $\Sigma\Delta$ bit-stream is inverted on the negative lobe of the OBT output signal. At the end of an evaluation period the result is the same as (6.4) but with the addition of the quantization noise term

$$\text{Amplitude Count} = \sum_{1}^{N_1} d(n) - \sum_{N_1+1}^{N} d(n) \approx \frac{2}{\pi} A \frac{f_{clock}}{f_{osc}} \pm 4 \tag{6.8}$$

where now A is the waveform amplitude with respect to the modulator full-scale.

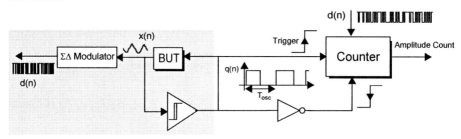

Figure 6.39: Amplitude Evaluation

Likewise, the DC-level evaluation scheme is converted to the scheme in Fig. 6.40 where the integrator driven by the OBT signal is replaced by a counter driven by the $\Sigma\Delta$ modulator bit-stream. The result is the same as (6.8) but with the addition of the quantization noise term whose worst case is expressed as

$$\text{DC Count} = \sum_{1}^{N_1} d(n) \approx BN \pm 2 \tag{6.9}$$

where B is the DC-level value, with respect to the modulator full-scale, and N is the oversampling ratio as defined above. It can be seen that the noise term is higher than in the case of the amplitude because, when the count is inverted during half of the period, the modulator error can not be cancelled out.

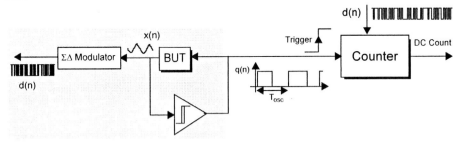

Figure 6.40: DC Evaluation

In summary, we propose the on-chip evaluation mechanism displayed in Fig. 6.41. This scheme makes use of a first-order sigma-delta modulator and some simple counters. The first-order sigma-delta modulator encodes the oscillation waveform into a digital bitstream, and the non-linear feedback loop around the BUT provides a digital signal giving its sign. The oscillation parameters (both the oscillation frequency and the oscillation amplitude) are entirely characterized in the digital domain by the bitstream and the square wave, with the corresponding benefits of robustness, ease of implementation and technology scaling.

The so-called *Digital Evaluation Unit* (Fig. 6.41) consists of three counters which provide a coded digital version of the frequency, the amplitude and the DC level of the oscillatory signal, respectively. The counters work as was explained above. The DC-level and the amplitude counters principle of operation is based on a synchronous integration. The square wave $q(n)$ (which gives the sign of the oscillation waveform) is used to trigger the counters which perform the sum (integration) of the sigma-delta bitstream. A "1" increments the counter and a "0" decrements the counter.

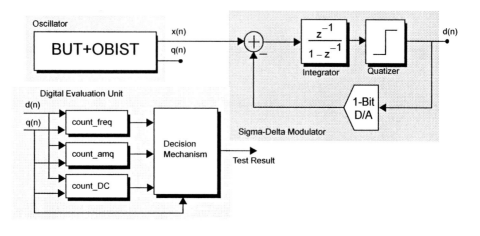

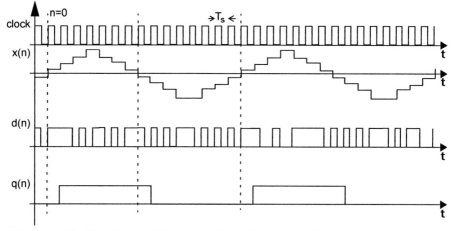

Figure 6.41: Simple on-chip evaluation of test signals

In short, the counter outputs are thus directly proportional to the oscillation parameters with a precision that is mainly driven by the oversampling ratio. It has been shown in [90] that it is possible to improve the precision by a factor of two at a very low cost by averaging out the bitstream values around the zero crossing instants defined by $q(n)$. Then, the interpretation unit would only consist of a digital comparison with a nominal window stored on-chip. More details about this strategy are shown in [85], [87]-[90].

Finally, we can go a step farther when taking into account some features of our demonstrator. In particular, we have studied a two-channel filter and we have described (Fig. 6.18) an OBIST strategy which simultaneously provides two sine-waves (one for each channel). Then, to jointly codify both test signals we propose, for instance, adding a *Premodulation Block* which allows us to obtain only a digital test output signal (see Fig. 6.42). Such a premodulation block involves the use of separate product modulators that are supplied with one of the waveforms coming from the OBIST scheme and the other one but differing in phase by -90 degrees (using a Hilbert transformer which causes a phase shift of -90 degrees). The resulting signal consists of the sum of these two product modulator outputs as shown in Fig. 6.42. In short, it is the scheme of a quadrature multiplexing.

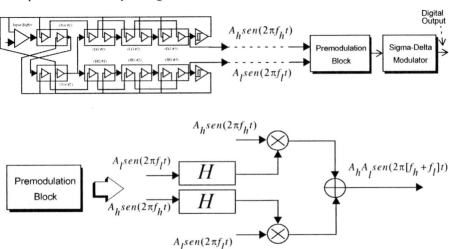

Figure 6.42: Premodulation Block to jointly codify the two test output signal

The digital signal resulting of the premodulation block will be then inserted in the sigma-delta modulator and likewise, its output will be inserted in the proposed Digital Evaluation Unit. Moreover, a square-version of the signal in the output of the premodulation block must be available. Such a square-version will be the reference signal in the counters of the Digital Evaluation Unit (Fig. 6.43).

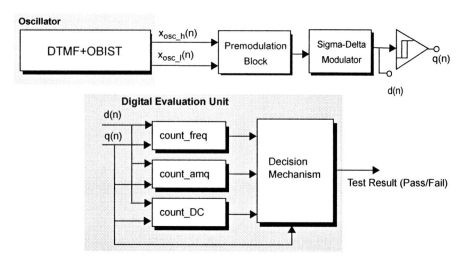

Figure 6.43: Proposed on-chip evaluation scheme

A problem derived from this scheme (Fig. 6.43) is that the Decision Mechanism will detect if there is a deviation in the output signal of the premodulation block, but it will be impossible to determine where the fault is. Therefore, with this evaluation scheme we lost the observability of the fault locations.

6.4 ELECTRICAL SIMULATION RESULTS IN THE OBIST MODE

In previous sections the design of different parts of our OBIST structure was described. The required modifications for the main blocks of the macro-cell were also reported. Now, simulation results utilizing the prototype implemented using a CMOS 0.6 μm technology are also presented.

The entire analog circuitry has been used to perform electrical simulations using SPECTRE. Bearing in mind the simulations, the expected results for the oscillation parameters are those given in Table 6.6 where we have compiled all the data coming from different means of decision. For example, to determine the oscillation amplitude we have used the proposed peak detector. However, to determine the frequency basically two ways have been used: 1) FFT and 2) number of pulses (frequency counter).

Therefore, the first row of Table 6.6 reports the feedback gain (that is, the reference level of the voltage limiter, V_{ref}) required for each biquad reconfigured as an oscillator. On the other hand, the second and the third rows collect the amplitude results obtained with the peak detector for both available biquad outputs. Finally, the remaining rows include many oscillation frequency results obtained by different mathematical methods (different windows have been used for evaluating the FFT).

Theoretical Results		Not #1	Not #2	LG #1	LG #2	LG #3	HG #1	HG #2	HG #3
Feedback Gain (V_{ref})		0.532	0.367	-0.355	-0.246	-0.109	-0.519	0.420	-0.162
$A_{osc}\_V_{o1\_peak\_detector}$		-0.450	-0.650	-0.451	-0.451	-0.450	-0.650	-0.650	-0.734
$A_{osc}\_V_{o2\_peak\_detector}$		-1.086	-0.998	-0.448	-0.511	-0.450	-0.649	-0.775	-0.650
FFT	f_{osc_FFT1}	699.1	586.7	822.5	966.9	594.9	1300.7	1177.5	1928.6
	f_{osc_FFT2}	710.1	573.6	819.4	955.9	600.9	1311.0	1174.4	1939.1
	$n_{\_pulse}$	78.8	97.5	68.3	58.5	93.1	42.7	47.6	28.9
Counter	$n_{\_pulse}$	80	95.3	68	57.813	94	43	47.5	29
	f_{osc}	699.2	586.9	822.6	967.5	595.1	1300.8	1177.6	1928.8

Table 6.6: Simulation OBIST results (fosc in Hz and Aosc in V)

All the oscillations were defined in such a way that the amplitude values of the OBT output signals (V_{o1} for all the biquad with the exception of HG #3) were very similar: either 0.45 V or 0.65 V (see Table 6.6).This is achieved by adequately adjusting the reference levels. Obviously, this tries to make the test interpretation process easier. On the other hand, we have to consider the remaining output signals (V_{o2} for all the biquad with the exception of HG #3) as well. We have to guarantee that the amplitude of such signals are below the saturation levels of the involved operational amplifiers in the biquad structures.

All these theoretical results (Table 6.6) will be used in Chapter 7 to validate the experimental results obtained by the IC prototype used as a circuit demonstrator.

6.5 DIGITAL PROCESSING PART OF THE DTMF

Although the digital algorithm is complex, the resulting circuitry is very conventional. It means that any of the well-accepted DfT strategies developed for digital circuits can be applied. However, they have not been considered in this case. Only a very simple programming of the existing circuitry has been introduced to cope with the frequency measurements of the oscillation waves.

6.5.1 Digital Detection algorithm

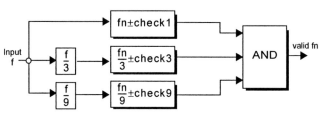

Figure 6.44: Digital Detection Algorithm for a given nominal frequency f_n

The detection algorithm is responsible for frequency tolerance specifications concerning maximum accepted frequency offset and minimum rejected frequency offset. Also, both the talk-off and noise immunity performance depend on the implemented detection algorithm. It receives the low frequency input and the high frequency input coming from the zero-crossing detectors, and processes the two signals separately by means of two similar subsystems. Each one determines whether one of the four DTMF frequencies is being received and, if so, which one. Frequency discrimination is performed by digital counting techniques which calculate the number of cycles of a given reference clock contained in a period of the input signal. A sophisticated detection algorithm involving many periods of the input signal is actually applied in order to tolerate frequency deviations required by DTMF specification and frequency variations resulting from the presence of interference frequencies and noise (jitter of the zero-crossing). A signal called *EST* is active while a valid frequency is being detected. In order to maintain *EST* high, the algorithm performs three different checks involving frequency measurements over one period, over three periods and over nine periods as it is sketched in Fig. 6.44. Algorithm design parameters *check1*, *check2* and *check3* have been chosen in order to achieve a good talk-off and SNR performance while satisfying frequency tolerance and timing specifications.

6.5.2 Steering logic

On the other hand, the steering logic must satisfy the timing specifications concerning minimum accepted tone length ($t_{\overline{REC}}$), maximum rejected tone length (t_{REC}), pause time (t_{ID}) and tone interrupt (t_{DO}). It activates *DST* when a valid tone pair has been present for a minimum amount of time and deactivates this after this is absent for given time. That is, it controls that a) too short (duration less than $t_{\overline{REC}}$) DTMF signals are never accepted, b) enough long (duration equal or greater than t_{REC}) DTMF signals are always detected, c) a DTMF signal with a too short interruption (interruption smaller than t_{DO}) are never validated as two distinct signals, and d) two DTMF signals with an enough long pause between them (interruption equal or greater than t_{ID}) are always accepted as two distinct signals. Steering logic measures the valid tone pair duration and pause duration and compares then to parameters *REC* and *GTA* respectively. These numbers are mainly constrained by the performance of the analogue part.

The digital part has been synthesized from a VHDL description. The layout has been generated using place and route automatic tools.

6.5.3 Simple Frequency Measurement Counter Block

Because there are many counters in the digital part whose mission is to check the frequency of the present tones, it is easy to understand that such counters can be re-used for the frequency measurements of the generated oscillation waves during BIST modes. So, a minor change in the digital algorithm have been introduced to adjust the counters during BIST modes in order to provide an internal measurement of such a parameter.

The module responsible for the frequency measurements is represented in Fig. 6.45. Basically, this system performs as a counter where one has stored two count numbers for every test configuration (for every involved frequency).

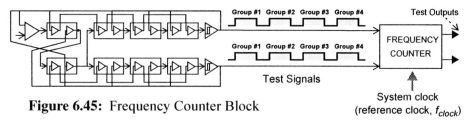

Figure 6.45: Frequency Counter Block

The count numbers are called N_{cmin} and N_{cmax} in Fig. 6.46 (with the j subscript if we are referring to the j-th group of oscillators). They correspond to a minimum value and a maximum value, respectively, what shows the tolerance margin for every obtained frequency. This tolerance margin must be determined by an exhaustive simulation (Monte Carlo analysis and fault simulations) and its fixing is very important for a correct fault detection.

Fig. 6.46 illustrates the counter operation. The idea was to design a mechanism which provides a go/no-go decision. Therefore, the counter measures the number of clock frequencies existing in the square-version of an obtained oscillation frequency. Such a number is called N_{count} in Fig. 6.46. If the corresponding N_{count} belongs to the interval $[N_{cmin}, N_{cmax}]$, then, the output signal of the counter circuit activates. Otherwise, if $N_{count} \notin [N_{cmin}, N_{cmax}]$, the counter output stays low and we can state that the corresponding biquad-under-test has failed. Fig. 6.46-(a) shows the case where all the biquads (of a channel) pass the go/no-go test and Fig. 6.46-(b) displays the case where a biquad does not go (to be precise a biquad from the Group #2). A significant fact is that the implementation of this block was completely embedded in the digital processing part without adding any extra area. Moreover, such a circuit is very simple and its implementation does not mean a considerable design effort.

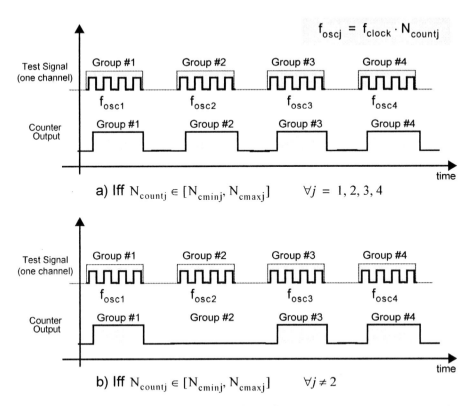

Figure 6.46: Frequency Counter Block Performance

6.6 DTMF/OBIST OPERATION MODES DESCRIPTION

Fig. 6.47 shows a simplified block diagram where the peripheral circuitry and the control signals of the different DTMF/OBIST operating modes are displayed. For the sake of clarity, these operating modes and their use are summarized in Table 6.7 and Table 6.8 where we have highlighted the particular case of the OBIST Mode.

The existing control and I/O pins available in the DTMF chip (Fig. 6.47) have been exploited to be able to set the different circuit operating modes. Moreover, a simple multiplexing strategy of the analog and the digital output signals has been applied to be able to read out the internal signals. Tri-state buffers for the digital signals and analog muxs have been disposed to reach this objective.

The extra pin named TEST has been used to change the meaning of the control signals PD and TOE to generate internally different control signals (MUX, TOE1, TOE2, PD) to control five different operating modes: a *Normal Mode*, three functional test modes (modes called, in Table 6.8, *Test Analog Part*, *Test Digital Part-I* and *Test Digital Part-II*) and an *OBIST Mode*.

In summary, we have 4 configurations where the system performs in its normal operation. They are: (000, 001, 010, 011). Notice, moreover, that the analog part is only readable in the Test Analog Part mode (100). In this mode, the filter outputs and the x10 gain stages as well as the comparators are externally monitored in parallel. It allows verifying all the analog part of the system at the same time. On the other hand, in the called OBIST Mode (101), we have access to the OBIST output signals as well as to the outputs of the proposed frequency counter. Finally, the test of the digital part is divided in two: first, an external test pattern is injected (mode 110) while the steering logic is externally monitored and verified. Once the steering logic has been activated, PD should change from 0 to 1 (mode 111) so the output of the decoder as well as the reaction of the steering logic to the injected test pattern is readable through Q0*-Q3*. Take into account that the change to the read mode is understood for the digital circuitry as if the input signal ceases.

EXTERNAL SIGNALS			REAL OUTPUTS / INPUTS					
TEST	TOE	PD	Q_0^*	Q_1^*	Q_2^*	Q_3^*	STD*	EST*
0	0	0	Q0	Q1	Q2	Q3	STD	EST
0	0	1	Q0	Q1	Q2	Q3	STD	EST
0	1	0	H	H	H	H	STD	EST
0	1	1	H	H	H	H	STD	EST
1	0	0	voflx10	vofl	vofh	vofhx10	vocl	voch
1	0	1	COUNT1	vofl	vofh	COUNT2	vocl	voch
1	1	0	vil	H	H	vih	STD	EST
1	1	1	Q0	Q1	Q2	Q3	STD	EST

Table 6.7: Operating modes summary (I)

EXTERNAL SIGNALS			OPERATING MODES
TEST	TOE	PD	
0	0	0	Normal Mode. Power Down OFF
0	0	1	Normal Mode. Power Down ON
0	1	0	Normal Mode. Q0-4 in High Impedance Mode. Power Down OFF.
0	1	1	Normal Mode. Q0-4 in High Impedance Mode. Power Down ON.
1	0	0	Test Analog Part- Analog circuitry. Outputs externally monitored as indicated through Q0*-Q3* and STD,EST.
1	0	1	OBIST Mode - Analog circuitry. Outputs externally monitored as indicated through Q0*-Q3* and ST_D,E. Positive edges in PD are used to change the configuration (reset is obtained whith TEST=0)
1	1	0	Test Digital Part-I. Signal injection through Q0 and Q3 . The steering logic indicates the recognition and validation (or not) of the test patterns
1	1	1	Test Digital Part-II. The outputs of the decoder (Q0-Q3) and the steering logic are all observable through Q0*-Q3* and STD, EST

Table 6.8: Operating modes summary (II)

Of course, pads corresponding to the digital outputs must be analog pads (only with protection diodes) to allow observing the analog outputs and injecting (bidirectionally) the test pattern in the test mode of the digital part. More details about these operation modes as well as their experimental results will be given in the next chapter. Now let us focus on the OBIST Mode.

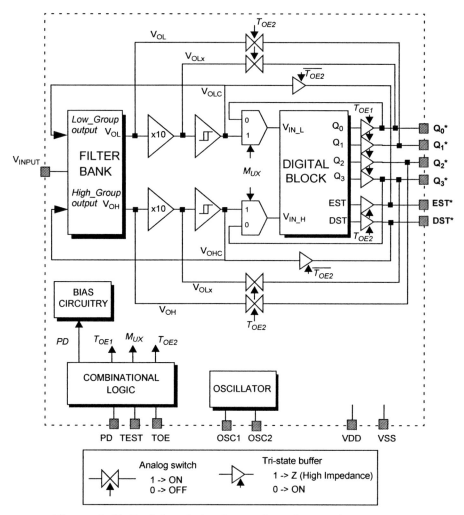

Figure 6.47: Peripheral circuitry and used control strategy

6.6.1 OBIST Mode description

As was said in the previous section, we can configure the DTMF system in different operation modes. Among them, we can distinguish the OBIST Mode where we have available the oscillation outputs (labelled V_{ofl} and V_{ofh} in Table 6.7), their square versions (labelled V_{ocl} and V_{och} in Table 6.7) and the signals coming from the interpretation circuit called Frequency Measurement Counter (labelled *COUNT1* and *COUNT2* in Table 6.7).

In this operation mode the control signals TEST, TOE and PD must be 1, 0 and 1 respectively. In fact, the test procedure is as indicated in Table 6.9 and Fig. 6.48. Notice that the configuration 0 0 0 is used once to start this mode. Observe, moreover, that when the configuration 1 0 1 is used, only one test group is activated. To change to another test group, the system has to move the control signals to 0 1 0 as represented in Fig. 6.48. That means, in fact, that the signal PD is used to change the involved test configuration whereas the signal TEST is set to low to reset the OBIST mode.

OBIST Mode			
TEST	TOE	PD	Test Configuration
0	0	0	
1	0	0	
1	0	1	Group #1
1	0	0	
1	0	1	Group #2
1	0	0	
1	0	1	Group #3
1	0	0	
1	0	1	Group #4

Table 6.9: OBIST Configuration

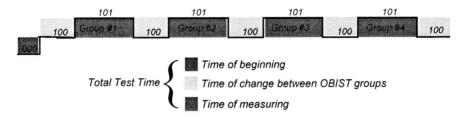

Figure 6.48: OBIST procedure Control.

We can, then, estimate the test time required to this proposed OBIST approach. Table 6.10 shows the time necessary for measuring each group the oscillations. We have computed two periods of the signal with minimal frequency for every group (the number of necessary periods depends on the required measurement accuracy, relative to the number of measured cycles). The total test time equals the total time of measuring plus the time needed to start the OBIST mode and the time required to change the OBIST configurations (this time is governed not only by the interval required to switch the control signals but also by the space required to interrupt the innate evolution of oscillations).

Group #1	Group #2	Group #3	Group #4	Total Time
~3.5ms	~2.5ms	~2.5ms	~2.9ms	~11.4ms

Table 6.10: Estimated measurement time

6.6.2 Test Strategy Comparison

To qualify the OBIST approach we need:

a) to compare it with the functional test strategies, and

b) to define a way to use both basic strategies (structural OBIST and functional) in a cooperative way.

Our intention is to discuss how the designer community can take the best of both for any particular case. To accomplish this task, we need to consider the testing cost for each approach in terms of required resources and test application time, as well as its suitability for BIST.

As a first consideration for comparison, we need to describe typical test routines for the functional testing of a stand-alone DTMF chip. Besides a complete set of measurements giving the transfer functions associated to the twin filter channels, there is an extensive characterization test used in industry [113], based on determining how the 16 tone combinations (high- and low-band) can be separated by the circuit (Comprehensive Functional Test described in Appendix 6.A). This is functionally sufficient, but the time required to apply it renders this test prohibitive for mass-produced ICs (see Appendix 6.A for details).

For production testing, a simpler alternative consists of detecting the separation between a subset of these combinations. A possible subset can be tone pairs in one of the diagonals in the keyboard, or even a two-tone signal formed by the highest of the low-frequency group and the lowest of the high-frequency group. Because of transient effects, a burst of periods is needed; the burst time duration depends on the settling time of the filter.

As explained earlier, we have two alternatives- parallel and sequential- for applying OBIST. One is to convert all biquads simultaneously into oscillators (parallel test) and another is to convert them sequentially (one after the other).

For the former we will need one comparator per biquad. On the other hand, the latter approach (sequential test) only requires one comparator, with significant savings in area overhead. However, test time differs for both cases. In the parallel case, all measurements are simultaneously carried out as opposed to the sequential case where the total test time is the sum of the test time for the eight biquads in the DTMF.

Oscillation start-up might be a problem if no special care is taken. We saw in previous sections that, in practice, using swopamps it is simple to force a start-up condition and to shorten the oscillation build-up time. In any case, this extra time may have an influence on the overall test time. But again, in the parallel method this time is equal to the larger start-up time of the eight oscillators, whereas for the sequential method, the oscillation settling is the sum of the start-up times of every oscillator. The difference between both cases depends on the frequencies to be measured, the biquad structures, and the required measurement accuracy (relative to the number of cycles to be measured).

From the point of view of a practical implementation, it is worth considering the use of OBT in combination with a simplified functional counterpart. We should make both options available, so that the designer can choose either one of them (or a combination) depending on the particular application. The main limitations from which a trade-off must be established between external and internal test options are:

a) number of pins,
b) external tester demands,
c) internal memory,
d) internal (extra) circuitry, and
e) testing time.

Test	Type	Coverage	Time	Signal Generation	Signal Interpretation
Comprehensive Test	Functional	Very High	Very High	External/Complex	External
Functionality Test	Functional	Very High	High	External/16-tone	External
Two-tone Test	Functional	Reasonable.	Medium	External/2-tone	Ext./Internal
Serial OBT	Structural	High/Very high	Medium	Internal	Ext./Internal
Parallel OBT	Structural	High/Very high	Low	Internal	Ext./Internal

Table 6.11: Test feature comparison.

Concerning tester demands, the situation can vary greatly, but, typically, there are scenarios in which a mixed-signal tester required for just a small part of a chip should be avoided. Another important constraint is the number of available pins. Normally, this is a scarce resource and has to be shared when several cores are used. The problem is that sharing pins increases test time proportionally, and then it should be appealing to devise methodologies with low-cost internal generation of test stimuli. Table 6.11 must be understood as a complete set of possibilities from which the user can select a test strategy to apply. Depending on pin availability, intended test time, external equipment cost, internally existing resources, etc, the system customer can choose one or combine two procedures.

Therefore, all said previously show that OBT-OBIST is a potential candidate to be used in combination with functional test techniques.

6.7 SUMMARY

In this chapter we have established a set of Design Decisions or rules to follow when a designer is interested in implementing the OBT-OBIST approach to a complex mixed-signal system. The analog part of a DTMF receiver has been used as vehicle to show all the requirements necessary to incorporate the on-chip test facilities.

In summary, let us express the view that the insertion of the OBIST approach in the DTMF analog core does not mean a lot of design effort. On the contrary, this test strategy can be easily embedded in the system without big problems in terms of performance degradation, design cost or area overhead.

On the other hand, we can use this implementation example (together with its obtained experimental results compiled in next chapter) not only to try to convince the mixed-signal community that commercial BIST solutions for analog circuits are already feasible but, also to encourage designers to adapt OBIST concept to other complex commercial mixed-signal systems.

Let us highlight some achievements:

-the wise use of a modified opamp (called swopamp) to ensure start-up and to provide accessing to internal blocks.

-the smart reuse of the existing zero-crossing detectors (in the original DTMF core) as the required non-linear feedback blocks in the closed-loop OBT scheme.

-the existence of several low-cost alternatives for testing the feedback path (sequential or parallel test) and the easy choice of an optimum solution based on a sequential-parallel test.

-the possibility of testing the normal signal path as well as the extra circuitry during the test mode.

-the chance of verifying the functionality of the OBIST structure itself.

-the availability of obtaining a one-pin digital signature, with low cost.

-the feasibility of coding the non-frequency data.

-the minimum cost of such an on-chip test approach: just one extra pin and an area overhead of around 7%.

-the compatibility of this technique with functional test counterparts.

Chapter 7

OBT-OBIST silicon validation
Experimental Integrated Prototypes

THIS CHAPTER reports experimental results corresponding to the application of OBT-OBIST to Switched-Capacitor integrated filters. Silicon prototypes of an universal programmable biquad as well as a Dual Tone Multifrequency Detector (DTMF) system are used as exploratory examples to demonstrate the feasibility of the OBT-OBIST approach. This Chapter is, perhaps, the most complete compendium of empirical results presented until date on the practical use of OBT-OBIST for mixed-signal and analog ICs.

7.1 INTRODUCTION

All the preceding chapters of this book were conceived to define the frame of a general OBT-OBIST mathematical theory and to detail the main guidelines that must be considered when OBT-OBIST is being applied to a particular system. However, an exhaustive proof based on a meaningful empirical evidence on practical integrated circuits was needed. Consequently, there is a need for providing empirical data confirming both the practicality of incorporating this new test technique to the regular design flow and the quality of its results. Consequently, this Chapter aims to fill in this lack through providing extensive experimental measurements on silicon demonstrators.

In short, all the mathematical results studied in previous chapters pave the way for employing the OBT-OBIST strategy from a practical point of view. Therefore, in this moment, we feel willingness to prove OBT-OBIST in practical and even industrial circuits.

Two ICs have been chosen for this purpose. The first device was specifically built to justify and give a firm evidence of our previous theoretical results. It incorporates abundant additional circuitry and numerous access points to different internal nodes. The idea is to carefully examine the power and efficiency of the OBT method by checking every step and every critical feature of this test approach. Basically, the device is a programmable circuit capable of performing three different biquadratic filters and/or of including several potential faults.

On the other hand, the second circuit is an industrial prototype with a well-defined structure whose main characteristics were described in the previous chapter. It is, in fact, the DTMF system which allowed us to study OBT-OBIST in a realistic context. In general lines, the objective with this experimental circuit is to prove that the OBT-OBIST strategy is in such an advanced developing stage than can be even applied to an industrial system.

7.2 FIRST EXPERIMENTAL DEMONSTRATOR

The first demonstrator used to support experimentally all said until now concerning the OBT approach, is based on a programmable version of the biquad displayed in Fig. 7.1. This particular structure was selected in Chapter 3 as a general OBT theoretical validation vehicle since it incorporates most of the common features in discrete-time analog circuits, allowing to validate OBT in many alternative filter configurations. An additional motivation for choosing such a cell was, of course, the experience that has been acquired in the course of these last years exploiting a similar biquadratic section in complex testable filters [25]-[36].

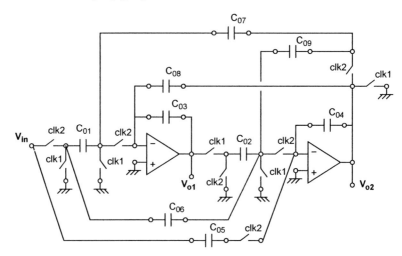

Figure 7.1: Studied SC Biquadratic Structure

Making use of the programmable nature of the biquad, some compatible functionality to simulate faults has been added at electrical level. Then, both soft and hard faults can be programmed (through external electrical variables) and experiments can be carried out to check the effectiveness of the OBT technique in this demonstrator. The way we can re-program the set of faults used for validating the OBT strategy in the programmable device, is also worthwhile since it resorts to circuit components compatible with the implementation technique.

Changes in capacitor values, stuck-on and -off for switches, and active component faults can be experimentally emulated by the circuit and many measurements on how the OBT architecture reacts to those faults can be performed. In fact, the versatility of this demonstrator allows us to authenticate the feasibility of the OBT method in an actual circuit.

7.2.1 Programmable biquad and fault programming

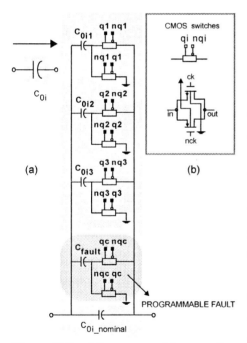

Figure 7.2: a) Programmable capacitor implementation b) CMOS switches

From the scheme in Fig. 7.1, a programmable biquad can be implemented by substituting every capacitor in this structure by a digital programmable capacitor array (DPCA) formed by some capacitors and some switches, as shown in Fig. 7.2. Each value of capacitor is conveniently altered to perform an individual type of biquadratic filter by directly using CMOS switches (Fig. 7.2-b). After selecting the set of different filters that will be realized with this biquadratic structure, a method to select the minimal capacitor value ($C_{0i\_nominal}$) required for any filter in this set [25] is used. Then, the increments in capacitance needed to realize all filters in the set are calculated. Using the minimal value as a seed, every capacitor in the programmable biquad is implemented by adding the incremental capacitors (capacitors C_{0i1}, C_{0i2} and C_{0i3}) under the command of switches.

The same principle can be used for implementing soft faults. Adding extra capacitors (or removing of existing ones) under the control of switches that can be externally manipulated, allow us to simulate a change in capacitor ratios that can slightly modify the overall filter transfer function. Rather than increasing the silicon area with capacitors, those dummy capacitors already placed in the filter for compensating layout dissymmetries are employed. It is shown in Fig. 7.3, where the extra capacitor introduced to emulate a faulty behaviour is represented by C_{fault}. Moreover, since these kind of soft faults implementations take advantage of the existing dummy capacitors, they can be considered relatively realistic. On the other hand, hard faults are easier to introduce [25], since there are many switches whose operation can be controlled in order to fix an internal node to a specific voltage.

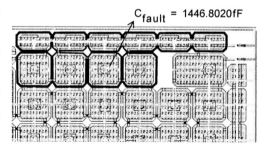

Figure 7.3: Programmable Fault

7.2.2 Experimental results

Fig. 7.4 shows the general diagram of the implemented chip with the different blocks and the required I/O pins. This circuit is composed of the programmable biquad, an amplifier stage, a non-linear block which consists of a feedback comparator with a controllable voltage limiter, a first order $\Sigma\Delta$ modulator (for encoding the test outcomes), a digital control block (for commanding the different operation modes) and all the necessary current and voltage reference generators. Moreover, we have added some extra I/O pins in order to observe many internal system nodes.

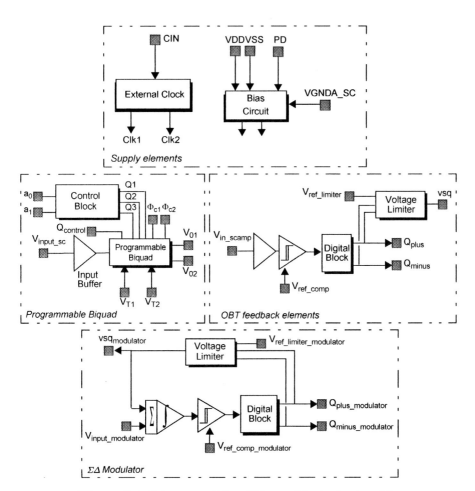

Figure 7.4: Diagram of the different blocks in the chip

On the other hand, Table 7.1 shows the accessible pines and all their functions.

Block	Pin Name	Simbol Explanation
Supply Elements	CIN	External clock to provide the system clocks: Clk1 and Clk2
	VDD	Supply Voltages
	VSS	
	PD	
Programmable Biquad	V_{input_sc}	Primary circuit input
	a_0	Signals to control the programmable biquad configuration
	a_1	
	V_{o1}	Signals to monitor the OBT configuration
	V_{o2}	
	Φ_{c1}	Signals to supervise the use of swopamps
	Φ_{c2}	
	V_{T1}	
	V_{T2}	
	$Q_{control}$	Signal to charge the programmable fault
OBT Feedback Elements	V_{in_scamp}	Input to the feedback elements
	$V_{ref\_limiter}$	Signal to guide the sign or/and the value of the feedback loop
	V_{ref_comp}	Signal to define the value of the comparator hysteresis
	Q_{plus}	Signals to visualize and process the chip outputs
	Q_{minus}	
	vsq	
SD Modulator	$Q_{plus\_modulator}$	
	$Q_{minus\_modulator}$	
	$vsq_{modulator}$	
	$V_{ref\_comp\_modulator}$	Signal to define the value of the comparator hysteresis

Table 7.1: Pin Description

The prototype was designed in a 0.6μm double-poly double-metal technology. The active area is $917.70\mu m \times 1808.20\mu m$, and the complete die size is $1492.10\mu m \times 2404.00\mu m$. In fact, two chips were implemented: one including only normal opamps and the other one in where the two opamps belonging to the programmable biquad were replaced by swopamps.

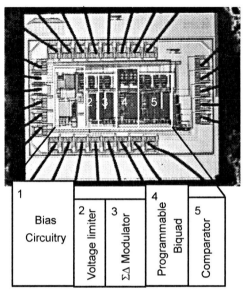

In Fig. 7.5 a microphotograph of the second prototype is displayed. The goal is to compare the results coming from each one of the chips and to verify experimentally not only the practicality of the swopamp but also its negligible impact on the system performance.

We can define different system configurations allowing the systems to operate in different operation modes (given by different connections of signals and different values of the control signals). Among them, we can highlight the *normal operation*

Figure 7.5: Filter Microphotograph

mode and *the test operation mode*. These operation modes will be studied in detail in next sections underlining their main features. Experimental results for both modes will be gathered as well.

-Normal Operation Mode:

In this operation mode the circuit works as a biquad (a second-order filter) and depending on the considered configuration (the values of the signals (a_0, a_1)), a particular input tone will be filtered or not. The programmable biquad is capable to perform as any of three different filters (a High-Pass and two Band-Pass, represented in Table 7.2 by BQ1, BQ2 and BQ3, respectively).The device can also incorporate several potential soft faults, managed by the external signals ($Q_{control}$, a_0 and a_1), as summarized in Table 7.2.

Configuration	$a_1=0\ a_0=0$		$a_1=1\ a_0=0$		$a_1=0\ a_0=1$	
Control	Qcontrol=0	Qcontrol=1	Qcontrol=0	Qcontrol=1	Qcontrol=0	Qcontrol=1
	BQ1	BQ1_Fault	BQ2	BQ2_Fault	BQ3	BQ3_Fault
C_{03}	6777.729 fF		2566.510 fF		10030.995 fF	
$C_{03\,fault}=$ $=C_{03}+C_{fault}$		8224.531 fF		4013.312 fF		11477.797 fF
Deviation	21.4%		56.4%		14.4%	

Table 7.2: Programmable Biquad Configuration

For the sake of convenience, the faults are inserted in the capacitor called C_{03} which is situated in the layout next to the dummy capacitors (observe again Fig. 7.3). This feedback capacitor C_{03} affects the feedback loops formed by the capacitors C_{01}, C_{07} and C_{08} (red dotted lines in Fig. 7.6).

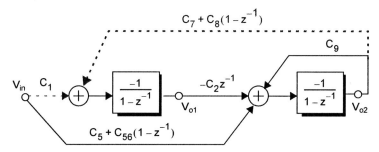

Figure 7.6: Programmable biquad z-domain equivalent circuit

Therefore we are, in some way, checking simultaneously different kind of deviations considering faults in C_{03}: faults in the branch controlled by $C_1 = C_{01}/C_{03}$ and faults in the branch controlled by $C_7 = C_{07}/C_{03}$ and $C_8 = C_{08}/C_{03}$. Obviously, we are not talking about a multiple fault because only a capacitor is strayed from its nominal value. We are talking, however, about a capacitor with a strong influence in the biquad structure. Just consider the z-equivalent circuit for the programmable biquad shown in Fig. 7.6, where we can see how two essential branches of the biquad are affected by C_{03} through the capacitors C_1, C_7 and C_8. Therefore, studying faults in this capacitor is useful to estimate the practicality of the OBT strategy applied to this device.

Fig. 7.7 displays the experimental transfer functions corresponding to the fault-free filters and, overimposed in gray, the corresponding faulty counterparts given in Table 7.2.[1] At first glance, it could seem that the faults do not have a prominent impact on the corresponding transfer functions. Notice, however, from Fig. 7.8, that the effect of the faults is very significant mainly in the case BQ2 (where 56.4 % deviation in C_{03} has been inserted).

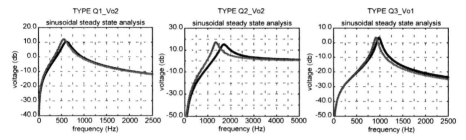

Figure 7.7: Transfer Functions for each SC- second order component

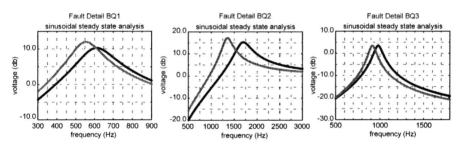

Figure 7.8: Details of the transfer functions

-Test Operation Mode:

This Test Operation Mode refers to the operation mode in which the OBT technique is applied to the programmable biquad. In this mode, the circuit works as an oscillator (since it is reconfigured following the OBT scheme). Again, depending on the considered configuration (the values of the signals (a_0, a_1)), a particular oscillator will be obtained.

---

[1]. The experimental conditions were VDD=3/2.7V, VSS=0V, PD=0V, VGNDA_SC=0V and CIN=55.934kHz.

The main features of this OBT mode are summarized in Table 7.3.

$a_1=0$ $a_0=0$		$a_1=1$ $a_0=0$		$a_1=0$ $a_0=1$	
Qcontrol=0	Qcontrol=1	Qcontrol=0	Qcontrol=1	Qcontrol=0	Qcontrol=1
BQ1	BQ1_Fault	BQ2	BQ2_Fault	BQ3	BQ3_Fault
Feedback Vo2		Feedback Vo2		Feedback Vo1	
Vref_limiter = 110mV		Vref_limiter = 161mV		Vref_limiter = 252mV	

Table 7.3: SC Programmable Biquad OBT Characteristics.

All experimental results obtained in this Section to validate the OBT methodology were measured making a by-hand evaluation of the oscillation parameters. We use different strategies to examine the waveforms: an oscilloscope and/or a network analizer (Fig. 7.9). Obviously, this is not a good alternative to be used in production, because it would need a very long test time. But, the goal herein is to experimentally show the feasibility of OBT, no matter the required test time.

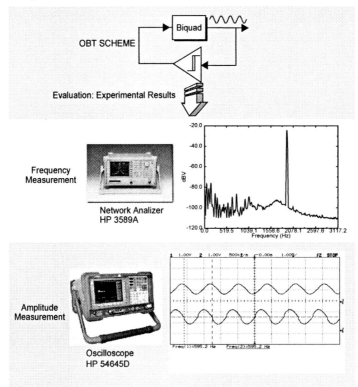

Figure 7.9: Evaluation of the oscillation parameters

However, in the next Section, we will work out an on-chip evaluation solution (see Chapter 6) which uses the involved $\Sigma\Delta$ modulator (Fig. 7.4). The practicality of the proposed on-chip scheme will be demonstrated by means of the obtained on-chip experimental results.

-Prototype including normal opamp:

Although the impact of the faults on the transfer functions is relatively small (see again Fig. 7.7), when translated into the oscillation parameters lead to a simple discrimination, as shown in Fig. 7.10, Fig. 7.11 and Fig. 7.12 where changes in amplitude and frequency for the faulty circuits have been represented on every sub-figure. The clusterized points (diamonds in green) in the lower left-side corner correspond to the experimentally measured values for the fault-free circuits (5 samples were characterized) as well as to the values predicted by Monte Carlo analysis for the corresponding filter (circles in black). Notice that, with the exception of the case BQ2, in the cases BQ1 and BQ3 the fault-free circuits deviates slightly from the Monte Carlo predictions. The reason is that Monte Carlo evaluations are not completely tuned. We have not checked all possible component deviations such as, for example, mismatching in the operational amplifier elements. On the contrary, only capacitor deviations in the programmable biquad were considered.

On the other hand, we have also marked off the zone in the frequency and amplitude space where the circuits with acceptable deviations are placed. This tolerance window is drawn in the last graphs of Fig. 7.10, Fig. 7.11 and Fig. 7.12. Its shape is a rectangular gray area in the left-side corner.

The starred points (in blue, see again Fig. 7.10, Fig. 7.11 and Fig. 7.12) correspond to the experimental values obtained for the experimental filters when oscillating under the influence of the soft faults. And finally, the red squares depict the simulated hard faults. For the sake of clarity, only faults which cause deviations smaller than a 50% are displayed. From the referred figures it should be clear that, assuming a 5% measurement accuracy, faults like the ones simulated are easily detectable with the exception of BQ2 where one finds a fault even inside the Monte Carlo region. Such a fault would be an undetectable fault with this OBT strategy. But it has not a strong repercussion, it would only mean that the percentage of detected faults (from those injected by simulation) would be 98.3 instead of 100% (see the third chapter for details). On the other hand, it can be observed from Fig. 7.10, Fig. 7.11 and

Fig. 7.12 that for the cases BQ1 and BQ2 (where the introduced soft faults cause a considerable alteration of their transfer functions, mostly in the case BQ2) the faulty points (in blue) are out of the tolerance window, that means inside the faulty zone determined by simulation. However, some points of the case BQ3 are borderline to both regions (the fault-free area and the faulty area). Obviously, a 14.4% deviation in C_{03} is not enough to assert that the circuit is malfunctioning.

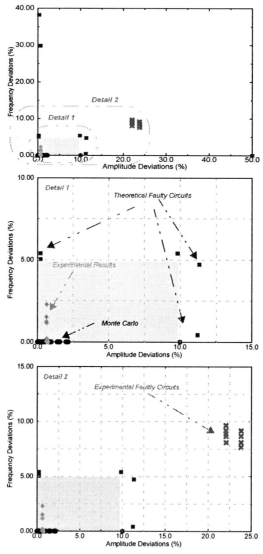

Figure 7.10: Oscillation parameter experimental measurements (BQ1)

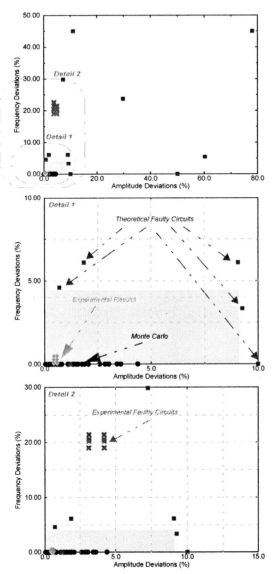

Figure 7.11: Oscillation parameter experimental measurements (BQ2)

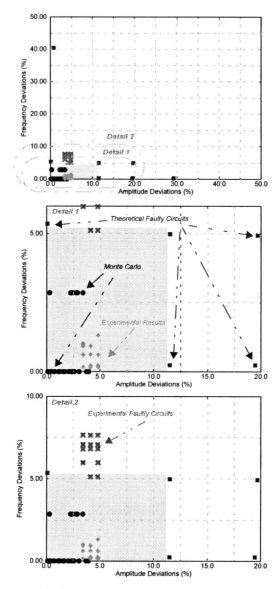

Figure 7.12: Oscillation parameter experimental measurements (BQ3)

To check the accuracy of the OBT method, Fig. 7.13, Fig. 7.14 and Fig. 7.15 show the experimental spectra for the three particular fault-free filter structures (the three different configuration of the programmable biquad displayed in black in Fig. 7.7). On every diagram we have represented data corresponding to the five tested samples. Moreover, all the spectra are displayed together in a last subgraphic. From these Figures, it should be clear that *a priori* prediction gives a good diagnosis of the experimental oscillation values.

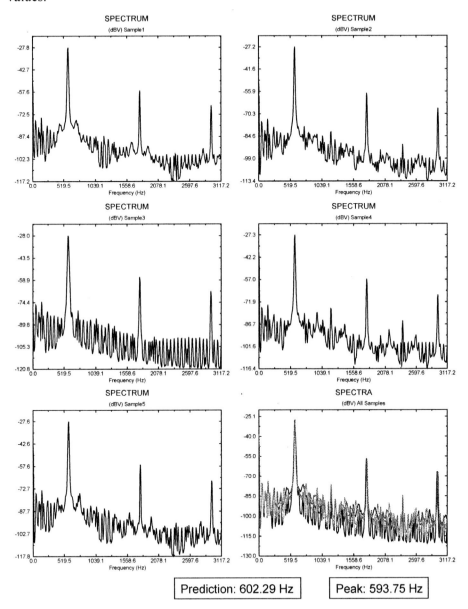

| Prediction: 602.29 Hz | Peak: 593.75 Hz |

Figure 7.13: Experimental frequency results from five different samples (BQ1)

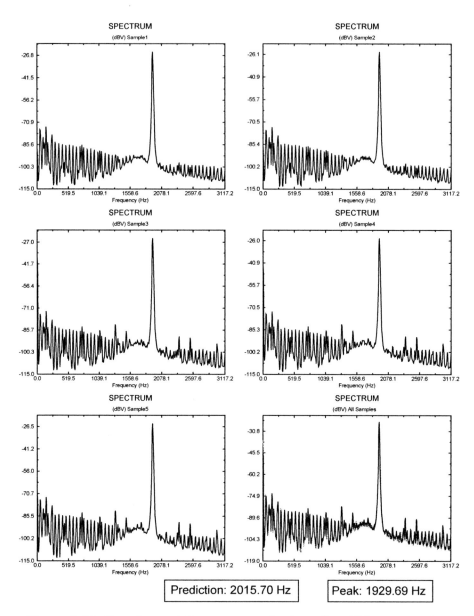

Figure 7.14: Experimental frequency results from five different samples (BQ2)

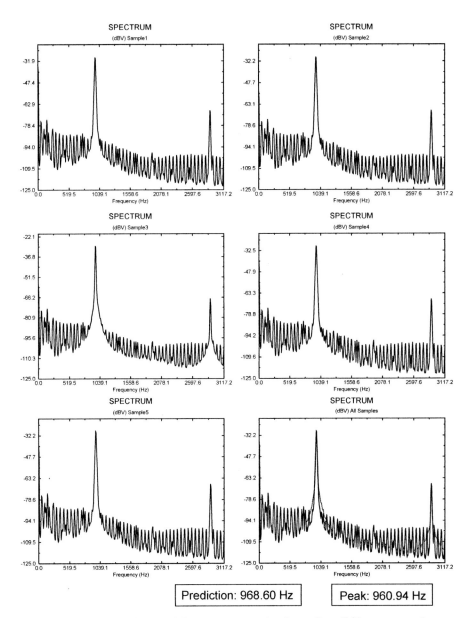

| Prediction: 968.60 Hz | Peak: 960.94 Hz |

Figure 7.15: Experimental frequency results from five different samples
(BQ3)

In addition, Fig. 7.16, Fig. 7.17 and Fig. 7.18 give an overall view of the
actual waveforms as predicted by simulation (MATLAB has been used, since
its results appear to be quite satisfactory) and as obtained empirically on the

actual silicon. Again, a good agreement has been observed. Notice how the experimental results (on the lower side) present the typical peaks for a SC circuit whereas the profile of the simulation waveforms (on the upper side) is smoother. On the other hand, experimental waves exhibit a DC level given by the analog ground (provided by the signal called above VGNDA_SC).

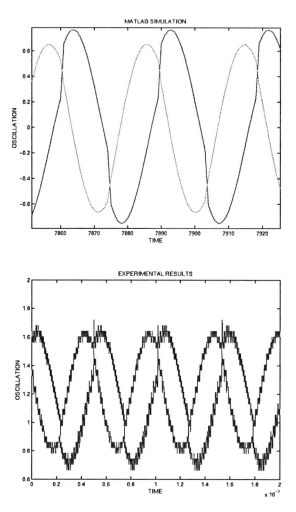

Figure 7.16: Comparing simulations and experimental measurements (BQ1)

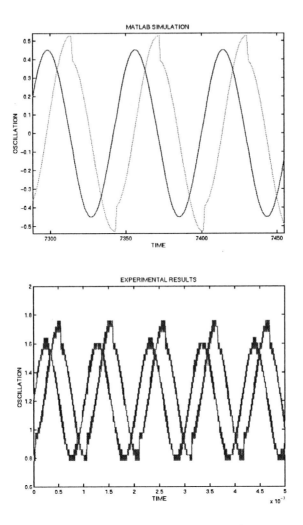

Figure 7.17: Comparing simulations and experimental measurements (BQ2)

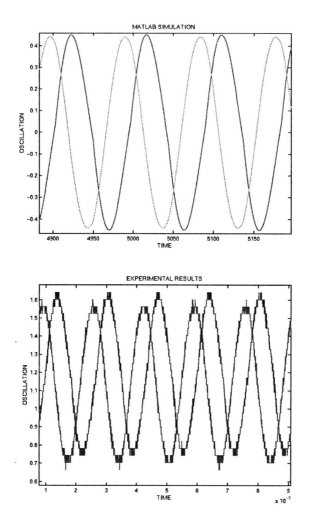

Figure 7.18: Comparing simulations and experimental measurements (BQ3)

On the other hand, we can also gather together all these data and check at the same time both the accuracy of the oscillator linearized model and the fault deviation (see tables in Appendix 7.A). Therefore, Table 7.4 provides with a summary of the predicted values for both the fault-free biquads and their corresponding faulty counterparts (considering as accepted values those coming from the SWITCAP simulations).

Parameters	BQ1	BQ1_fault	BQ2	BQ2_fault	BQ3	BQ3_fault
Amplitude	0.912 V	1.099 V	1.475 V	1.550 V	0.899 V	0.872 V
Frequency	595.01 Hz	543.18 Hz	1928.70 Hz	1532.40 Hz	966.89 Hz	901.26 Hz
Amplitude Deviation	20.50%		5.08%		3.00%	
Frequency Deviation	8.71%		20.55%		6.79%	
Fault Magnitude	21.4% of deviation		56.4% of deviation		14.4% of deviation	

Table 7.4: Predicted values for fault-free and corresponding faulty filters

We can observe from Table 7.4 that the value of a fault magnitude is not translated into exactly the same percentage in the value of the amplitude and frequency deviations. It depends on the expression of the oscillation parameters as a function of C_{03} (see Chapter 5).

On the other hand, the experimental results from the Tables given in Appendix 7.A (comparing only with the SWITCAP simulation data) are shown in Table 7.5.

Parameters	BQ1	BQ1_fault	BQ2	BQ2_fault	BQ3	BQ3_fault
Amplitude Deviation	21-23%		4-5%		3-5%	
Frequency Deviation	8-9%		19-21%		5-8%	
Fault Magnitude	21.4% of deviation		56.4% of deviation		14.4% of deviation	

Table 7.5: Experimental values for fault-free and corresponding faulty filters

In conclusion, these experiments (in both the time and frequency domains) we have performed with this programmable biquad, contribute to reinforce the results described elsewhere [33],[36] based on fault simulation. We have emulated physical defects through the use of switches and dummy capacitors. When these faulty circuits were characterized, all available samples were spot as "problematic" through the measurement of the oscillation frequency and amplitude. In this sense, this work can be considered as a point of reference for OBT in integrated analog circuits.

-Prototype including swopamps:

Until now only experimental results coming from the experimental IC without swopamps have been provided. The swopamp device was studied, from a theoretical viewpoint, in the fifth chapter. No essential disparities, in term of performance degradation or power consumption, was observed by simulation comparing the designed swopamp with its corresponding related opamp.

The main difference between both circuits (swopamp and its related opamp) is in the occupied area. Obviously, owing to the fact that the swopamp is built replicating several parts of the normal opamp (the differential pair to be precise), it must be bigger than the opamp. In fact, in our demonstrator if we estimate the discrepancy of areas between the IC containing swopamps and the IC containing opamps, we find from the layouts in Fig. 7.19 that the additional area is only the shaded rectangle on the right figure (approximately $64\mu m \times 240\mu m$) which means exclusively a 0.93% of the active area. This extra area is due to the two swopamps employed in the programmable biquad.

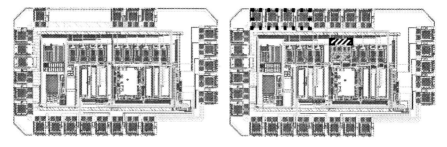

Figure 7.19: Layouts of the both IC used as demonstrators

Notice, likewise, from Fig. 7.19 that the second prototype (that one containing the swopamps) has more external pins than the first one (see area in dotted black lines). It is because every swopamp requires two extra external pins (one to control the operation mode, called before Φ_C, and the other one to supply the signal introduced in the buffer mode, called before V_T). But, the extra pin V_T, however, is usually used in the OBT-OBIST strategy to bypass the test signals from an intermediate point of the circuit to the circuit output or to the point where they must be interpreted. Therefore, this kind of pin usually performances without an external connection when OBT-OBIST is being applied to a more complex system. For this reason the only inevitable external pin derived from the use of the swopamp would be Φ_C.

In order to evaluate the impact of the swopamp on the experimental OBT results, experimental proofs shown above have been repeated. It is: the experimental spectra for the three particular fault-free filter structures and all those oscillation data given by the oscilloscope and the network analizer, again for five samples (whose tables are given in Appendix 7.A).

If we compare all the experimental results obtained with both prototypes (with and without swopamp), we achieve essentially the same. A good agreement is observed in Fig. 7.20, Fig. 7.21, Fig. 7.22 and Fig. 7.23 with the theoretical predictions.

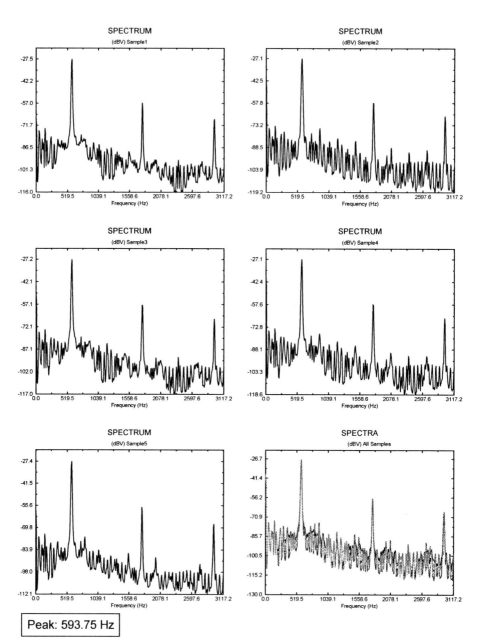

Peak: 593.75 Hz

Figure 7.20: Experimental frequency results from five different samples (BQ1_SW)

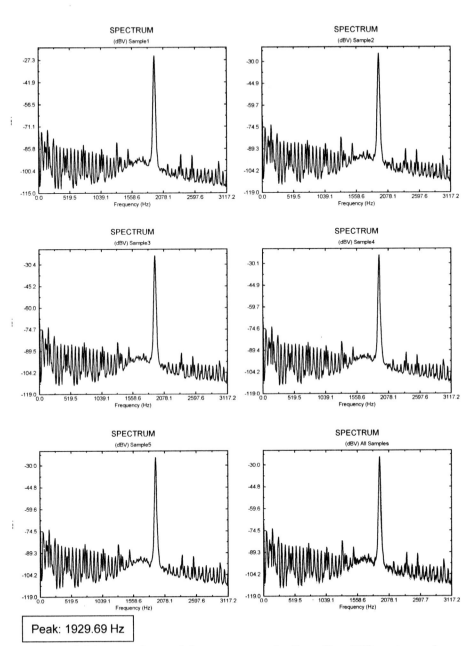

Peak: 1929.69 Hz

Figure 7.21: Experimental frequency results from five different samples (BQ2_SW)

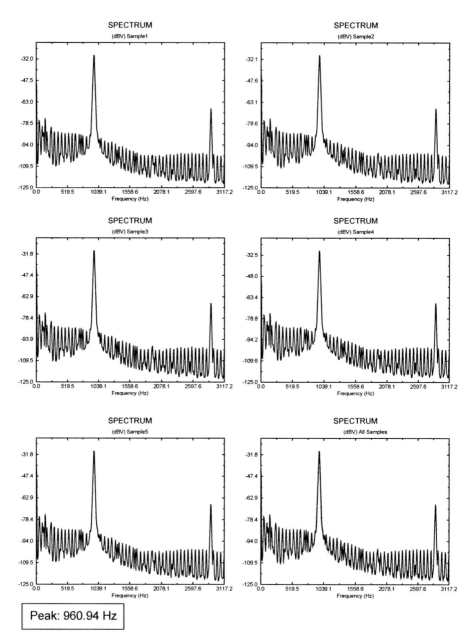

Peak: 960.94 Hz

Figure 7.22: Experimental frequency results from five different samples
(BQ3_SW)

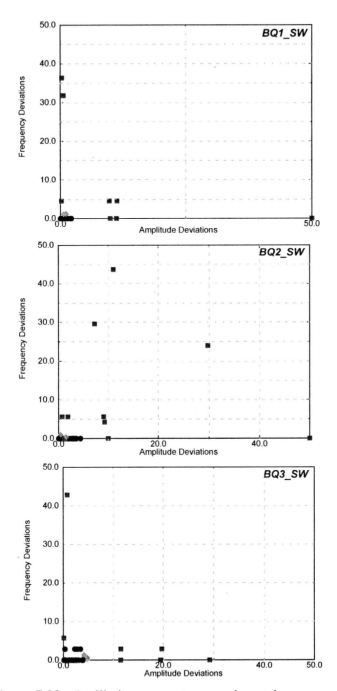

Figure 7.23: Oscillation parameter experimental measurements

7.2.3 On-chip evaluation

As was explained in Fig. 7.4 we have a $\Sigma\Delta$ modulator at our disposal embedded in the circuit. Therefore, we can employ the general scheme of Fig. 7.24 as a practical and low cost digital on-chip evaluation of the OBT output signals. On the left, one can see the BUT (in this case the programmable biquad) that has been reconfigured as an oscillator employing a comparator (the comparator which will give the square version of the sine-signal). The resulting sine-wave feeds a first order sigma-delta modulator. Finally, on the rigth, the sigma delta output bit-stream and the square wave signal are digitally processed with very simple hardware to compute the oscillation parameters: the DC-level, the amplitude and the frequency. We do not need any precise analog block and the extra area overhead due to the incorporation of this hardware should be minimum.

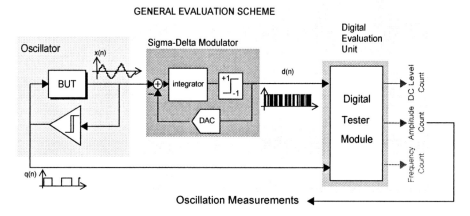

Figure 7.24: General Evaluation Scheme

In Fig. 7.25 we can see the test setup. On the left, we find the chip. In the middle of the scheme, we can see a digital tester to acquire the square wave and the modulator bit-stream, over around 500 signal periods. The results are processed by a workstation with MATLAB achieving the three different counts. Notice, however, that the work of the digital tester as well as the workstation can be perfectly substituted by an on-chip unit evaluation composed of three counters and some simple control circuitry.

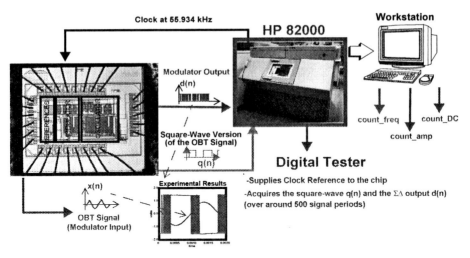

Figure 7.25: Test Set-up

In order to validate the experimental results, we used a clock frequency of 55.934 kHz and the modulator full-scale was set up 900 mV peak to peak. The OBT signal parameters were measured and shown in Table 7.6 and Table 7.7.

Oscillation Example	value
Frequency	598Hz
Amplitude	800 mVpp
3rd harmonic	-30 dB
5th harmonic	-40 dB

Table 7.6: Experimental Oscillation Parameters

count-freq	[93;94]
count-amp	$53 \pm 4 = [49;57]$

Table 7.7: Count results

The frequency is 598 Hz. So, the oversampling ratio is 93.5 and the frequency counter varies between 93 and 94 ($N = f_{clock}/f_{osc}$ = count_freq). The amplitude is $800 mV_{pp}$ so the normalized value is 0.89 which gives an amplitude counter value between 49 and 57 ($2\dfrac{A}{\pi}\dfrac{f_{clock}}{f_{osc}} \pm 4$ = count_amp).

In Fig. 7.26 we can see the experimental results for the frequency. The expected count_freq result is [93:94] ∼ 598*Hz*. It is clearly observed that they perfectly match predictions. Moreover, we can see that the mean counter value (93.43) tends to converge to the exact value of the oversampling ratio.

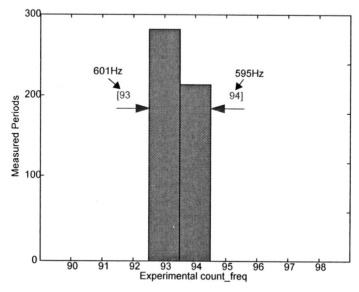

A sample measured in 50 periods with 10 different initial conditions

Figure 7.26: Experimental Frequency Results

On the other hand in Fig. 7.27 we can see the amplitude results. The expected count_amp result is [49:57] ∼ 800*mV*. Again the experimental results are in good agreement with predictions. Notice, moreover, that the extreme count values are hardly reached. This is because the worst-case quantization noise is unlikely. Observe also that the mean value of these results (52.9, ∼802*mV*) converges towards the exact value of the experimental oscillation amplitude (see Table 7.6).

In summary, we have an experimental basis which shows the feasibility and robustness of the proposed on-chip technique to measure and interpret the oscillation frequency and amplitude of the OBT output signals. We simply use a first-order sigma-delta modulator and several digital counters. The precision of these measurements is mostly driven by the oversampling ratio. However, other simple extraction algorithms have been developed that can double this precision [90].

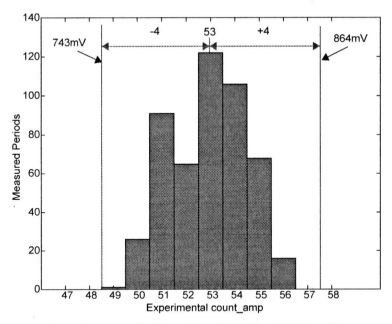

A sample measured in 50 periods with 10 different initial conditions

Figure 7.27: Experimental Amplitude Results

7.3 SECOND CIRCUIT DEMONSTRATOR: DTMF RECEIVER

Again, this part of the chapter focus on the validation of the OBT-OBIST technique proposed in previous chapters but now applying it to the SC filter macrocell used in the DTMF demonstrator considered in Chapter 6 [25]-[36], [121]-[122]. This silicon demonstrator, that will be described herein, is a two-channel filter bank. As was explained, it is an integrated SC circuit to which OBT-OBIST has been applied through minimal modifications and wise re-use of available components on chip. In fact, this filter structure was the complex core where we incorporated an elaborate OBIST strategy which was widely explained throughout Chapter 6.

The main goal of this section is to illustrate how the overall filter bank performance can be tested under many operating conditions. OBIST approach will be also reported comparing the design data obtained through extensive

Monte Carlo simulations and the test experimental results under different OBIST test configurations.

7.3.1 Floor-Planning and Chip

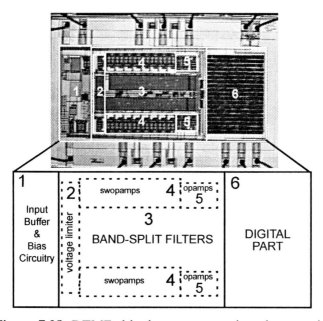

Figure 7.28: DTMF chip demonstrator microphotograph

Fig. 7.28 shows the chip microphotograph of the DTMF receiver. All the cells of the chip were designed in a 0.6µm double-poly technology. It has 14 pins and the total occupied area is of approximately 1x3mm^2. From Fig. 7.28, the extra area overhead (this area is only due to the area took up by the voltage limiters and the additional area due to the increase caused by the swopamp implementation) is very small, and can be quantified about 7%, which is a pretty good result. Consequently, it can be concluded that the re-use of the existing circuitry has been done efficiently. Moreover, power consumption is not penalized because all the extra circuitry for testing purpose is powered-off during normal mode.

Fig. 7.29 shows the general scheme displaying the chip blocks and the required I/O pins. We have 14 I/O pins, however, as explained in Chapter 6, only the pin called TEST is incorporated for test purposes.

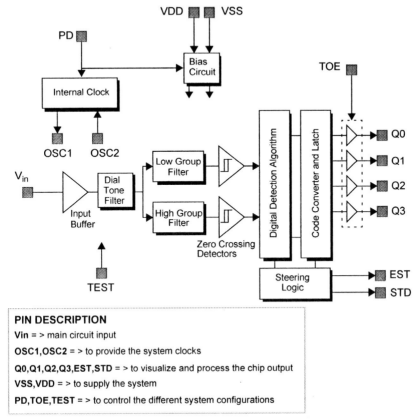

Figure 7.29: Block diagram of the DTMF receiver

7.3.2 DTMF Operation Modes

As was discussed in other chapters, the OBT-OBIST concept requires to divide the operation of the system under test in, at least, two operation modes. That is, at least, in a normal operation mode and a test operation mode. This is the case of the implemented DTMF system where we have introduced, by means of many control signals, programmable connections and some multiplexers, the possibility of working in several ways. In reality, more than the two basic operation modes has been incorporated in order to prove the feasibility and practicality of not only the DTMF prototype designed by us, but also the OBT-OBIST test technique proposed in previous chapters. In fact, five different operation modes can be distinguished. The values of their control signals are shown in Table 7.8.

Control Signals	Normal Operation Mode	Test Mode			
		Analog Test Mode		Digital Test Mode	
		Test Analog Part	OBIST Mode	Test Digital Part-I	Test Digital Part-II
TOE	0	0	0	1	1
TEST	0	1	1	1	1
PD	1	0	1	0	1

Table 7.8: DTMF/OBIST Operation Modes

For the sake of simplicity, the peripheral circuitry has been omitted in Fig. 7.29. We already gave a detailed description of it in Chapter 6 where we highlighted the basic usefulness of this circuitry and its involvement in the DTMF performance.

-Normal Operation Mode:

When the pin called TEST=LOW, the system is set to digital normal operation mode. Table 7.9 describes the meaning of each I/O pin while Table 7.10 explains briefly how the circuit works. Obviously, in this mode, the extra circuitry for testing purposes is not active.

Pin Name	Digital Normal Operation Mode (TEST="0")
VIN	DTMF input. Signal must be AC coupled via 100nF capacitor.
OSC1&2	Oscillator Input and Output respectively. A crystal or ceramic oscillator (3.58MHz) connected between them completes the internal oscillator circuit.
VDD, VSS	Positive and Negative Power Supply respectively.
TOE	Tri-State Output Enable.
Q0-3	Tri-State Coded output. When enabled by TOE, provide the last valid tone pair received. Otherwise are high-Impedance nodes.
EST	Early Steering. A logic high means that the digital algorithm has detected a valid tone-pair.
STD	Delayed Steering. A logic high means that a new tone-pair has been registered and latched.
PD	Power-Down mode select. It powers down the analog part and inhibits the oscillator.

Table 7.9: Pin description

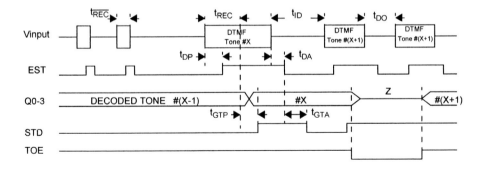

SIMBOL EXPLANATION	
Vinput	DTMF input signal
EST	Early steering. Indicates detection of valid tone
Q0-3	4-bit decoded output
STD	Delayed steering. Indicates that valid tones have been latched
TOE	Tone output enable. Shifts Q0-3 to high impedance (Z)
t_{REC}	Minimum tone duration required
t_{ID}	Minimum time between valid tones
t_{DO}	Maximum allowable drop out during valid tone
t_{DP}	Time to detect the presence of valid tones
t_{DA}	Time to detect the absence of valid tones
t_{GTP}	Guard time, tone present
t_{GTA}	Guard time, tone absent

Table 7.10: Explanation of events in the normal operating mode of the circuit

-Digital Test Mode:

The characterization of the circuit in its digital normal mode has been performed by applying in the lab the industrial audio test tape [113]. To illustrate the behaviour of the circuit, the obtained experimental oscillograms corresponding to the recognition and validation of the DTMF different tones whenever the time schedule is accomplished are shown in Fig. 7.30. The test executed is called *Decoder Check* (see the previous chapter for more details) where all pairs associated with standard 4x4 keypad digits are pulsed sequentially using 50ms bursts at 100mV per frequency. Each tone pair is pulsed once and a group of four pairs are sent consecutively. The receiver should respond to all tone pairs. Therefore, STD is activated only once per group of tones as expected whereas EST must be activated once each change of tone pairs.

Table 7.11 summarizes the main characteristics of the DTMF receiver as the result of testing 10 samples. As can be seen, they correspond to a high performance DTMF receiver, with only 5 hits in the speech testing.

Current Consumption	Operating	1 mA
Current Consumption	Power-Down ON	<1μA
Valid Input Signal levels		< 10mV to 350 mV
Freq. deviation	accept	±1.5%± 2Hz
	reject	±3.5%
Tone duration	accept	40ms
	reject	20ms
Tone pause	accept	40ms
	reject	20ms
Talk-off test		5 Hits
Voltage Supply		2.7V to 5.0V

Table 7.11: Main characteristics of the receiver

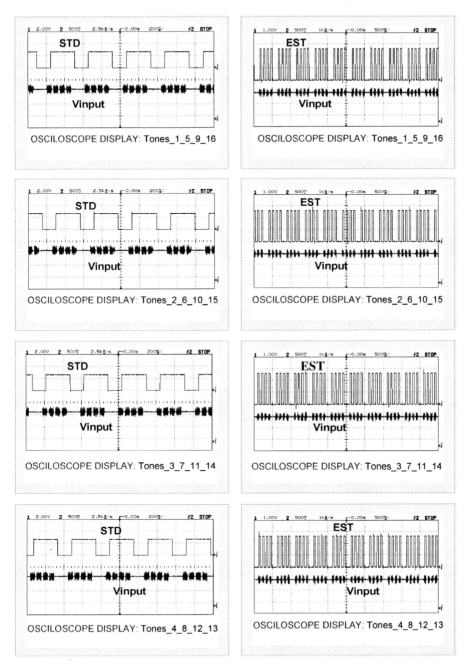

Figure 7.30: Tone pairs recognition

-Analog Test Modes:

-Analog Mode (TEST=1 TOE=0 PD=0):

In this operating mode, the digital circuitry and the extra one for testing purpose in BIST mode is not active. The available outputs at the I/O pins correspond to the main outputs for the analog filter banks as described in Table 7.12. It means that the filtering section can be fully characterized.

Pin Name	Analog Mode (TEST="1" and PD="0")
Q0-3	Output waveforms. $Q_0 = V_{olx10}$, $Q_1 = V_{ol}$, $Q_2 = V_{ohx10}$ and $Q_3 = V_{oh}$
EST	Zero crossing detector output for the high filter group
STD	Zero crossing detector output for the low filter group

Table 7.12: Output Pin Description

The following criteria and strategies have been used for the characterization of the filtering sections (Table 7.13 and Table 7.14).

IDDQA-ON	Operating Supply Current for the ANALOG Circuitry when PWDN mode is NOT ACTIVE (PWDN=GND)
IDDQA-OFF	Operating Supply Current for the ANALOG Circuitry when PWDN mode is ACTIVE (PWDN=VDD)
VOFF-L	Offset voltage measured at the output of the Low-Group Filter (Vol)
VOFF-H	Offset voltage measured at the output of the High-Group Filter (Vof)

Table 7.13: DC Electrical Characterization

A_{minl}	Minimum rejection in the lower stop-band
A_{minh}	Minimum rejection in the higher stop-band
A_{max}	Maximum ripple in the passband
f_{sl}	Lower stop-band frequency
f_{sh}	Higher stop-band frequency
f_{pl}	Lower passband frequency
f_{ph}	Higher passband frequency

Table 7.14: AC Electrical Characterization

The experimental results are shown in Table 7.15-Table 7.18 for 5V and 2.7V, respectively. It can be seen how both, the offset voltage and the current consumption have been kept as low as the application requires.

	Current Supply		LOW-GROUP							
	IddqA -ON	IddqA -OFF	Voff-L	A_{minl}	A_{minh}	A_{max}	f_{sl}	f_{pl}	f_{ph}	f_{sh}
M1	600	0.900	4			0.29				
M2	620	0.914	4			0.29				
M3	682	0.897	3			0.28				
M4	682	0.908	5			0.27				
M5	661	0.901	4	40.8	38.8	0.29	452	635	963	1172
M6	636	0.887	3			0.29				
M7	600	0.923	1			0.28				
M8	585	0.909	3			0.27				
M9	616	0.895	2			0.26				
M0	701	0.880	5			0.29				
Max	701	0.923	5			0.29		648	964	
Min	585	0.880	1			0.27		646	963	
Units	uA		mV	dB			Hz			
Test Cond	Vin=GND	PWDN = VDD		Vin(AC) = 15dBm						

Table 7.15: DC and AC electrical characteristics (VDD @5V)

	Current Supply		HIGH-GROUP						
	IddqA -ON	IddqA -OFF	Voff-H	A_{min}	A_{max}	f_{sl}	f_{pl}	f_{ph}	f_{sh}
M1	600	0.900	-1		0.44				
M2	620	0.914	-1		0.37				
M3	682	0.897	-1		0.42				
M4	682	0.908	-1		0.39				
M5	661	0.901	-1	38.8	0.38	963	1188	1662	1172
M6	636	0.887	-2		0.35				
M7	600	0.923	1		0.40				
M8	585	0.909	-1		0.41				
M9	616	0.895	-3		0.40				
M0	701	0.880	2		0.40				
Max	701	0.923	2		0.44		1187	1664	
Min	585	0.880	-3		0.35		1188	1668	
Units	uA		mV	dB		Hz			
Test Cond.	Vin=GND		Vin(AC) = 15dBm						
		PWDN = VDD							

Table 7.16: DC and AC electrical characteristics (VDD @5V)

	Current Supply		LOW-GROUP							
	IddqA -ON	IddqA -OFF	Voff-L	A_{minl}	A_{minh}	A_{max}	f_{sl}	f_{pl}	f_{ph}	f_{sh}
M1	488	0.207	- 0			0.30				
M2	477	0.210	-1			0.32				
M3	549	0.204	-1			0.30				
M4	544	0.208	0			0.28				
M5	547	0.204	-1	40.8	38.7	0.31	452	635	963	1172
M6	527	0.200	-1			0.30				
M7	470	0.216	-3			0.30				
M8	437	0.209	-1			0.30				
M9	˙502	0.199	-2			0.30				
M0	558	0.197	0			0.31				
Max	558	0.209	0			0.31		635	963	
Min	437	0.197	-3			0.28		634	962	
Units	uA		mV	dB			Hz			
Test Cond.	Vin=GND		Vin(AC) = 15dBm							
		PWDN = VDD								

Table 7.17: DC and AC electrical characteristics (VDD @2.7V)

	Current Supply		HIGH-GROUP					
	IddqA -ON	IddqA -OFF	Voff-H	A_{min}	A_{max}	f_{sl}	f_{pl}	f_{ph}
M1	488	0.207	-2		0.44			
M2	477	0.210	-2		0.39			
M3	549	0.204	-2		0.44			
M4	544	0.208	-2		0.41			
M5	547	0.204	-1	38.8	0.41	963	1188	1662
M6	527	0.200	-3		0.37			
M7	470	0.216	1		0.42			
M8	437	0.209	-1		0.43			
M9	502	0.199	-4		0.42			
M0	558	0.197	0		0.41			
Max	558	0.209	1		0.44		1188	1662
Min	437	0.197	-4		0.37		1187	1661
Units	uA		mV	dB		Hz		
Test Cond.	Vin=GND		Vin(AC) = 15dBm					
	PWDN= VDD							

Table 7.18: DC and AC electrical characteristics (VDD @2.7V)

On the other hand, the corresponding magnitude frequency response, given by a network analyzer, for the ten samples are shown in Fig. 7.31 with VDD=5V and Fig. 7.32 with VDD=2.7V. On the left-side of these figures, the entire experimental Bode's plots are represented whereas on the right-side details on the ripples of the corresponding passbands are displayed.

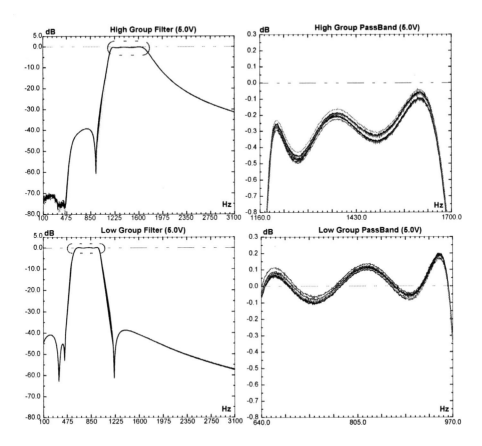

Figure 7.31: Bode diagrams (magnitude) of the filter paths for VDD=5.0V

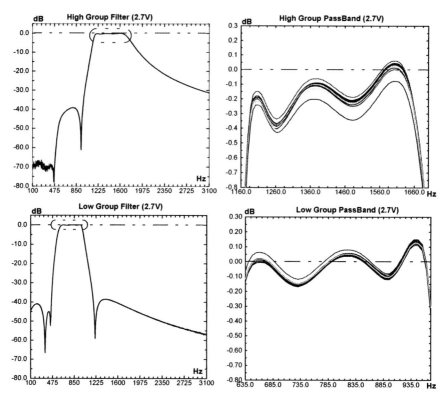

Figure 7.32: Bode diagrams (magnitude) of the filter paths for VDD=2.7V

Observe from this Figures, all measurements agree with the simulations and specifications at least at room temperature (27 °C.). Despite of the fact that only ten samples have been used, the circuit shows a very stable behaviour (critical frequency points, ripple, magnitude of the rejection, current supply, etc....).

Finally, and to illustrate the time-domain behaviour of the filtering section, Fig. 7.33 shows the response at the output of the zero-crossing detector when the input is a composite signal of 1300HZ and 835Hz (one dial tone). It can be seen that the frequency of the square-shaped signal agrees very well with the expected one. The screen of a digital oscilloscope has been captured to display the resulting signals.

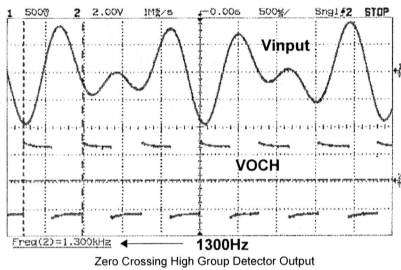

Zero Crossing High Group Detector Output

Zero Crossing Low Group Detector Output

Figure 7.33: Time-domain response at the output of the
zero-crossing detector for a composite signal of 835Hz
and 1300Hz and equal amplitude

-OBIST Mode (TEST=1 TOE=0 PD=1):

In the BIST mode both the digital circuitry and the extra elements added for testing purpose are active. As explained in Chapter 6, our DfT goal was to split the filter bank into biquads and feeding-back each one until achieving the sustained oscillations needed for applying the OBT technique. The available outputs at the I/O pins correspond to the oscillation outputs. It was seen in Chapter 6 that in the test mode only two biquads can be simultaneously tested (Fig. 7.34).

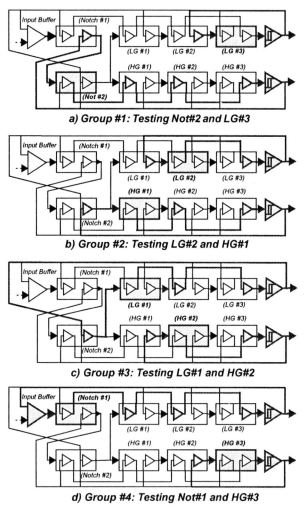

a) Group #1: Testing Not#2 and LG#3

b) Group #2: Testing LG#2 and HG#1

c) Group #3: Testing LG#1 and HG#2

d) Group #4: Testing Not#1 and HG#3

Figure 7.34: Configurations to test the whole DTMF Filter using the OBIST technique

Each configuration is realized changing the external pin called PD which is used to multiplex the corresponding outputs. Therefore, it is possible to find four different groups of oscillators depending on those biquads which are oscillating. The Group #1 is formed by the biquads called Not #2 and LG #3, the Group #2 by the biquads called LG #2 and HG #1, the Group #3 by the biquads called LG#1 and HG#2 and, finally, the Group #4 by the biquads called Not #1 and HG #3. Hence, the whole filter is tested in four phases (see Fig. 7.34). The corresponding theoretical results for each oscillation group are displayed in Tables 7.19 and 7.20 where the feedback sign and the extra existing delays (n) are also shown.

Biquad	Output / Feed-back Sign	extra delay				
		n=1				
		A (Vp)	fosc (FFT)	fosc ($1/T_{square-wave}$)		
				max	mean	min
Group #1						
LG #3	Vo2 / -					
Not #2	Vo1 / +	1.83	601	589	589	589
Group #2						
HG #3	Vo2 / -					
Not #1	Vo1 / +					
Group #3						
LG #1	Vo1 / -					
HG #2	Vo1 / +	1.55	1174	1190	1165	1179
Group #4						
LG #2	Vo1 / -	1.83	956	981	966	964
HG #1	Vo1 / -					
		V	Hz			

Table 7.19: Groups of obtained oscillations

Biquad	Output / Feed-back Sign	extra delay				
		n=2				
		A (Vp)	fosc (FFT)	fosc ($1/T_{square\text{-}wave}$)		
				max	mean	min
Group #1						
LG #3	Vo2 / -	4.15	601	601	595	599
Not #2	Vo1 / +					
Group #2						
HG #3	Vo2 / -	4.00	1939	1928	1928	1928
Not #1	Vo1 / +	0.85	710	699	699	699
Group #3						
LG #1	Vo1 / -	1.27	819	823	823	823
HG #2	Vo1 / +					
Group #4						
LG #2	Vo1 / -					
HG #1	Vo1 / -	1.25	1311	1300	1300	1300
		V	Hz			

Table 7.20: Groups of obtained oscillations

For one experimental sample, making use of a digital oscilloscope, the waveforms of the groups of oscillators obtained in the laboratory are shown in Fig. 7.35 and Fig. 7.36. Other experimental measurements were made for five samples using a network analyzer (for the FFT, see Fig. 7.37, Fig. 7.38 , Fig. 7.39 and Fig. 7.40). In these Figures, five subgraphics separately display the results of the five samples and a sixth subgraphic shows all the information together. Data coming from a virtual oscilloscope are also displayed in Fig. 7.41. Finally, we have collected all the experimental results obtained in this OBIST Mode in the second section of the Appendix 7.A.

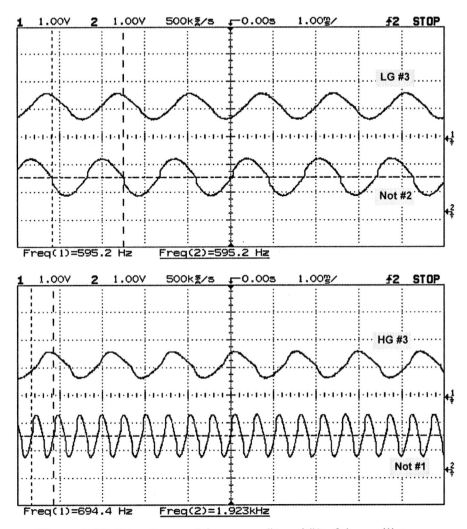

Figure 7.35: Waveforms of the groups #1 and #2 of the oscillators.
Digital Oscilloscope

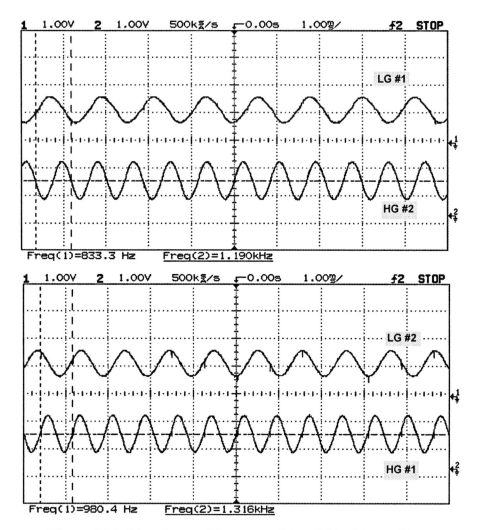

Figure 7.36: Waveforms of the groups # 3 and #4 of the oscillators. Digital Oscilloscope

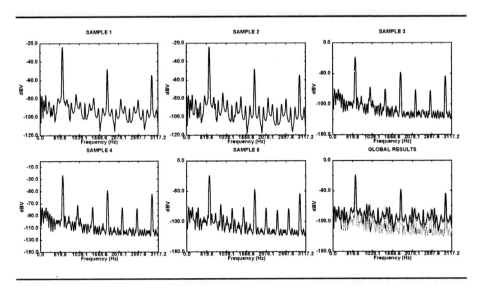

Not#2

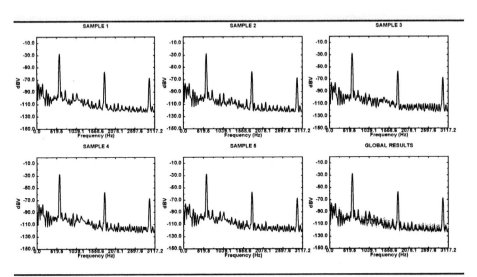

LG#3

Figure 7.37: Experimental Spectra for the group #1 of oscillators.
Five samples

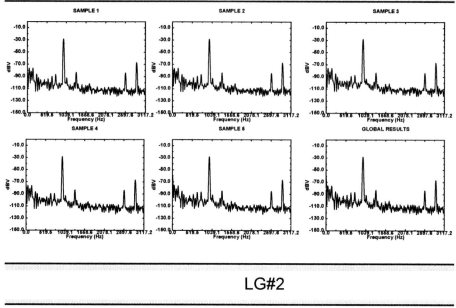

LG#2

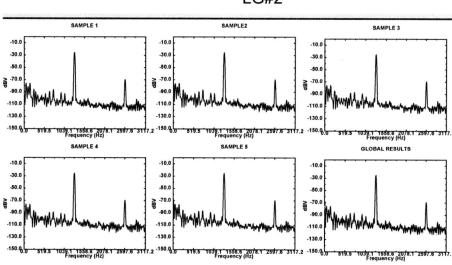

HG#1

Figure 7.38: Experimental Spectra for the group #2 of oscillators.
Five samples

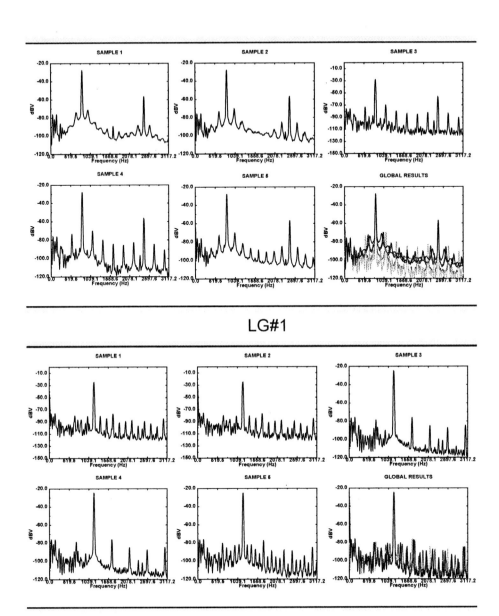

Figure 7.39: Experimental Spectra for the group #3 of oscillators.
Five samples

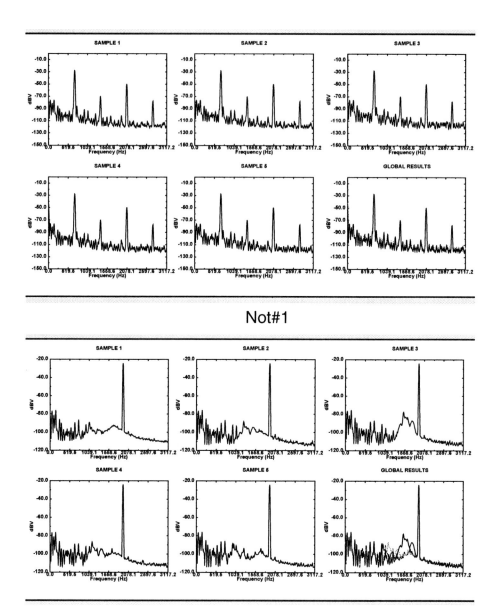

Not#1

HG#3

Figure 7.40: Experimental Spectra for the group #4 of oscillators.
Five samples

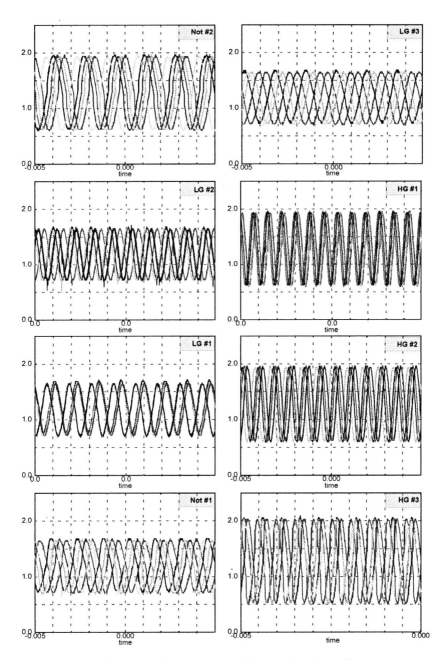

Figure 7.41: Waveforms captured by a virtual oscilloscope.
Five samples

The robustness of the OBIST approach an be also discussed comparing the design data gathered through extensive Monte Carlo simulations and the test experimental results under different test configurations. Fig. 7.42, depicts a typical case for one of the filter biquadratic sections representing the design acceptability region and several faulty situations in the oscillator parameter space.

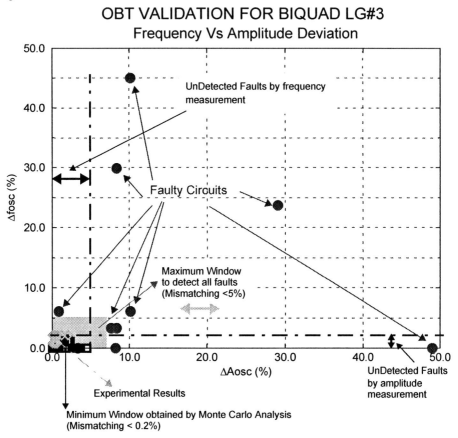

Figure 7.42: Details about simulation data and actual experimental data

On the diagram in Fig. 7.42, we have represented (in green diamonds) data corresponding to the five tested samples for only a biquad (LG#3). From this Figure it should be clear that a priori simulation prediction give a good esti-mation of the experimental oscillation values. As it is displayed, data in green (taken from the experimental results), are placed in or closed to the minimum window obtained by Monte Carlo Analysis (in the worst case never out of the maximum window to detect faults). In addition, Fig. 7.43, Fig. 7.44, Fig. 7.45

y Fig. 7.46 give an overall view of the actual oscillation parameters as predicted by simulation (SWITCAP has been used, since its results appear to be quite satisfactory) and as obtained empirically on the actual silicon. A good agreement has been observed for every biquad.

In general, all the experimental results obtained with this validation demonstrator exhibit a good agreement with the predicted oscillations. Physical defects were emulated through the use of switches models and changes in the capacitors. When these fault-free circuits were characterized, all available samples were spot as "no problematic" through the measurement of the oscillation frequency and amplitude. Likewise, in the lab, all the resulting experimental data corroborated the predicted theoretical results.

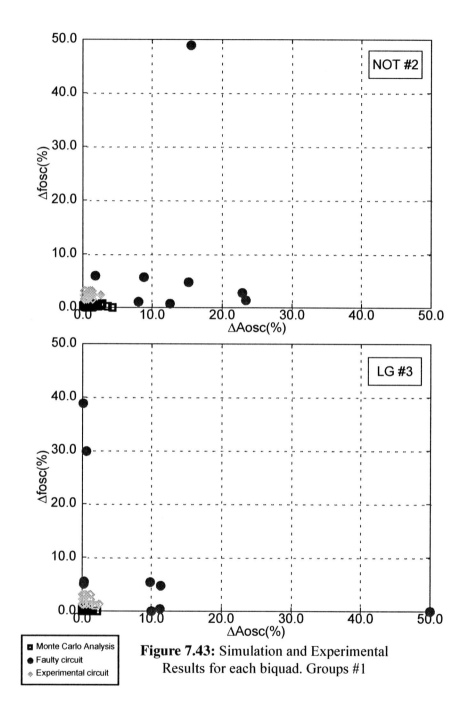

Figure 7.43: Simulation and Experimental
Results for each biquad. Groups #1

Legend:
- Monte Carlo Analysis
- Faulty circuit
- Experimental circuit

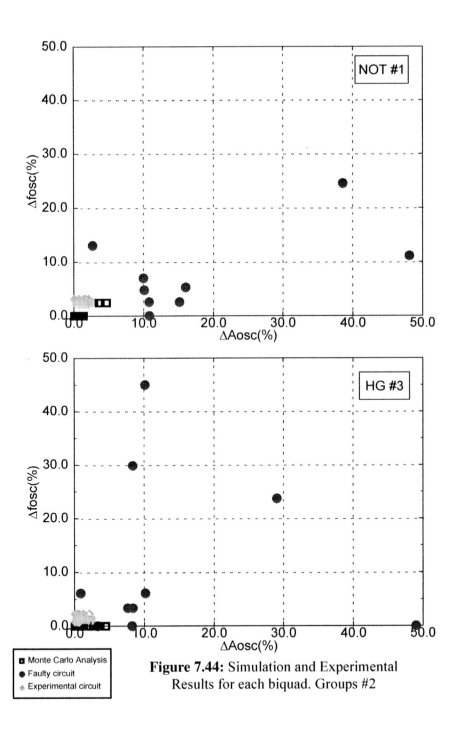

Figure 7.44: Simulation and Experimental Results for each biquad. Groups #2

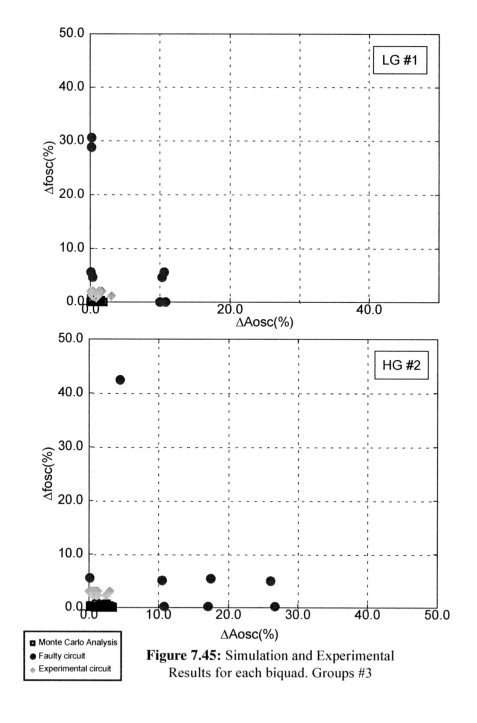

Figure 7.45: Simulation and Experimental
Results for each biquad. Groups #3

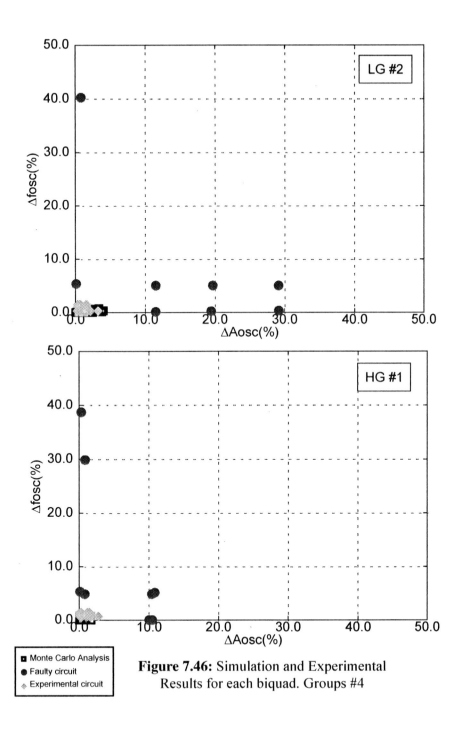

Figure 7.46: Simulation and Experimental Results for each biquad. Groups #4

7.4 SUMMARY

This Chapter provides a clear experimental evidence of the OBT-OBIST approach feasibility. Many experimental results, extracted from two demonstrator circuits, reinforce it. All the resulting data obtained herein practically agree with the theoretical predictions obtained in previous chapters.

Up to day, no experimental proof validated OBT-OBIST in a circuit implemented in silicon. In this thesis, however, we have designed two exploratory devices. One of them is a programmable biquad specifically built to empirically explore the characteristics of the proposed OBT-OBIST approach: the accuracy in determining the oscillation parameters, the start-up strategy, the impact of the swopamp use, the proposed on-chip evaluation mechanism (use of a $\Sigma\Delta$ modulator to digitally encode the oscillation signal), etc. On the other hand, the second implemented system is an industrial prototype of a DTMF receiver which allows us to prove the feasibility of integrating OBT-OBIST into the frame of an industrial environment.

In fact, this work can be considered as a benchmark for OBIST in integrated analog circuits which provides empirical data confirming both the practicality of incorporating this test technique to the regular design flow and the quality of its results.

Appendix 2.A

Error bound calculation

THE GOAL IN THIS APPENDIX is to introduce a systematic way to fence in the error bounds of the DF methodology. As discussed, the DF method is approximate. Then, it is crucial to develop any kind of error analysis. A lot of work has been made in this sense [99]-[100], [109]-[112], [114]. However, we have selected a particular reference [99], not only because the proposed method is relatively easy to use by nonexperts, but also because that method is not restricted to a particular set of nonlinear elements. On the contrary, a wide range of nonlinear elements can be included. In this Appendix we will follow [99], our intention being to facilitate the designer the practical use of a set of concepts available elsewhere at a rather theoretical level.

For the sake of clarity, only the example of Fig. 2.A.1 will be herein considered. It is a closed-loop system composed of a generic second-order structure followed by a comparator. Conclusions from this example will guide us to establish a methodological approach to decide when the DF approach can be considered valid and when not.

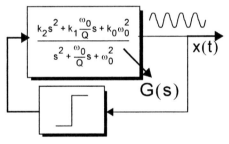

Figure 2.A.1: Studied Oscillator

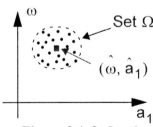

Figure 2.A.2: Set Ω

We are interested in determining how exact is the oscillation result obtained by the DF method. The idea is to find a *"confidence interval"* called Ω. This interval (see Fig. 2.A.2) is a set of fundamental frequency and amplitude values, (ω, a_1), containing the fundamental frequency and amplitude of a true oscillatory solution, $(\omega, \hat{a}_1)$, for the system in Fig. 2.A.1. Ω will be found after defining several error functions called p, q, r and an auxiliary variable, σ. Such functions will be needed to state a key inequality at all points in Ω. The steps needed to do it will be described, introducing notation as required [99].

The error functions as well as the auxiliary variable σ are related to some fundamental issues that intuitively may ensure the validity of the hypothesis supporting the DF approach. In essence, with these functions we try to compare the effect of higher-order harmonics to the effect of the fundamental on different points of the involved closed-loop system.

Thus, ρ determines whether or not the effect of the linear block on the higher-order components is negligible. Function p measures the difference between the actual output of the non-linear block and the assumed sinusoidal output (fundamental frequency). Function q is an estimate of the error due to disregarding non-fundamental harmonics at the input of the non-linear block. Function r is an upper bound to the nonlinearity output. Finally, with the variable σ, we define error discs around the $N(a_1)$ and $-1/G(j\omega)$ loci which allow us to find out the set Ω.

Step 0: Find $(\hat{\omega}, a_1)$ satisfying $N(a_1) + 1/G(j\hat{\omega}) = 0$ (see Section 2.3.1).

Step 1: Defining the error function, $\rho(\omega)$, which measures the effect of the system's response, $G(j\omega)$ to high-order <u>undesired</u> harmonics.

Since the non-linear element in Fig. 2.A.1 is an odd function, let us look for π-symmetric solutions[1]. We define $K = \{0, 1, 2, 3, ...\}$ and $K^* = \{0, 2, 3, ...\}$.

On the other hand, if $x(t)$ is a periodic function of period π/ω, we can write:

$$x(t) = Re \sum_{k \in K} a_k e^{jk\omega t}$$

and

$$x^*(t) = Re \sum_{k \in K^*} a_k e^{jk\omega t}.$$

---

[1] A π-symmetric solution is one with the property $x(\omega t + \pi) = x(\omega t)$, and, consequently, containing only odd harmonics.

Thus, x^* is the part of the solution that is neglected when we use the describing function to approximate the solution. Therefore, we will assume reliable results only if x^* is sufficiently small.

The filtering effect is the main reason for assuming x^* is small. Let us thus define

$$\rho(\omega) = \sqrt{\sum_{k \in K^*} |G(jk\omega)|^2}$$

Observe that the sum will converge if $|G(s)|$ is $o(|1/s|^m)$ with m>1/2, as $|1/s|$->0. In practice, G is usually strictly proper and the convergence is very fast, so only a few terms are needed to get a very good evaluation of $\rho(\omega)$. Small values of $\rho(\omega)$ are desirable; the smaller $\rho(\omega)$, the better the eventual error estimate.

On the other hand, we can write

$$G(jk\omega) = \frac{-k_2 k^2 \omega^2 + jk_1 \dfrac{\omega_0}{Q} k\omega + k_0 \omega_0^2}{-k^2 \omega^2 + j\dfrac{\omega_0}{Q} k\omega + \omega_0^2}$$

And then

$$|G(jk\omega)|^2 = \frac{k_2^2 k^4 \omega^4 - \left(2k_2 k^2 k_0 \omega_0^2 - k_1^2 \dfrac{\omega_0^2}{Q^2}\right) k^2 \omega^2 + k_0^2 \omega_0^4}{k^4 \omega^4 - k^2 \omega^2 \left(2\omega_0^2 - \dfrac{\omega_0^2}{Q^2}\right) + \omega_0^4}$$

In this case we have

$$\rho(\omega) = \sqrt{\sum_{k = 0, 2, 3, \ldots} \frac{k_2^2 k^4 \omega^4 - \left(2k_2 k_0 \omega_0^2 - k_1^2 \dfrac{\omega_0^2}{Q^2}\right) k^2 \omega^2 + k_0^2 \omega_0^4}{k^4 \omega^4 - k^2 \omega^2 \left(2\omega_0^2 - \dfrac{\omega_0^2}{Q^2}\right) + \omega_0^4}} \quad \text{"""Step1}$$

It has to be guaranteed that $\rho(\omega)$ has a small value in order to obtain a good solution with the DF approach. Notice that actually, as was discussed above, if $k_0 = k_2 = 0$ (filter transfer function, $G(s)$, is a bandpass function) then

$$\rho(\omega) = \sqrt{\sum_{k = 0, 2, 3, \ldots} \frac{k_1^2 \dfrac{\omega_0^2}{Q^2} k^2 \omega^2}{k^2 \omega^2 - \left(2\omega_0^2 - \dfrac{\omega_0^2}{Q^2}\right) + \dfrac{\omega_0^4}{k^2 \omega^2}}} \rightarrow 0$$

when $Q \rightarrow \infty$. This means that, in the case of a bandpass transfer function, the main requirement to successfully apply the DF method is $Q \gg 1$ as was discussed in Chapter 2.

However, if G is not strictly proper, $\rho(\omega)$ is infinite. One way to circumvent this problem is to remove the constant part of G and absorb it into n, i.e, we poleshift n [99]. We can rewrite

$$G(s) = \frac{k_2 s^2 + k_1 \dfrac{\omega_0}{Q} s + k_0 \omega_0^2}{s^2 + \dfrac{\omega_0}{Q} s + \omega_0^2} = k_2 \left[\frac{s^2 + \dfrac{k_1}{k_2}\dfrac{\omega_0}{Q} s + \dfrac{k_0}{k_2}\omega_0^2}{s^2 + \dfrac{\omega_0}{Q} s + \omega_0^2} \right]$$

Instead of the model shown in Fig. 2.A.1, we can use the model shown in Fig. 2.A.3 where the gain k_2 is accounted for by the first feedback loop.

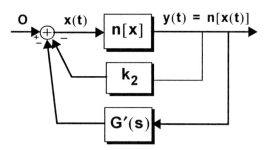

Figure 2.A.3: Poleshifting the nonlinearity to obtain a strictly proper G·(s)

A simple analysis shows that Fig. 2.A.3 can further be reduced to the equivalent system of Fig. 2.A.4.

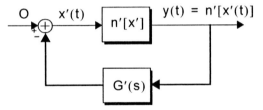

Figure 2.A.4: New oscillator scheme

where

$$G'(s) = G(s) - k_2 = k_2 \left[\frac{s^2 + \dfrac{k_1 \omega_0}{k_2 Q}s + \dfrac{k_0}{k_2}\omega_0^2}{s^2 + \dfrac{\omega_0}{Q}s + \omega_0^2} - 1 \right]$$

$$G'(s) = \frac{\dfrac{\omega_0}{Q}(k_1 - k_2)s + \omega_0^2(k_0 - k_2)}{s^2 + \dfrac{\omega_0}{Q}s + \omega_0^2} = \frac{As + B}{s^2 + z_{21}s + z_{31}}$$

and $n'(x') = n(x) - k_2 x$ (Fig. 2.A.5) is the saturation characteristic of Fig. 2.A.4.

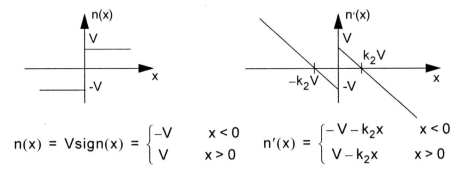

$$n(x) = V\text{sign}(x) = \begin{cases} -V & x < 0 \\ V & x > 0 \end{cases} \qquad n'(x) = \begin{cases} -V - k_2 x & x < 0 \\ V - k_2 x & x > 0 \end{cases}$$

Figure 2.A.5: New nonlinear characteristic

The describing function will be in this case

$$N'(a_1) = -k_2 + \frac{4V}{\pi a_1}$$

And the linear element, $G'(j\omega)$, has the following generic transfer function in the frequency domain

$$G'(j\omega) = \frac{\frac{\omega_0}{Q}(k_1 - k_2)s + \omega_0^2(k_0 - k_2)}{s^2 + \frac{\omega_0}{Q}s + \omega_0^2}$$

The aim would be to repeat the *Step 0* with these new expressions, the describing-function $N'(a_1)$ as well as the linear element transfer function, $G'(s)$

$$N'(a_1) - \frac{1}{G'(s)} = 0$$

$$N'(a_1) = -k_2 + \frac{4V}{\pi a_1}$$

The new solution will be given by

$$s^2 + bs + \omega^2 = 0$$

with

$$b = \frac{\omega_0}{Q}[1 - (k_1 - k_2)N'(a_1)]$$

$$\omega^2 = \omega_0^2[1 - (k_0 - k_2)N'(a_1)]$$

Then, the solution will be obtained from

$$b(\hat{a}_1) = \frac{\omega_0}{Q}[1 - (k_1 - k_2)N'(\hat{a}_1)] = 0 \Rightarrow -k_2 + \frac{4V}{\pi \hat{a}_1} = \frac{1}{(k_1 - k_2)}$$

$$"" \Rightarrow \hat{a}_1 = \frac{4V}{\pi}\left[\frac{k_2 - k_1}{1 - k_2(k_2 - k_1)}\right]$$

$$\hat{\omega}^2 = \omega_0^2[1 - (k_0 - k_2)N'(\hat{a}_1)] = \omega_0^2\left[1 - \frac{k_0 - k_2}{k_1 - k_2}\right]$$

And, after poleshifting the nonlinearity, we would have

$$|G'(jk\omega)|^2 = \frac{\dfrac{\omega_0^2}{Q^2}(k_1 - k_2)^2 k^2 \omega^2 + \omega_0^4 (k_0 - k_2)^2}{k^4 \omega^4 - \left(2\omega_0^2 - \dfrac{\omega_0^2}{Q^2}\right) k^2 \omega^2 + \omega_0^4}$$

which is strictly proper.

Step 2: Defining the describing function output error, $p(a_1)$.

This value takes into account the error of assuming that the output of the nonlinear element, n, is sinusoidal when its input is sinusoidal. Such a function is

$$p^2(a_1) = \|n(a_1 \cos(\omega t))\|_2^2 - |a_1 N(a_1)|^2$$

where the L_2 norm on $[0, 2\pi]$ is $\|\circ\|_2^2$, defined by

$$\|f(t)\|_2^2 = \frac{1}{\pi} \int_0^{2\pi} f(t)^2 dt$$

The function $p(a_1)$ can always be calculated explicitly, but if n has a finite gain β (i.e., $|n(x)| \le \beta |x|$ for all x in the region of interest) then, with some loss of accuracy in eventual error estimate, we can replace $p(a_1)$ by an upper bound βa_1. Incidentally, that it is sometimes possible to get by it without calculating p.

In our specific case (if we do not need to poleshift the nonlinearity), we have

$$\|n(a_1 \cos t)\|_2^2 = \frac{1}{\pi} \int_0^{2\pi} n^2(a_1 \cos(\omega t)) dt = 2V^2$$

since $n^2(x) = V^2 \ \forall x$.

Therefore, $p(a_1)$ is

$$p^2(a_1) = \left\| n(a_1 \cos(\omega t)) \right\|_2^2 - \left| a_1 N(a_1) \right|^2 \Rightarrow \text{""}$$

$$p^2(a_1) = \frac{2V^2\pi^2 - 16V^2}{\pi^2} = 2V^2\left(\frac{\pi^2 - 8}{\pi^2}\right) \quad \forall a_1$$

$$p(a_1) = V\sqrt{2\left(\frac{\pi^2 - 8}{\pi^2}\right)} \approx 0.62V \qquad \textbf{Step2}$$

Step 3: Defining the function $q(a_1, \varepsilon)$.

A decisive step is to compute a function that measures *the error introduced by disregarding high harmonics at the input of n*. The function is defined employing the supreme norm

$$\|f(t)\|_\infty = sup|f(t)| \qquad t \in [0, 2\pi]$$

and uses an upper bound ε on $\|x^*\|_\infty$, the supreme norm of the neglected (usually higher) harmonics.

This function is defined as

$$q(a_1, \varepsilon) = sup\left\| n(a_1 \cos(\omega t) + x^*(t)) - n(a_1 \cos(\omega t)) \right\|_2$$

$$\|x^*\|_\infty \leq \varepsilon$$

Take careful note of the *two different norms* used here. The actual calculation of q is by a worst-case analysis of the integral involved in the L_2 norm. If n is single-valued, we can define

$$m(x, \varepsilon) = sup|n(y) - n(x)| \qquad |y - x| \leq \varepsilon$$

where $m(x, \varepsilon) = max\{|n(x + \varepsilon) - n(x)|, |n(x - \varepsilon) - n(x)|\}$.

So that

$$q(a_1, \varepsilon) = \sqrt{\frac{1}{\pi}\int_0^{2\pi} m(a_1 \cos(\omega t), \varepsilon)^2 dt}$$

Now, supposing the proposed nonlinearity, $n(x) = Vsign(x)$ (shown in the next Fig. 2.A.6), then $m(x, \varepsilon) = 2V$ when $|x| \leq \varepsilon$ and $m(x, \varepsilon) = 0$ when $|x| > \varepsilon$.

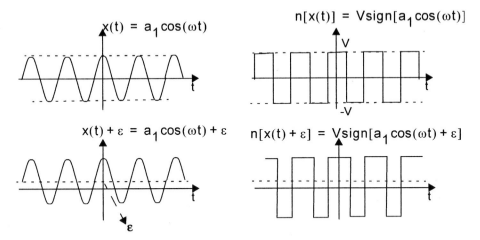

Figure 2.A.6: Calculating the function $q(a_1, \varepsilon)$

Analysing the function $x(t)$ we can find the value of $q(a_1, \varepsilon)$. Thus, let us draw in Fig. 2.A.7 the signal x with respect to the time.

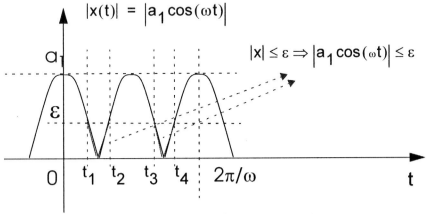

Figure 2.A.7: x with respect to the time

Then, we have

$$t_1 = \frac{x_1}{\omega} = acos\left(\frac{\varepsilon}{a_1}\right) = \frac{\pi}{2} - asen\left(\frac{\varepsilon}{a_1}\right)$$

$$t_2 = \frac{x_2}{\omega} = \frac{\pi}{2} - \frac{x_1}{\omega}$$

$$t_3 = \frac{x_3}{\omega} = \pi + \frac{x_1}{\omega} = 2\pi - \frac{x_2}{\omega}$$

$$t_4 = \frac{x_4}{\omega} = \pi + \frac{x_2}{\omega} = 2\pi - \frac{x_1}{\omega}$$

Finally, we can find the q function for this case

$$q(a_1, \varepsilon) = \sqrt{\frac{1}{\pi} \int_0^{2\pi} m(a_1 \cos(\omega t), \varepsilon)^2 dt}$$

$$\int_0^{\frac{2\pi}{\omega}} m(a_1 \cos(\omega t), \varepsilon)^2 dt = \int_{\frac{x_1}{\omega}}^{\frac{x_2}{\omega}} 4V^2 dt + \int_{\frac{x_3}{\omega}}^{\frac{x_4}{\omega}} 4V^2 dt = ''''$$

$$'''' = 4V^2 \left[\frac{x_2 + x_4 - x_1 - x_3}{\omega}\right] = 4V^2 \left[2\pi - 4\frac{x_1}{\omega}\right] = 16V^2 asen\left(\frac{\varepsilon}{a_1}\right)$$

$$q(a_1, \varepsilon) = 4V \sqrt{\frac{1}{\pi} asen\left(\frac{\varepsilon}{a_1}\right)} \qquad \textbf{Step3}$$

Steps to find the Set Ω:

With ρ, p and q known, the rest of the process entails resolving an equation to find an upper bound on the higher harmonic error, ε, then finding a set Ω of (ω, a_1) values satisfying a key inequality, and finally checking a non-degeneracy condition which is nearly always trivially fulfilled. In order to find an appropriate value of ε, we have to satisfy the inequality

$$\rho(\omega) min\{q(a_1, \varepsilon) + p(a_1), r(a_1, \varepsilon)\} \le \varepsilon$$

with $r(a_1, \varepsilon) = \sqrt{2} sup|n(y)| \qquad |y| \le a_1 + \varepsilon$.

In our example

$$\rho(\omega)min\left\{\left[4\sqrt{\frac{1}{\pi}asen\left(\frac{\varepsilon}{a_1}\right)}+\sqrt{2\left(\frac{\pi^2-8}{2}\right)}\right]V,\sqrt{2}V\right\}\le\varepsilon$$

where $r(a_1,\varepsilon) = \sqrt{2}V$.

This inequality has to be fulfilled for some $\varepsilon(a_1,\omega)>0$ for all $(a_1,\omega)\in\Omega$. In fact, we do not know Ω yet. All we know at this stage is that $(\hat{a}_1,\hat{\omega})\in\Omega$, where $(\hat{a}_1,\hat{\omega})$ is the DF solution. The method is as following:

Step 4: Find the smallest ε which satisfies the inequality for $(a_1,\omega) = (\hat{a}_1,\hat{\omega})$:

$$\rho(\hat{\omega})min\left\{\left[4\sqrt{\frac{1}{\pi}asen\left(\frac{\pi\varepsilon}{4Vk_1}\right)}+\sqrt{2\left(\frac{\pi^2-8}{\pi^2}\right)}\right]V,\sqrt{2}V\right\}\le\varepsilon$$

$$\rho(\hat{\omega}) = \sqrt{\sum_{k=3,5,\dots}\frac{k_2^2k^4\omega_0^4\left(\frac{k_1-k_0}{k_1-k_2}\right)^2-\left(2k_2k_0\omega_0^2-k_1^2\frac{\omega_0^2}{Q^2}\right)k^2\omega_0^2\left(\frac{k_1-k_0}{k_1-k_2}\right)+k_0^2\omega_0^4}{k^4\omega_0^4\left(\frac{k_1-k_0}{k_1-k_2}\right)^2-k^2\omega_0^2\left(\frac{k_1-k_0}{k_1-k_2}\right)\left(2\omega_0^2-\frac{\omega_0^2}{Q^2}\right)+\omega_0^4}}$$

We can then guess a larger value of ε, check that it still satisfies the inequality for $(\hat{a}_1,\hat{\omega})$, and later, complete the check when Ω is known.

Step 5

Otherwise, we can try to solve the inequality as an implicit equation: for each given pair of values of ω and a_1, we look for the smaller positive ε for which there is an intersection between the line $y = \varepsilon$ and the curve

$$y = f(\varepsilon) = \rho(\omega)min\left\{4V\sqrt{\frac{1}{\pi}asen\left(\frac{\varepsilon}{a_1}\right)}+V\sqrt{2\left(\frac{\pi^2-8}{\pi^2}\right)},\sqrt{2}V\right\}$$

Since q and r are both monotone increasing functions of ε, if there are intersections between the line and the curve, the first intersection will be found by applying the contraction mapping theorem to find a fixed point of $f(\varepsilon)$, starting from $\varepsilon = 0$. In this way we can find the smallest ε for each ω and a_1.

Because of the term $r(a_1, \varepsilon)$, any nonlinear element that saturates (or, more generally, any nonlinear element that eventually grows slower than linearly) will produce a finite value of ε. In general, though, it is possible that no solution exists to the inequality, either because the linear part is not a good-enough filter or the nonlinear part is badly behaved. In such a case, we can go no further with this method. Poleshifting may be employed to reduce the values of q and p as was discussed in **Step 1**.

We now try to find a closed bounded set Ω that contains $(\hat{a}_1, \hat{\omega})$ and all nearby points that satisfy the key inequality

$$\left| N(a_1) - \frac{1}{G(j\omega)} \right| \le \sigma(\omega, a_1) \tag{8.1}$$

where

$$\sigma(\omega, a_1) = \frac{q(a_1, \varepsilon(\omega, a_1))}{a_1}$$

with $\varepsilon > 0$ satisfying the inequality; the tightest bounds are found using the smaller values of ε. Let us explain two ways of finding Ω; the first one provides the smaller set, but requires more work than the other one. If no bounded Ω can be found, the error analysis has failed.

Let us evaluate this secondary variable. We start from the error function $q(a_1, \varepsilon)$

$$q(a_1, \varepsilon) = 4V\sqrt{\frac{1}{\pi} asen\left(\frac{\varepsilon}{a_1}\right)} \Rightarrow q(a_1, \varepsilon)^2 = \frac{16V^2}{\pi}\left[asen\left(\frac{\varepsilon}{a_1}\right)\right]$$

Let us suppose a_1 is large compared with ε. We can rewrite

$$q(a_1, \varepsilon)^2 = \frac{16V^2}{\pi}\left[asen\left(\frac{\varepsilon}{a_1}\right)\right] \approx \frac{16V^2}{\pi}\frac{\varepsilon}{a_1}$$

And then

$$\sigma(\omega, a_1) = \frac{q(a_1, \varepsilon(\omega, a_1))}{a_1} \approx \left(\frac{16V^2}{\pi} \frac{\varepsilon}{a_1}\right)^{\frac{1}{2}} \frac{1}{a_1} = \left(\frac{16V^2}{\pi}\varepsilon\right)^{\frac{1}{2}} a_1^{-\frac{3}{2}} \quad (8.2)$$

Step 5_1:

We can find Ω straightforwardly encircling $(\hat{a}_1, \hat{\omega})$ by a grid of points and at each point, find ε and calculate the ratio

$$\frac{\left|N(a_1) - \frac{1}{G(j\omega)}\right|}{\sigma(\omega, a_1)}$$

The boundary of Ω consists of those points where the ratio is 1.

Step 5_2:

We can settle ε at some slightly pessimistic (large) value and employ the previous inequality (8.1). This is easy to implement graphically since it says that points (ω, a_1) inside Ω must be such that the distance between $N(a_1)$ and $-1/G(j\omega)$ is at most $q(a_1, \varepsilon(\omega, a_1))/a_1$. Consequently, we can choose a range of a_1 values, and draw discs centered on $N(a_1)$ and of radius $q(a_1, \varepsilon(\omega, a_1))/a_1$. The envelope of the discs cuts off a range of ω values and the first and last discs to intersect the $-1/G(j\omega)$ locus define the correct a_1 range. (See Fig. 2.A.8). We obtain a rectangle $\Omega = [\omega_{min}, \omega_{max}] \times [a_{1min}, a_{1max}]$ which will contain the set Ω found in method step 5_1 and we check that the ε we fixed is actually big enough over this rectangle.

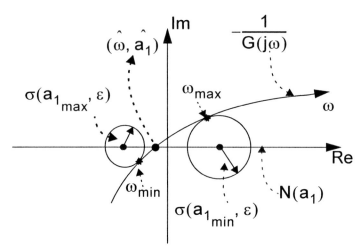

Figure 2.A.8: Error discs used in locating the set Ω, in
which the exact solution lies

Step 6: Check that Ω contains the one describing function solution
$(\omega, \hat{a}_1)$.

Step 7: There is at least one true oscillation solution in the system
whose frequency and first harmonic amplitude are within the above
ranges and $\|x*\| \leq \varepsilon$.

Step 8: A final step in the error analysis would be to check for nondegenerancy of the intersection between the loci of N and $-1/G$. The meaning of
"nondegenerancy" as well as the way to calculate it are explained in [99],
[109]-[112].

Getting to the point, the values of the involved error functions and
the variable σ are

$$\rho(\omega) = \sqrt{\sum_{k=3,5,\ldots} \frac{k_2^2 k^4 \omega^4 - \left(2k_2 k_0 \omega_0^2 - k_1^2 \frac{\omega_0^2}{Q^2}\right) k^2 \omega^2 + k_0^2 \omega_0^4}{k^4 \omega^4 - k^2 \omega^2 \left(2\omega_0^2 - \frac{\omega_0^2}{Q^2}\right) + \omega_0^4}} \quad (8.3)$$

$$p(a_1) = V \sqrt{2 \left(\frac{\pi^2 - 8}{\pi^2} \right)}$$

$$q(a_1, \varepsilon) = 4V \sqrt{\frac{1}{\pi} asen\left(\frac{\varepsilon}{a_1} \right)} \tag{8.4}$$

$$\sigma(\omega, a_1) = \frac{q(a_1, \varepsilon(\omega, a_1))}{a_1} \approx \left(\frac{16V^2}{\pi} \varepsilon \right)^{\frac{1}{2}} a_1^{-\frac{3}{2}}$$

Appendix 5.A

Characterizing the involved oscillators

We can differentiate several groups of oscillators in the DTMF core when it is reconfigured in the OBT mode. Table 5.A.1 summarizes the existing groups as well as the types of biquads involved in every group.

Group	BIQUAD	TYPE	FEEDBACK OUTPUT
A	LG #1	BP00-LP01	
	HG #1		
B	Not #1	GENERAL-HPNOTCH	
C	Not #2	GENERAL-HPNOTCH	V_{o1}
	LG #2		
	HG #2		
D	LG #3	GENERAL-BP00	
		GENERAL-BP00	V_{o2}
E	HG #3	GENERAL-HP	V_{o1}

Table 5.A.1: Groups of oscillator types

In what follows, we will separately study every oscillator group. The results achieved in this Appendix are used in Chapter 5 to obtain some main conclusions. The given steps are the following:

Step #1: we calculate the numerical values of the oscillation parameters by the theoretical expressions (using the DF approach) as well as by simulation (using Simulink/Matlab).

Step #2: we determine the oscillation frequency in relation to the resonant frequencies of the biquads. To do that, we draw the corresponding Bode Diagrams. We then estimate the value of all the frequencies of interest.

Step #3: we calculate the amplitude and the frequency sensitivities.

Step #4: we plot a set of graphics to show how every individual capacitor deviation influences on every oscillation parameters. We will remark in every graphic the deviation (in %) between the value of the involved nominal oscillation parameters and the value of such an oscillation parameter when a capacitor deviates a $\pm 10\%$ from its actual value.

-Group A (LG #1 and HG #1 Oscillators):

Starting from the first biquads in the DTMF structure, we examine the biquads called LG #1 and HG #1 which belong to the same group. Fig. 5.A.1 displays the corresponding oscillators.

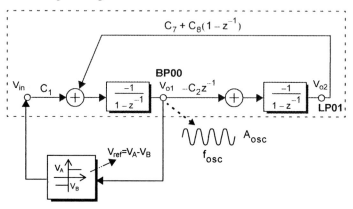

Figure 5.A.1: LG #1 and HG #1 oscillators

Observe from this Figure that the feedback loop is closed by the first output, V_{o1}. For this particular case, the oscillation conditions and the oscillation parameters are shown in Table 5.A.2. **Step #1** is summarized in Table 5.A.3, **Step #2** is represented in Fig. 5.A.2 and **Step #3** is given in Table 5.A.4.

$C_{05} = C_{09} = C_{56} = 0$	
Oscillation Conditions	$\text{sign}(V_{ref}) = \text{sign}(-C_1) = \text{sign}[-C_1(1 - C_2 C_8)] < 0$
Oscillation Frequency	$f_{osc} = \dfrac{1}{2\pi T_s} \cdot acos\left[\dfrac{1}{2} \cdot \dfrac{2 - C_2(C_7 + 2C_8)}{1 - C_2 C_8}\right]$
Oscillation Amplitude	$A_{osc} = \dfrac{-2V_{ref}}{\pi} \cdot \dfrac{C_1}{C_2 C_8}$

Table 5.A.2: LG #1 and HG #1 reconfigured as an oscillator: Steady Oscillation Mode

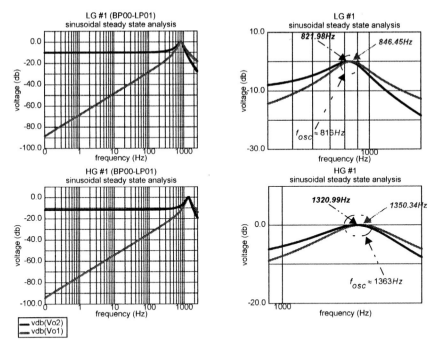

Figure 5.A.2: LG #1 and HG #1 Bode Diagrams

LG #1	HG #1
$C_1 = 0.03; C_2 = 0.09; C_7 = 0.09; C_8 = 0.32$	$C_1 = 0.04; C_2 = 0.15; C_7 = 0.15; C_8 = 0.26$
$k_2 = -0.03; k_1 = 0.03; k_0 = 0$ $b_1 = -1.96; b_0 = 0.97$	$k_2 = -0.04; k_1 = 0.04; k_0 = 0$ $b_1 = -1.94; b_0 = 0.96$
Negative feedback $\text{sign}(k_2) < 0$	
$f_{osc}^{theoretical} = 816.19\,\text{Hz}$ $\rightarrow \Delta f = 2.41\,\%$ $f_{osc}^{matlab} = 796.51\,\text{Hz}$	$f_{osc}^{theoretical} = 1363.12\,\text{Hz}$ $\rightarrow \Delta f = 1.29\,\%$ $f_{osc}^{matlab} = 1345.49\,\text{Hz}$
$V_{ref} = -1\,\text{V}$	
$A_{osc}^{theoretical} = 1.272\,\text{V}$ $\rightarrow \Lambda A = 0.31\,\%$ $A_{osc}^{matlab} = 1.276\,\text{V}$	$A_{osc}^{theoretical} = 1.306\,\text{V}$ $\rightarrow \Lambda A = 0.31\,\%$ $A_{osc}^{matlab} = 1.310\,\text{V}$

Table 5.A.3: LG #1 and HG #1 reconfigured as an oscillator: main oscillation results

Amplitude Sensitivities (%)	Frequency Sensitivities (%)
$S_{C_1}^{A_{osc}} = \dfrac{100}{C_1}$	$S_{C_1}^{\cos(\Theta_{osc})} = 0$
$S_{C_2}^{A_{osc}} = -\dfrac{100}{C_2}$	$S_{C_2}^{\cos(\Theta_{osc})} = -100\dfrac{C_7}{(-2+C_2C_7+2C_2C_8)(-1+C_2C_8)}$
$S_{C_8}^{A_{osc}} = -\dfrac{100}{C_8}$	$S_{C_8}^{\cos(\Theta_{osc})} = -100\dfrac{C_2^2C_7}{(-2+C_2C_7+2C_2C_8)(-1+C_2C_8)}$
$S_{C_7}^{A_{osc}} = 0$	$S_{C_7}^{\cos(\Theta_{osc})} = -100\dfrac{C_2}{(-2+C_2C_7+2C_2C_8)}$

Table 5.A.4: Sensitivities for LG #1 and HG #1

From this last table we have, for instance, that the oscillation frequency does not depend on capacitor C_1 and that the oscillation amplitude does not depend on capacitor C_7. Both cases are a clear proof of this statement:

---

measuring both oscillation parameters is necessary to guarantee a good fault coverage

---

Otherwise, if one of the oscillation parameters is not considered, an unacceptable test quality is achieved. But, from the expressions of Table 5.A.4, it is very complicated to guess how each capacitor deviation affects the oscillation parameters. **Step #4**, however, lets us to straightforward determine it. The involved graphics are Fig. 5.A.3 for LG #1 and Fig. 5.A.4 for HG #1. Such Figures display, on the left side, the oscillation frequency cosine versus every capacitor deviation and, on the right side, the oscillation amplitude versus every capacitor deviation.

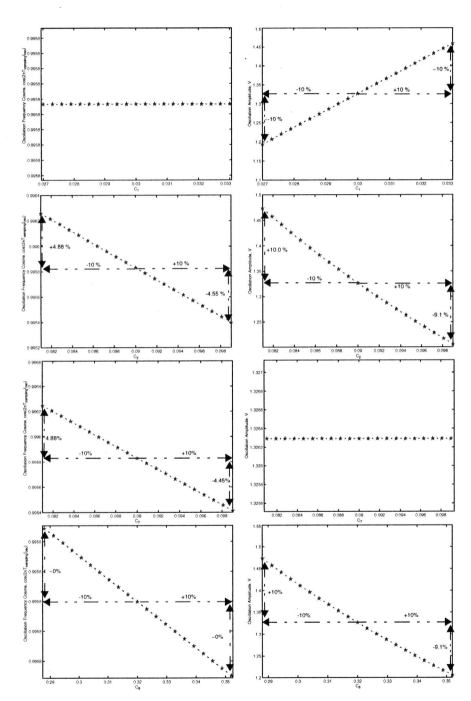

Figure 5.A.3: LG #1 (BP00-LP01)

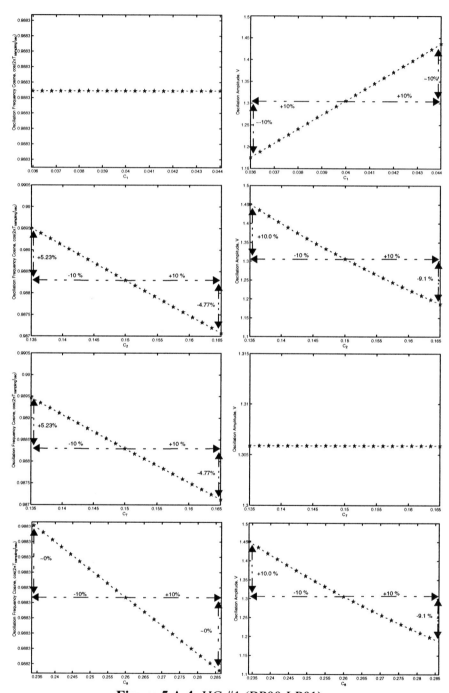

Figure 5.A.4: HG #1 (BP00-LP01)

From Fig. 5.A.3 and Fig. 5.A.4, it can be observed that a $\pm 10\%$ deviation in C_1 is reflected in a $\sim \pm 10\%$ deviation in the oscillation amplitude, while oscillation frequency is not affected by this kind of deviation. Similarly, if a $\pm 10\%$ in C_8 is being considered, only the oscillation amplitude deviates visibly from its nominal value (more specifically $\sim \pm 10\%$). On the contrary, if a $\pm 10\%$ deviation in C_7 is being considered, only the oscillation frequency deviates from its nominal value (to be precise a $\sim \pm 4.5\%$). Finally, when a $\pm 10\%$ deviation in C_2 is being considered, both oscillation parameters are affected (to be precise, a $\sim \pm 4.5\%$ deviation in the frequency and a $\sim \pm 10\%$ in the amplitude). Undoubtedly, in view of these figures (Fig. 5.A.3 and Fig. 5.A.4), we are faced with a particular case where oscillation amplitude is of vital importance as even provides more fault coverage than oscillation frequency.

-Group B (Not #1 Oscillator):

Let us, on the other hand, examine the biquad called Not #1. Fig. 5.A.5 shows the corresponding oscillator.

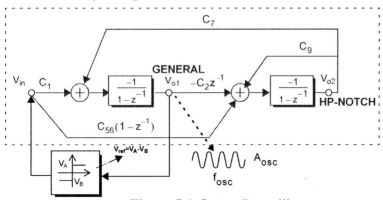

Figure 5.A.5: Not #1 oscillator

Now, the oscillation conditions and the oscillation parameters are shown in Table 5.A.5. **Step #1** is summarized in Table 5.A.6, **Step #2** is represented in Fig. 5.A.6 and **Step #3** is given in Table 5.A.7.

$C_{05} = C_{08} = 0$	
Oscillation Conditions	$\text{sign}(V_{ref}) = \text{sign}(-C_1(1+C_9)+C_{56}C_7)$
Oscillation Frequency	$f_{osc} = \dfrac{1}{2\pi T_s} \cdot \text{acos}\left[\dfrac{-2C_1 + 2C_{56}C_7 - C_1 C_9^2 - 2C_1 C_9 + C_2 C_7 C_1(1+C_9) - C_2 C_7^2 C_{56}}{2[-C_1(1+C_9)+C_{56}C_7]}\right]$
Oscillation Amplitude	$A_{osc} = \dfrac{2V_{ref}}{\Pi} \cdot \left[\dfrac{-C_1(1+C_9)+C_{56}C_7}{C_9}\right]$

Table 5.A.5: Not #1 reconfigured as an oscillator: Steady Oscillation Mode

$C_1 = 0.039; C_2 = 0.041; C_7 = 0.077; C_9 = 0.076; C_{56} = 1.061$
$k_2 = 0.037; k_1 = -0.040; k_0 = 0$ $b_1 = -1.927; b_0 = 0.930$
Positive feedback $\text{sign}(k_2) > 0$
$f_{osc}^{theoretical} = 503.66 \text{Hz}$ $\qquad\qquad\qquad\qquad \to \Delta f = 0\ \%$ $f_{osc}^{matlab} = 503.66 \text{Hz}$
$V_{ref} = 1V \to \begin{array}{l} A_{osc}^{theoretical} = 6.523\,V \\[4pt] \qquad\qquad\qquad\qquad \to \Lambda A = 0.34\ \% \\[4pt] A_{osc}^{matlab} = 6.545\,V \end{array}$

Table 5.A.6: Not #1 reconfigured as an oscillator: main oscillation results

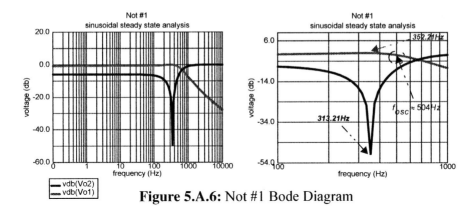

Figure 5.A.6: Not #1 Bode Diagram

Amplitude Sensitivities (%)	Frequency Sensitivities (%)
$S_{C_1}^{A_{osc}} = 100\left[\dfrac{1+C_9}{C_1(1+C_9)-C_7C_{56}}\right]$	$S_{C_1}^{\cos(\Theta_{osc})} = 100\left[\dfrac{C_9^2C_7C_{56}}{\left[(-2C_1+C_2C_7C_1)(1+C_9)-C_1C_9^2+2C_7C_{56}-C_2C_7^2C_{56}\right][C_1(1+C_9)-C_7C_{56}]}\right]$
$S_{C_2}^{A_{osc}} = 0$	$S_{C_2}^{\cos(\Theta_{osc})} = 100\dfrac{C_7(C_1(1+C_9)-C_7C_{56})}{(-2C_1+C_2C_7C_1)(1+C_9)-C_1C_9^2+2C_7C_{56}-C_2C_7^2C_{56}}$
$S_{C_7}^{A_{osc}} = -100\left[\dfrac{C_{56}}{C_1(1+C_9)-C_7C_{56}}\right]$	$S_{C_7}^{\cos(\Theta_{osc})} = 100\dfrac{C_2C_1^2\left(1+2C_9+C_9^2\right)-C_1C_{56}\left(2C_2C_7(1+C_9)+C_9^2\right)+C_2C_7^2C_{56}^2}{\left[(-2C_1+C_2C_7C_1)(1+C_9)-C_1C_9^2+2C_7C_{56}-C_2C_7^2C_{56}\right][C_1(1+C_9)-C_7C_{56}]}$
$S_{C_9}^{A_{osc}} = -100\left[\dfrac{C_1-C_7C_{56}}{C_9[C_1(1+C_9)-C_7C_{56}]}\right]$	$S_{C_9}^{\cos(\Theta_{osc})} = -100\dfrac{C_1C_9[C_1(2+C_9)-2C_7C_{56}]}{\left[(-2C_1+C_2C_7C_1)(1+C_9)-C_1C_9^2+2C_7C_{56}-C_2C_7^2C_{56}\right][C_1(1+C_9)-C_7C_{56}]}$
$S_{C_{56}}^{A_{osc}} = -100\left[\dfrac{C_7}{C_1(1+C_9)-C_7C_{56}}\right]$	$S_{C_{56}}^{\cos(\Theta_{osc})} = -100\dfrac{C_1C_7C_9^2}{\left[(-2C_1+C_2C_7C_1)(1+C_9)-C_1C_9^2+2C_7C_{56}-C_2C_7^2C_{56}\right][C_1(1+C_9)-C_7C_{56}]}$

Table 5.A.7: Sensitivities for Not #1

Again, from Table 5.A.7, we can not extract general conclusions except the necessity of both oscillation parameters to yield an acceptable test quality. Fig. 5.A.7 and Fig. 5.A.8 show the results of **Step #4**.

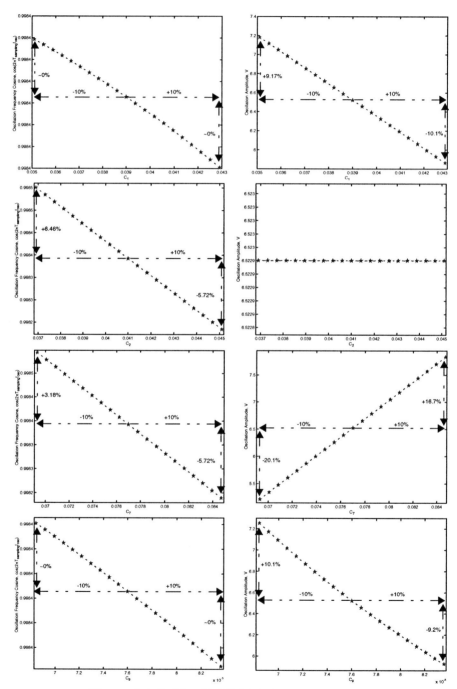

Figure 5.A.7: Not #1 (GENERAL - HPNOTCH)

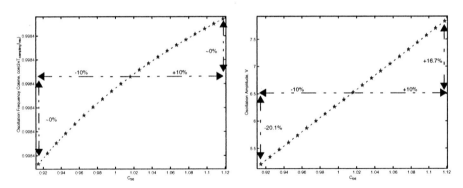

Figure 5.A.8: Not #1 (GENERAL - HPNOTCH)- Cont

-Group C (Not #2, LG #2 and HG #2 Oscillators):

Let us also examine the biquads called Not #2, LG #2 and HG #2. Fig. 5.A.9 shows the resulting oscillators.

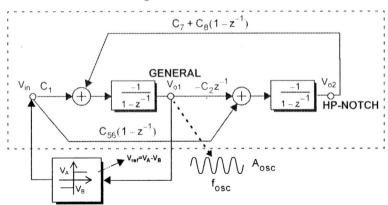

Figure 5.A.9: Not #2, LG #2 and HG #2 oscillators

The oscillation conditions and the oscillation parameters are shown in Table 5.A.8. **Step #1** is summarized in Table 5.A.9, **Step #2** is represented in Fig. 5.A.10 and **Step #3** is given in Table 5.A.10 and Table 5.A.11.

$C_{05} = C_{09} = 0$	
Oscillation Conditions	$\text{sign}(V_{ref}) = \text{sign}(-C_1 + C_{56}C_7)$
Oscillation Frequency	$f_{osc} = \dfrac{1}{2\pi T_s} \cdot \text{acos}\left[\dfrac{-2C_1 + 2C_{56}C_7 + (C_2C_7 + 2C_2C_8)(C_1 - C_{56}C_7) - 2C_2C_8^2C_{56}}{2\left[(C_1 - C_{56}C_7)(C_2C_8 - 1) - C_2C_8^2C_{56}\right]}\right]$
Oscillation Amplitude	$A_{osc} = \dfrac{2V_{ref}}{\pi} \cdot \left[\dfrac{-C_1 + C_{56}C_7}{C_2C_8}\right]$

Table 5.A.8: Not #2, LG #2 and HG #2 reconfigured as an oscillator: Steady Oscillation Mode

Not #2 $V_{ref} = 1\,V$	LG #2 $V_{ref} = 1\,V$ (Hysteresis=$15e^{-3}V$)	HG #2 $V_{ref} = 1\,V$
$C_1 = 0.036;$ $C_2 = 0.073;$ $C_7 = 0.058;$ $C_8 = 0.252;$ $C_{56} = 0.964;$	$C_1 = 0.044;$ $C_2 = 0.109;$ $C_7 = 0.110;$ $C_8 = 0.094;$ $C_{56} = 0.259;$	$C_1 = 0.024;$ $C_2 = 0.133;$ $C_7 = 0.129;$ $C_8 = 0.089;$ $C_{56} = 0.296;$
$k_2 = 0.264; k_1 = -0.507; k_0 = 0.243$ $b_1 = -1.977; b_0 = 0.982$	$k_2 = 0.009; k_1 = -0.034; k_0 = 0.0240$ $b_1 = -1.978; b_0 = 0.989$	$k_2 = 0.040; k_1 = -0.067; k_0 = 0.0264$ $b_1 = -1.971; b_0 = 0.988$
Positive feedback $\text{sing}(k_2 - k_0) > 0$	Negative feedback $\text{sing}(k_2 - k_0) < 0$	Positive feedback $\text{sing}(k_2 - k_0) > 0$
$f_{osc}^{teorica} = 666.34\,Hz$ $f_{osc}^{matlab} = 629.62\,Hz$ $\rightarrow \Delta f = 5.51\,\%$	$f_{osc}^{teorica} = 975.68\,Hz$ $f_{osc}^{matlab} = 967.51\,Hz$ $\rightarrow \Delta f = 0.84\,\%$	$f_{osc}^{teorica} = 1188.59\,Hz$ $f_{osc}^{matlab} = 1181.89\,Hz$ $\rightarrow \Delta f = 0.56\,\%$
$A_{osc}^{teorica} = 1.378\,V$ $A_{osc}^{matlab} = 1.532\,V$ $\rightarrow \Delta A = 10.03\,\%$	$A_{osc}^{teorica} = 1.927\,V$ $A_{osc}^{matlab} = 1.926\,V$ $\rightarrow \Delta A = 0.06\,\%$	$A_{osc}^{teorica} = 1.526\,V$ $A_{osc}^{matlab} = 1.533\,V$ $\rightarrow \Delta A = 0.45\,\%$

Table 5.A.9: Not #2, LG #2 and HG #2 reconfigured as an oscillator: oscillation results

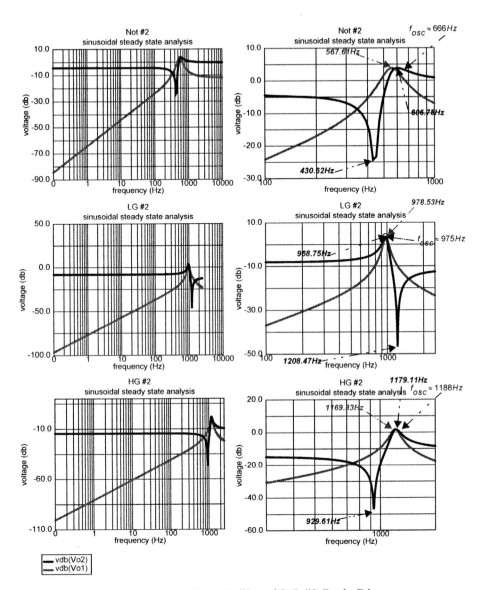

Figure 5.A.10: Not #2, HG #2 and LG #2 Bode Diagrams

Amplitude Sensitivities (%)
$S_{C_1}^{A_{osc}} = 100\left[\dfrac{1}{C_1 - C_7 C_{56}}\right]$
$S_{C_2}^{A_{osc}} = -\dfrac{100}{C_2}$
$S_{C_7}^{A_{osc}} = -100\left[\dfrac{C_{56}}{C_1 - C_7 C_{56}}\right]$
$S_{C_8}^{A_{osc}} = -\dfrac{100}{C_8}$
$S_{C_{56}}^{A_{osc}} = -100\left[\dfrac{C_7}{C_1 - C_7 C_{56}}\right]$

Table 5.A.10: Sensitivities for Not #2, HG #2 and LG #2 (I)

Frequency Sensitivities (%)
$S_{C_1}^{\cos(\Theta_{osc})} = -100\left[\dfrac{C_2^2 C_7 C_8^2 C_{56}}{[-C_1 + C_7 C_{56} + C_2 C_8(C_1 - C_7 C_{56} - C_8 C_{56})][2(-C_1 + C_7 C_{56} + C_2 C_8(C_1 - C_7 C_{56} - C_8 C_{56})) + C_2 C_7 (C_1 - C_7 C_{56})]}\right]$
$S_{C_2}^{\cos(\Theta_{osc})} = -100\dfrac{C_7\left(C_1(-2 C_7 C_{56} + C_1) + C_7^2 C_{56}^2\right)}{[-C_1 + C_7 C_{56} + C_2 C_8(C_1 - C_7 C_{56} - C_8 C_{56})][2(-C_1 + C_7 C_{56} + C_2 C_8(C_1 - C_7 C_{56} - C_8 C_{56})) + C_2 C_7 (C_1 - C_7 C_{56})]}$
$S_{C_7}^{\cos(\Theta_{osc})} = 100\dfrac{C_2\left[C_1 C_{56}(2 C_7 - C_2 C_8^2) + C_2 C_8(C_7^2 C_{56}^2 + 2 C_8 C_7 C_{56}^2 + C_1^2 - 2 C_1 C_{56}) - C_7^2 C_{56}^2 - C_1^2\right]}{[-C_1 + C_7 C_{56} + C_2 C_8(C_1 - C_7 C_{56} - C_8 C_{56})][2(-C_1 + C_7 C_{56} + C_2 C_8(C_1 - C_7 C_{56} - C_8 C_{56})) + C_2 C_7 (C_1 - C_7 C_{56})]}$
$S_{C_8}^{\cos(\Theta_{osc})} = -100\dfrac{C_2^2 C_7[(C_1 - C_7 C_{56} - 2 C_8 C_{56})(C_1 - C_7 C_{56})]}{[-C_1 + C_7 C_{56} + C_2 C_8(C_1 - C_7 C_{56} - C_8 C_{56})][2(-C_1 + C_7 C_{56} + C_2 C_8(C_1 - C_7 C_{56} - C_8 C_{56})) + C_2 C_7 (C_1 - C_7 C_{56})]}$
$S_{C_{56}}^{\cos(\Theta_{osc})} = 100\dfrac{C_1 C_2^2 C_7 C_8^2}{[-C_1 + C_7 C_{56} + C_2 C_8(C_1 - C_7 C_{56} - C_8 C_{56})][2(-C_1 + C_7 C_{56} + C_2 C_8(C_1 - C_7 C_{56} - C_8 C_{56})) + C_2 C_7 (C_1 - C_7 C_{56})]}$

Table 5.A.11: Sensitivities for Not #2, HG #2 and LG #2 (II)

Finally, **Step #4** is carried out in Fig. 5.A.11-Fig. 5.A.16.

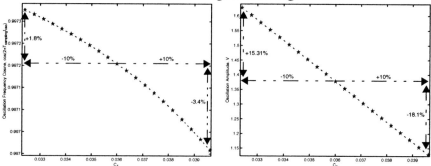

Figure 5.A.11: Not #2 (GENERAL-HP-NOTCH)

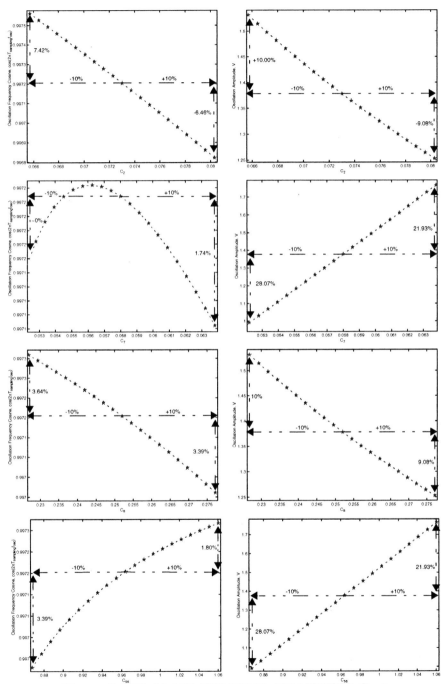

Figure 5.A.12: Not #2 (GENERAL-HP-NOTCH)-Cont

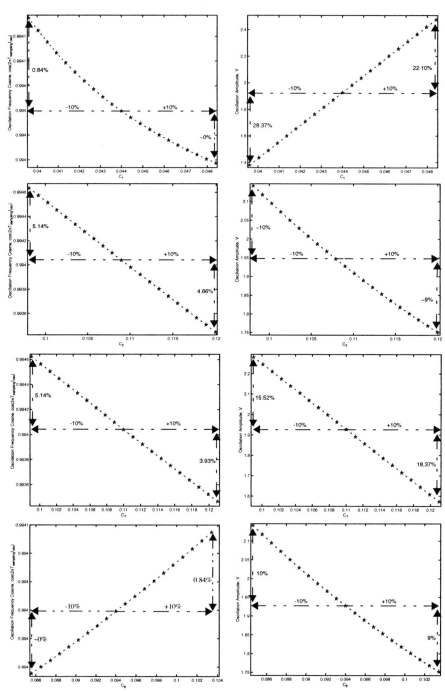

Figure 5.A.13: LG #2 (GENERAL-HP-NOTCH)

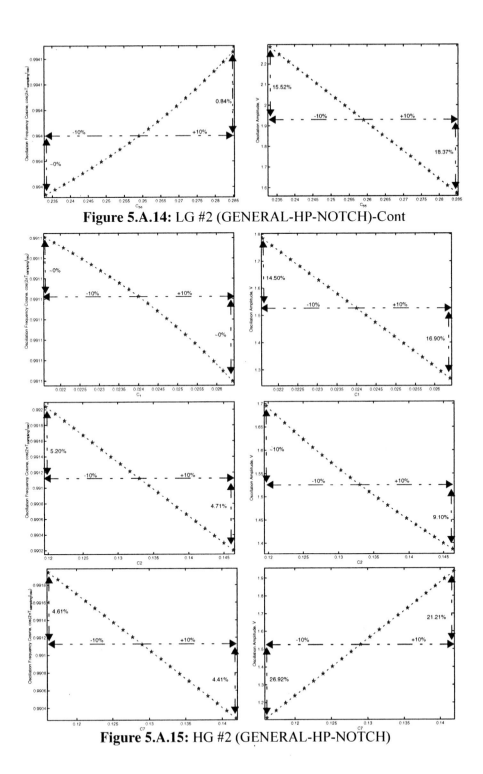

Figure 5.A.14: LG #2 (GENERAL-HP-NOTCH)-Cont

Figure 5.A.15: HG #2 (GENERAL-HP-NOTCH)

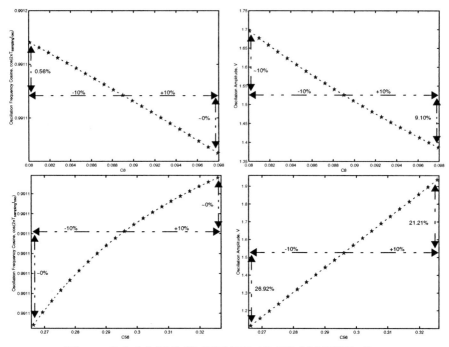

Figure 5.A.16: HG #2 (GENERAL-HP-NOTCH)-Cont

-Group D (LG #3 Oscillators):

Let us also examine the biquad called LG #3. Fig. 5.A.17 shows the resulting oscillator.

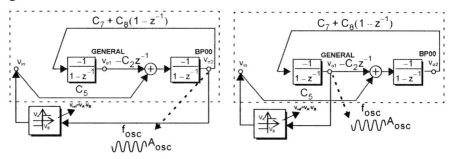

Figure 5.A.17: LG #3 possible oscillators

The oscillation conditions and the oscillation parameters for this particular case are shown in Table 5.A.12 and Table 5.A.13. On the other hand, **Step #1** is summarized in Table 5.A.14 and **Step #2** is represented in Fig. 5.A.18 and **Step #3** is given in Table 5.A.15.

$C_{01} = C_{56} = C_{09} = 0$	Ho1
Oscillation Conditions	$\text{sign}(V_{ref}) = \text{sign}(C_5(C_7 + C_8))$
Oscillation Frequency	$f_{osc} = \dfrac{1}{2\pi T_s} \cdot \text{acos}\left[\dfrac{-2C_7 - 2C_8 + C_2 C_7^2 + 2C_2 C_8 C_7 + 2C_2 C_8^2}{2[(C_7 + C_8)(C_2 C_8 - 1)]}\right]$
Oscillation Amplitude	$A_{osc} = \dfrac{2V_{ref}}{\Pi} \cdot \left[\dfrac{C_5(C_7 + C_8)}{C_2 C_8}\right]$

Table 5.A.12: LG #3 reconfigured as an oscillator (I): Steady Oscillation Mode

$C_{01} = C_{56} = C_{09} = 0$	Ho2
Oscillation Conditions	$\text{sign}(V_{ref}) = \text{sign}(C_5(C_7 + C_8))$
Oscillation Frequency	$f_{osc} = \dfrac{1}{2\pi T_s} \cdot \text{acos}\left[\dfrac{-2 + C_2 C_7 + 2C_2 C_8}{2(C_2 C_8 - 1)}\right]$
Oscillation Amplitude	$A_{osc} = \dfrac{2V_{ref}}{\Pi} \cdot \left[\dfrac{-C_5}{C_2 C_8}\right]$

Table 5.A.13: LG #3 reconfigured as an oscillator (II): Steady Oscillation Mode

Although it is theoretically feasible to obtain two kind of oscillators, there is, however, a problem to do so. If we use the first output, it is very difficult to force an adequate initial condition that allow to stimulate such an oscillator. A possible explanation of this is that in this particular case the value of oscillation frequency is not closed to the biquad resonant frequency, whereas for all the other cases that requisite is fulfilled (Fig. 5.A.18).

Ho1 (GENERAL)		Ho2 (BP00)	
$C_2 = 0.072; C_7 = 0.065; C_8 = 0.277; C_5 = 0.065$		$C_2 = 0.072; C_7 = 0.065; C_8 = 0.277; C_5 = 0.065$	
$k_2 = 0.022; k_1 = -0.018; k_0 = 0$ $b_1 = -1.975; b_0 = 0.980$		$k_2 = -0.065; k_1 = 0.065; k_0 = 0$ $b_1 = -1.975; b_0 = 0.980$	
Positive feedback $\text{sign}(k_2) > 0$		Negative feedback $\text{sign}(k_2) < 0$	
$f_{osc}^{theoretical} = 281.53 \text{Hz}$ $f_{osc}^{matlab} = \text{---Hz}$	$\rightarrow \Delta f = \text{--}\%$	$f_{osc}^{theoretical} = 616.89 \text{Hz}$ $f_{osc}^{matlab} = 603.90 \text{Hz}$	$\rightarrow \Delta f = 2.11\%$
$V_{ref} = 1V \rightarrow$	$A_{osc}^{theoretical} = 1.420 \text{V}$ $A_{osc}^{matlab} = \text{---V}$ $\rightarrow \Lambda A = \text{---}\%$	$V_{ref} = 1V \rightarrow$	$A_{osc}^{theoretical} = 5.0637 \text{V}$ $A_{osc}^{matlab} = 5.0721 \text{V}$ $\rightarrow \Lambda A = 0.17\%$

Table 5.A.14: LG #3 reconfigured as an oscillator: main oscillation results

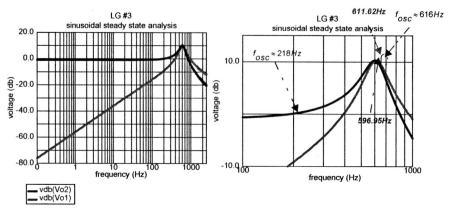

Figure 5.A.18: LG #3 Bode Diagram

Amplitude Sensitivities (%) Ho2	Frequency Sensitivities (%) Ho2		
$S_{C_2}^{A_{osc}} = -\dfrac{100}{C_2}$	$S_{C_2}^{\cos(\Theta_{osc})} = -100\dfrac{C_7}{(-1+C_2C_8)(-2+C_2C_7+2(C_2C_8))}$		
$S_{C_5}^{A_{osc}} = \dfrac{100}{C_5}$	$S_{C_5}^{\cos(\Theta_{osc})} = 0$		
$S_{C_7}^{A_{osc}} = 0$	$S_{C_7}^{\cos(\Theta_{osc})} = 100\dfrac{C_2}{-2+C_2C_7+2(C_2C_8)}$		
$S_{C_8}^{A_{osc}} = -\dfrac{100}{C_8}$	$S_{C_8}^{\cos(\Theta_{osc})} = -100\dfrac{C_2^2C_7}{(-1+C_2C_8)(-2+C_2C_7+2(C_2C_8))}$		

Table 5.A.15: Sensitivities for LG #3

Finally, **Step #4** is carried out in Fig. 5.A.19.

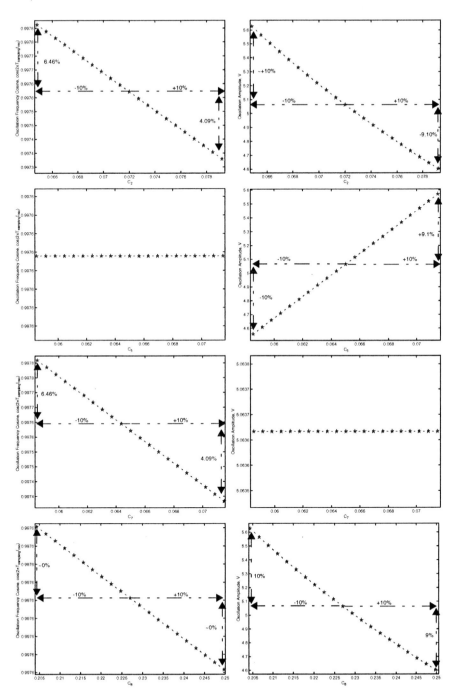

Figure 5.A.19: LG #3 (GENERAL-BP00)

-Group E (HG #3 Oscillator):

Let us also examine the biquad called HG #3. This is graphically as shown in Fig. 5.A.20.

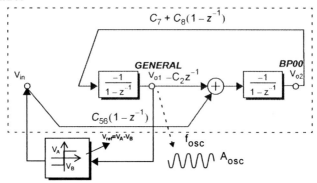

Figure 5.A.20: HG #3 oscillator

The oscillation conditions and the oscillation parameters for this particular case are shown in Table 5.A.16. On the other hand, **Step #1** is summarized in Table 5.A.17 and **Step #2** is represented in Fig. 5.A.21 and **Step #3** is given in Table 5.A.18.

$C_{01} = C_{05} = C_{09} = 0$	
Oscillation Conditions	$sign(V_{ref}) = sign(C_{56}C_7)$
Oscillation Frequency	$f_{osc} = \dfrac{1}{2\pi T_s} \cdot acos\left[\dfrac{-2C_7 + C_2 C_7^2 + 2C_2 C_7 C_8 + C_2 C_8^2}{-C_7 + C_2 C_7 C_8 + C_2 C_8^2}\right]$
Oscillation Amplitude	$A_{osc} = \dfrac{2V_{ref}}{\pi} \cdot \left[\dfrac{C_{56} C_7}{C_2 C_8}\right]$

Table 5.A.16: HG #3 reconfigured as an oscillator: Steady Oscillation Mode

$C_2 = 0.191; C_7 = 0.184; C_8 = 0.171; C_{56} = 0.994$
$k_2 = 0.353; k_1 = -0.353; k_0 = 0.170$ $b_1 = -1.932; b_0 = 0.967$
Positive feedback $sign(k_2 - k_0) > 0$
$f_{osc}^{theoretical} = 1728.93 Hz$ $\to \Delta f = 1.07\% \; V_{ref} = 1V \to$ $A_{osc}^{theoretical} = 7.130V$ $\to \Delta A = 1.36\%$ $f_{osc}^{matlab} = 1710.38 Hz$ $A_{osc}^{matlab} = 7.228V$

Table 5.A.17: HG #3 reconfigured as an oscillator: main oscillation results

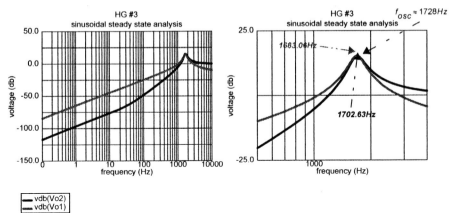

Figure 5.A.21: HG #3 Bode Diagram

Amplitude Sensitivities (%)	Frequency Sensitivities (%)
$S^{A_{osc}}_{C_2} = -\dfrac{100}{C_2}$	$S^{\cos(\Theta_{osc})}_{C_2} = -100 \dfrac{C_7\left(C_7^2 - C_8^2\right)}{\left(-2C_7 + C_2 C_7^2 + 2C_2 C_7 C_8 + C_2 C_8^2\right)\left(-C_7 + C_2 C_7 C_8 + C_2 C_8^2\right)}$
$S^{A_{osc}}_{C_8} = -\dfrac{100}{C_8}$	$S^{\cos(\Theta_{osc})}_{C_8} = -100 \dfrac{C_2 C_7\left(2C_2 C_7 C_8 + C_2 C_8^2 - 2C_8 + C_2 C_7^2\right)}{\left(-2C_7 + C_2 C_7^2 + 2C_2 C_7 C_8 + C_2 C_8^2\right)\left(-C_7 + C_2 C_7 C_8 + C_2 C_8^2\right)}$
$S^{A_{osc}}_{C_7} = \dfrac{100}{C_7}$	$S^{\cos(\Theta_{osc})}_{C_7} = 100 \dfrac{C_2\left(-C_8^2 - C_7^2 + C_2 C_7^2 C_8 + 2C_2 C_7 C_8^2 + C_2 C_8^3\right)}{\left(-2C_7 + C_2 C_7^2 + 2C_2 C_7 C_8 + C_2 C_8^2\right)\left(-C_7 + C_2 C_7 C_8 + C_2 C_8^2\right)}$
$S^{A_{osc}}_{C_{56}} = \dfrac{100}{C_{56}}$	$S^{\cos(\Theta_{osc})}_{C_{56}} = 0$

Table 5.A.18: Sensitivities for HG #3

Finally, **Step #4** is carried out in Fig. 5.A.22.

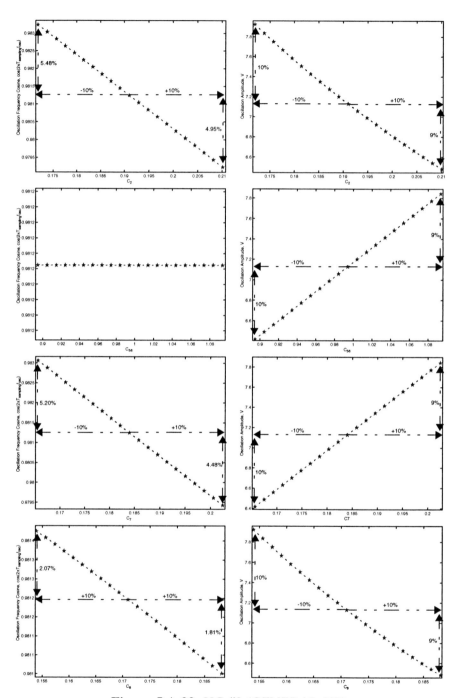

Figure 5.A.22: HG #3 (GENERAL-HP)

Appendix 5.B
Modelling Accuracy

All the studied biquads belonging to the DTMF core are examples where the predicted oscillations have a good agreement with more exact non-linear simulation results. This is true for the range of values of the involved capacitors at least. To see this, let us display for every biquad the curves obtained by both ways (Fig. 5.B.1-Fig. 5.B.12). Therefore, we have the evolution of the oscillation parameters (frequency and amplitude) when every involved capacitor is swept a ±10 % around its nominal value, as the DF approach predicts [99], [109]-[112]. And, on the other hand, we have the same evolution but now as Matlab/Simulink predicts [124], [126].

Observe from all the Figures that the evolution obtained by the DF approach is quasi-linear whereas, for most cases, the evolution obtained by a non-linear simulation presents ups and downs or is piece-wise linear. However, it must be clear from Fig. 5.B.1-Fig. 5.B.12, that the discrepancies between both methods of analysis are almost negligible for most of these particular examples.

Notice, on the other hand, we have marked in all the Figures the most significant deviation (in %) between the curves obtained by the DF approach and those obtained by Matlab/Simulink.

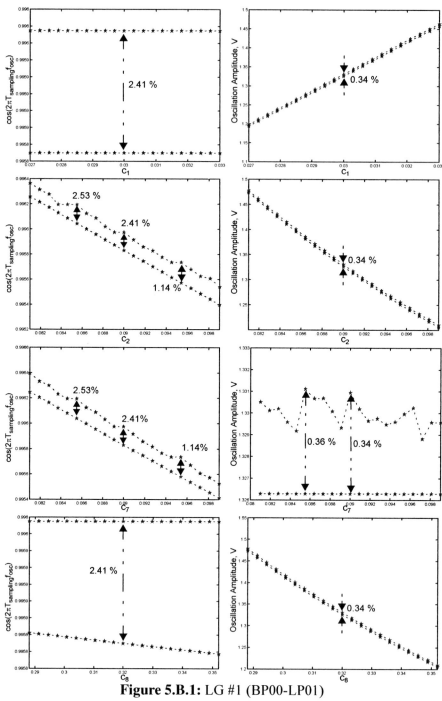

Figure 5.B.1: LG #1 (BP00-LP01)

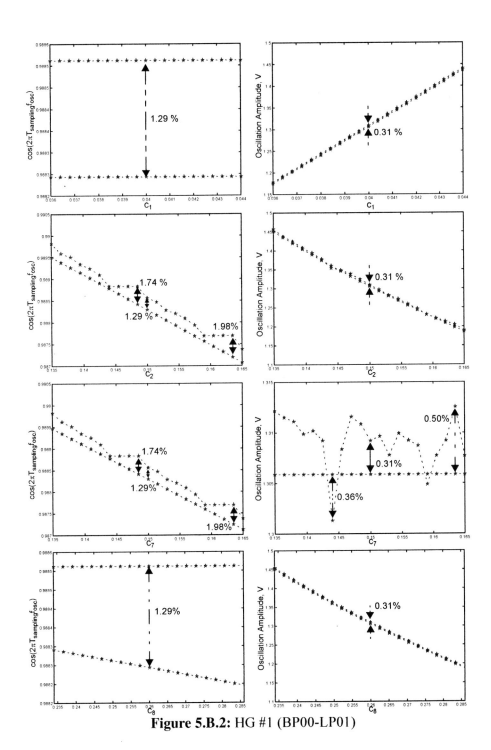

Figure 5.B.2: HG #1 (BP00-LP01)

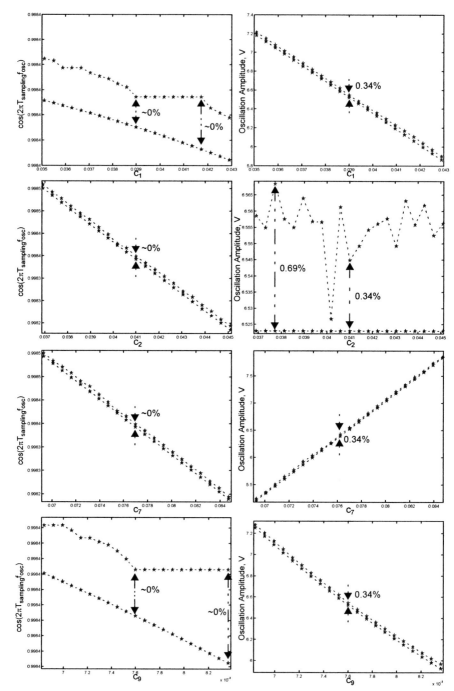

Figure 5.B.3: Not #1 (GENERAL - HPNOTCH)

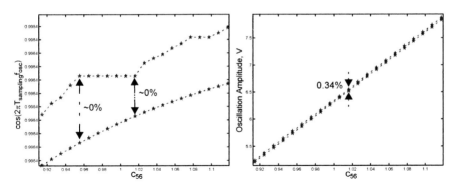

Figure 5.B.4: Not #1 (GENERAL - HPNOTCH) - Cont.

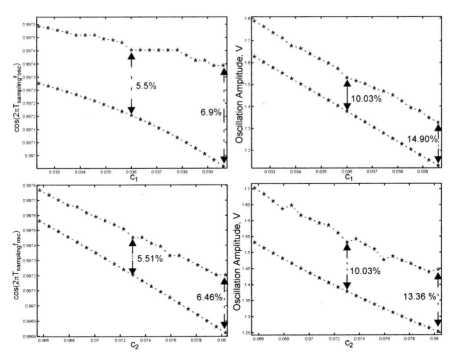

Figure 5.B.5: Not #2 (GENERAL-HP-NOTCH)

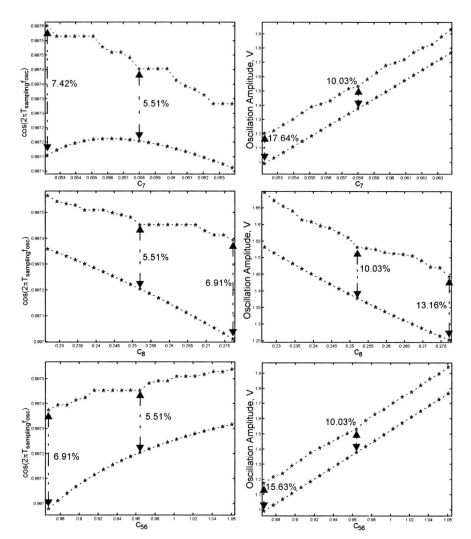

Figure 5.B.6: Not #2 (GENERAL-HP-NOTCH) - Cont.

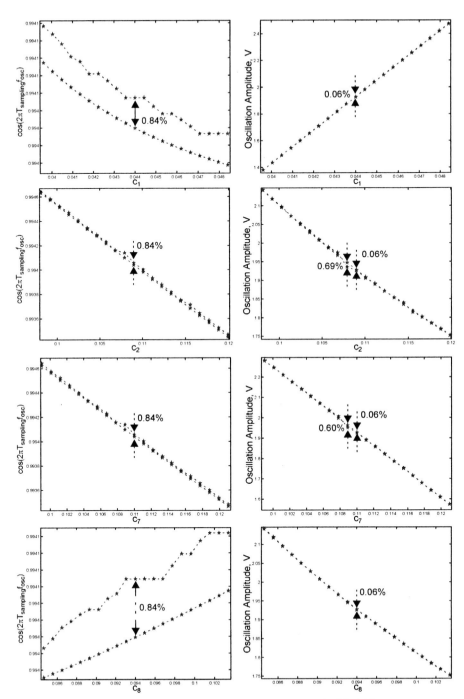

Figure 5.B.7: LG #2 (GENERAL-HP-NOTCH)

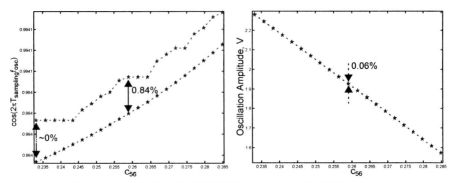

Figure 5.B.8: LG #2 (GENERAL-HP-NOTCH) - Cont.

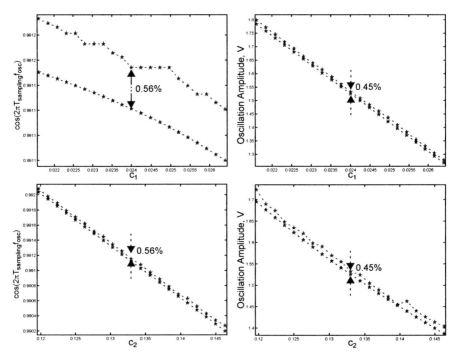

Figure 5.B.9: HG #2 (GENERAL-HP-NOTCH)

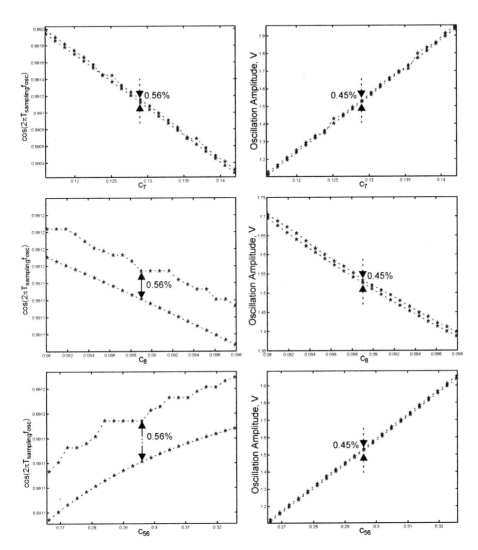

Figure 5.B.10: HG #2 (GENERAL-HP-NOTCH) - Cont.

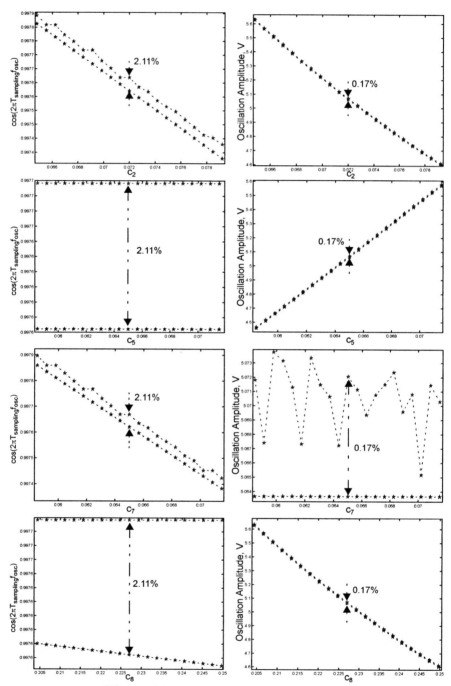

Figure 5.B.11: LG #3 (GENERAL-BP00)

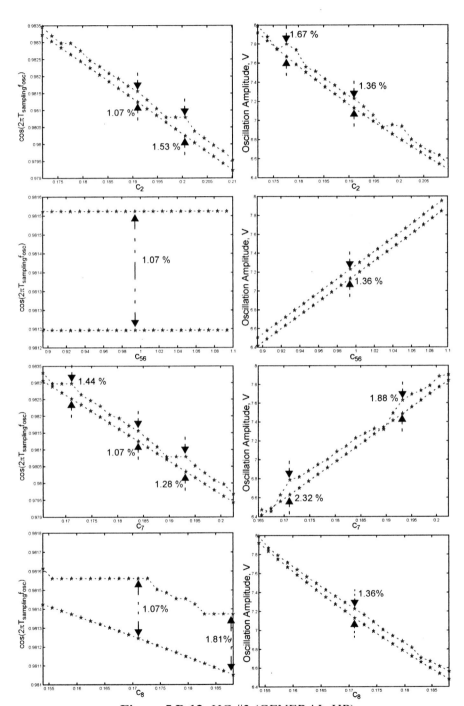

Figure 5.B.12: HG #3 (GENERAL-HP)

Appendix 5.C

Fault Models

In the switch-level fault analysis of the FL-SC-biquads, we have included both faults in switches and faults in capacitors. The employed SWITCAP model [74] for a faulty-free switch is shown in Fig. 5.C.1. Each switch is modeled as a voltage-controlled resistor with ON and OFF resistance values which incorporate the second-order effect of the actual ON and OFF resistances of the switch.

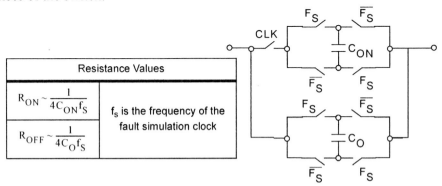

Resistance Values	
$R_{ON} \sim \dfrac{1}{4C_{ON}f_S}$	f_s is the frequency of the fault simulation clock
$R_{OFF} \sim \dfrac{1}{4C_O f_S}$	

Figure 5.C.1: Model for faulty-free switches

On the other one, the used SWITCAP fault modeling approach is displayed in Fig. 5.C.2 [74]. This model requires the ON and OFF resistance values of switches, the actual value of a resistive short and the capacitor deviation values.

We have injected the two general classes of analog faults, catastrophic and parametric faults. Amoung the so-called catastrophic faults we can distinguish:

a. stuck-on (**s_on**) and stuck-open (**s_open**) faults in switches

b. shorts between the analog terminals of a switch (**s_short**)

c. short in capacitors (**c_short**)

d. opens in capacitors (**c_open**).

Likewise, the so-called parametric faults take into account deviations on the value of the circuit capacitors (**c_dev(D)**, where D corresponds to a relative deviation of the capacitor nominal value).

Let us briefly explain how such faults are injected:

1. a **s_on** fault is injected by keeping the faulty switch permanently ON (thus equivalent to a resistor modeling the ON resistance of the switch).

2. a **s-open** fault is injected by keeping the faulty switch permanently OFF (thus equivalent to a resistor modeling the OFF resistance of the switch).

3. a **s_short** or **c_short** fault is represented as a resistive impedance between the shorted lines.

4. a **c_open** fault is modeled as a resistor in series with the capacitor (the default value of the switch OFF resistance is used as the value of a resistor open).

5. a **c_dev(.)** fault is injected by defining the deviation with respect the nominal value (e.g. a deviation of 50% is c_dev(0.5)). Switch ON and OFF resistances and voltage source output resistances are also considered.

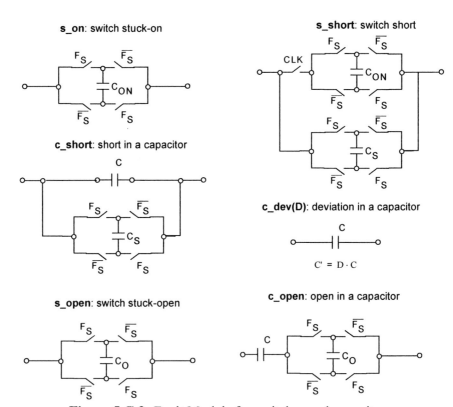

Figure 5.C.2: Fault Models for switches and capacitors

The values of the resistance involved in Fig. 5.C.2 are given in Table 5.C.1.

$R_{ON} \sim \dfrac{1}{4 \cdot C_{ON} \cdot f_S}$	
$R_{OFF} \sim \dfrac{1}{4 \cdot C_O \cdot f_S}$	f_s is the frequency of the fault simulation clock
$R_{SHORT} \sim \dfrac{1}{4 \cdot C_S \cdot f_S}$	

Table 5.C.1: Resistance Values

Appendix 6.A

DTMF Comprehensive Functional Test description

The DTMF functional test is a serie of tests involving tone bursts with their parameters varied in a number of different ways. Tests are performed by sending the tone bursts to the receiver, and counting the number of bursts to which the receiver responds. The results from these tests provide direct indications of receiver performance.

The following describes the tests carried out:

Decode Check: All tone pairs associated with standard 4x4 keypad digits (i.e. L1 H1 through L4 H4)(decode test digits 1 to 16) are pulsed sequentially using 50ms bursts at 1Vrms per frequency. Each tone pair is pulsed 10 times consecutively. The receiver should respond to all tone pairs that is designed to receive. As a whole this test requires 8000ms (4x4x10x50ms).

Recognition Bandwidth and Channel Center Frequency Check: This test utilizes the tone pairs L1 H1, L2 H2, L3 H3 and L4 H4 (i.e. digits 1, 5, 9 and 16). Each tone pair needs four test to complete the check, making 16 sections overall. Each section contains 40 pulses of 50ms duration, with an amplitude of 0.2Vrms per frequency. So, the total time will be 32000ms (16x40x50ms).

The four sections covering the tests for one tone (1 digit) are:

a. H frequency at 0% deviation from center, L frequency at +0.1%. L frequency is then incremented in +0.1% steps up to +4%. The number of tone bursts is noted and designated N^+.

b. H frequency at 0% deviation, L frequency at 0.1%. L frequency is then incremented in -0.1% steps, up to -4%. The number of tone bursts is noted and designated N^-.

c. The test in (a) is repeated with the L frequency at 0% and the H frequency varied up to +4%.

d. The test in (b) is repeated with the L frequency at 0% and the H frequency varied up to -4%.

Receiver Recognition Bandwidth (RRB(%)) is calculated as follows:

$RRB(\%)=(N^++N^-)/10$

Receiver Center Frequency Offset (RCFO) is calculated as follows:

$RCFO(\%)=(N^+-N^-)/20$

415

Acceptable Amplitude Ratio (Twist): This test utilizes the tone pairs L1 H1, L2 H2, L3 H3 and L4 H4 (i.e. digits 1, 5, 9 and 16). There are eight sections to the test. Each section contains 200 pulses with a 50ms duration for each pulse. Initially the amplitude of both tones is 1Vrms. Then, the total time will be 80000ms (8x200x50ms).

Two sections to test one tone pair are:

a. Standard Twist: H tone amplitude is maintained at 1Vrms, L tone amplitude is attenuated gradually until the amplitude ratio L/H is -20dB. Note the number of responses from the receiver.

b.Reverse Twist: L tone amplitude is maintained at 1Vrms, H tone amplitude is attenuated gradually until the amplitude ratio is 20dB. Note the number of responses from the receiver.

The Acceptable Amplitude Ratio in dB is equal to the number of responses registered in (a) or (b), divided by 10.

Dynamic Range: This test utilizes tone pair L1 H1 (digit 1). 35 tone pair pulses are transmitted, with both frequencies starting at 1 Vrms. The amplitude of each is gradually attenuated to -35dB at a rate of 1dB per pulse. The Dynamic Range in dB is equal to the number of responses from the receiver during the test. Then, in this case the time is 1750ms (35x50ms).

Guard Time: This test utilizes tone pair L1 H1 (digit 1). Four hundred pulses are transmitted at an amplitude of 1Vrms per frequency. Pulse duration starts at 49ms and is gradually reduced to 10ms. Guard time in ms is equal to (500-number of responses)/10. The time will be 20000ms (400x50ms)

Acceptable Signal to Noise Ration: This test utilizes tone pair L1 H1, transmitted on a noise background. The test consists of three sections in which the tone pair is transmitted 1000 times at an amplitude of 1Vrms per frequency, but with a different white noise level for each section. The first level is -24dBV, the second -18dBV and the third -12dBV. The Acceptable Signal to Noise Ratio is the lowest ratio of signal to noise in the test where the receiver responds to all 1000 pulses. Then, the total time will be 150000ms (3x1000x50ms).

Fig. 6.A.1 summarizes the different tests involved in the conventional DTMF functional test as well as the employed time.

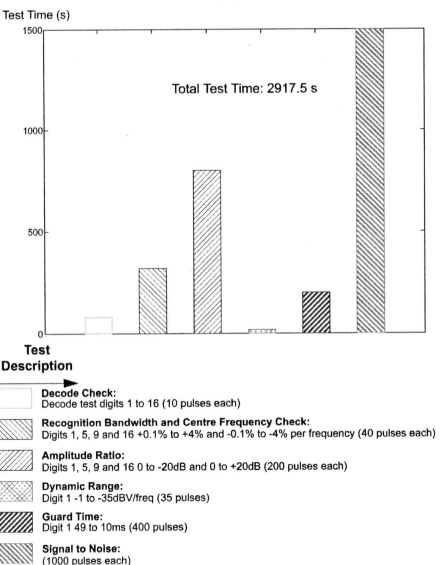

Figure 6.A.1: Summary of the conventional DTMF Functional Test

Therefore, the total time required for the conventional DTMF functional test would be 2917.5s (or 49 minutes). If we compare this result with the maximum test time to measure needed in the proposed DTMF OBIST strategy (that is 11.4 ms), we observe a very significant reduction. Obviously, we are talking about two very different types of testing. The functional testing includes the measurement of a lot of parameters because it pursues to characterize the perfomance of the SUT. However, the OBIST scheme presented in this chapter tries to be only a structural (or also called Defect-Oriented Test, DOT) testing approach. Consequently, making a comparison between both methodologies must be based on a lot of issues and not simply on the difference between the involved test times.

Appendix 7.A

Experimental Results

7.A.1. Experimental results for the programmable biquad

Biquad	Sample	Amplitude (V)	Deviation (%)	Frequency (fff) (Hz)	Deviation (%)	Frequency (Oscilloscope) (Hz)	Deviation (%)
BQ1	1	1.109 / 0.906	22.04 / 0.624	546.88 / 593.75	8.101 / 0.225	595.20 546.40 / 602.40 543.50 / 588.20 540.50	0.019 8.182 / 1.229 8.669 / 1.158 9.173
	2	1.125 / 0.906	23.80 / 0.624	546.88 / 593.75	8.101 / 0.225	595.20 546.40 / 602.40 549.50 / 588.20 540.50	0.019 8.182 / 1.229 7.661 / 1.158 9.173
	3	1.125 / 0.906	23.80 / 0.624	546.88 / 585.94	8.101 / 1.537	595.20 546.40 / 602.40 543.50 / 588.20 540.50	0.019 8.182 / 1.229 8.669 / 1.158 9.173
	4	1.109 / 0.906	22.04 / 0.624	546.88 / 593.75	8.101 / 0.225	595.20 546.40 / 602.40 537.60 / 588.20 540.50	0.019 8.182 / 1.229 9.661 / 1.158 9.173
	5	1.125 / 0.906	23.80 / 0.624	546.88 / 593.75	8.101 / 0.225	595.20 546.40 / 602.40 549.50 / 588.20 540.50	0.019 8.182 / 1.229 7.661 / 1.158 9.173

BQ1 Fault

Table 7.A.1: Experimental Results versus SPECTRE Simulation Results (0.909 V, 595.09 Hz)

Biquad	Sample	Amplitude (V)	Deviation (%)	Frequency (fff) (Hz)	Deviation (%)	Frequency (Oscilloscope) (Hz)		Deviation (%)	
BQ1 / BQ1 Fault	1	0.906	0.625	593.75	0.212	595.20	546.40	0.031	8.170
		1.109	21.60	546.88	8.090	602.40	543.50	1.241	8.658
						588.20	540.50	1.145	9.162
	2	0.906	0.625	593.75	0.212	595.20	546.40	0.031	8.170
		1.125	23.36	546.88	8.090	602.40	549.50	1.241	7.649
						588.20	540.50	1.145	9.162
	3	0.906	0.625	585.94	1.525	595.20	546.40	0.031	8.170
		1.125	21.60	546.88	8.090	602.40	543.50	1.241	8.658
						588.20	540.50	1.145	9.162
	4	0.906	0.625	593.75	0.212	595.20	546.40	0.031	8.170
		1.109	23.36	546.88	8.090	602.40	537.60	1.241	9.649
						588.20	540.50	1.145	9.162
	5	0.906	0.625	593.75	0.212	595.20	546.40	0.031	8.170
		1.125	21.60	546.88	8.090	602.40	549.50	1.241	7.649
						588.20	540.50	1.145	9.162

Table 7.A.2: Experimental Results versus SWITCAP Simulation Results (0.912 V, 595.01 Hz)

Biquad	Sample	Amplitude (V)	Deviation (%)	Frequency (fff) (Hz)	Deviation (%)	Frequency (Oscilloscope) (Hz)	Deviation (%)
BQ2 / BQ2 Fault	1	0.735	1.057	1929.7	0.077	1931.0 / 1515.0	0.145 / 21.43
		1.531	3.112	1531.5	20.57	1923.0 / 1538.0	0.270 / 20.24
						1938.0 / 1562.0	0.508 / 18.99
	2	0.735	1.057	1929.7	0.077	1931.0 / 1515.0	0.145 / 21.43
		1.531	3.112	1531.5	20.57	1923.0 / 1538.0	0.270 / 20.24
						1938.0 / 1562.0	0.508 / 18.99
	3	0.735	0.054	1929.7	0.077	1931.0 / 1515.0	0.145 / 21.43
		1.531	3.112	1531.5	20.57	1923.0 / 1538.0	0.270 / 20.24
						1938.0 / 1493.0	0.508 / 22.57
	4	0.735	1.057	1929.7	0.077	1931.0 / 1515.0	0.145 / 21.43
		1.531	3.112	1531.5	20.57	1923.0 / 1538.0	0.270 / 20.24
						1953.0 / 1562.0	1.286 / 18.99
	5	0.735	1.057	1929.7	0.077	1931.0 / 1515.0	0.145 / 21.43
		1.547	4.189	1531.5	20.57	1923.0 / 1538.0	0.270 / 20.24
						1938.0 / 1562.0	0.508 / 18.99

Table 7.A.3: Experimental Results versus SPECTRE Simulation Results (0.742 V, 1928.2 Hz)

Biquad	Sample	Amplitude (V)	Deviation (%)	Frequency (ff) (Hz)	Deviation (%)	Frequency (Oscilloscope) (Hz)		Deviation (%)	
BQ2 / BQ2 Fault	1	0.735	0.434	1929.7	0.051	1931.0	1515.0	0.119	21.45
		1.531	3.768	1531.5	20.59	1923.0	1538.0	0.296	20.26
						1938.0	1562.0	0.482	19.01
	2	0.735	0.434	1929.7	0.051	1931.0	1515.0	0.119	21.45
		1.531	3.768	1531.5	20.59	1923.0	1538.0	0.296	20.26
						1938.0	1562.0	0.482	19.01
	3	0.742	0.583	1929.7	0.051	1931.0	1515.0	0.119	21.45
		1.531	3.768	1531.5	20.59	1923.0	1538.0	0.296	20.26
						1938.0	1493.0	0.482	22.59
	4	0.735	0.434	1929.7	0.051	1931.0	1515.0	0.119	21.45
		1.531	3.768	1531.5	20.59	1923.0	1538.0	0.296	20.26
						1953.0	1562.0	1.260	19.01
	5	0.735	0.434	1929.7	0.051	1931.0	1515.0	0.119	21.45
		1.547	4.853	1531.5	20.59	1923.0	1538.0	0.296	20.26
						1938.0	1562.0	0.482	19.01

Table 7.A.4: Experimental Results versus SWITCAP Simulation Results (0.738 V, 1928.7 Hz)

Biquad	Sample	Amplitude (V)		Deviation (%)		Frequency (fff) (Hz)		Deviation (%)		Frequency (Oscilloscope) (Hz)		Deviation (%)	
BQ3 / BQ3 Fault	1	0.863	0.856	0.243	0.960	960.94	898.44	0.097	6.594	965.3	900.9	0.353	6.342
										969.0	909.1	0.738	5.489
										957.9	892.9	0.416	7.173
	2	0.869	0.869	0.486	0.486	960.94	898.44	0.097	6.594	965.3	900.9	0.353	6.342
										972.8	909.1	1.133	5.489
										957.9	892.9	0.416	7.173
	3	0.869	0.863	0.486	0.243	960.94	898.44	0.097	6.594	965.3	900.9	0.353	6.342
										976.6	917.4	1.528	4.626
										957.9	892.9	0.416	7.173
	4	0.856	0.863	0.972	0.243	960.94	898.44	0.097	6.594	965.3	900.9	0.353	6.342
										972.8	917.4	1.133	4.626
										969.0	892.9	0.738	7.173
	5	0.856	0.856	0.972	0.960	960.94	898.44	0.097	6.594	965.3	900.9	0.353	6.342
										954.2	909.1	0.800	5.489
										969.0	917.4	0.738	4.626

Table 7.A.5: Experimental Results versus SPECTRE Simulation Results (0.865 V, 961.9 Hz)

Biquad	Sample	Amplitude (V)	Deviation (%)	Frequency (fff) (Hz)	Deviation (%)	Frequency (Oscilloscope) (Hz)		Deviation (%)	
BQ3 / BQ3 Fault	1	0.863	4.113	960.94	0.625	965.3	900.9	0.165	6.826
						969.0	909.1	0.217	5.978
		0.856	4.803	898.44	7.080	957.9	892.9	0.931	7.653
	2	0.869	3.413	960.94	0.625	965.3	900.9	0.165	6.826
						972.8	909.1	0.610	5.978
		0.869	3.413	898.44	7.080	957.9	892.9	0.931	7.653
	3	0.869	3.413	960.94	0.625	965.3	900.9	0.165	6.826
						976.6	917.4	1.003	5.119
		0.863	4.113	898.44	7.080	957.9	892.9	0.931	7.653
	4	0.856	4.814	960.94	0.625	965.3	900.9	0.165	6.826
						972.8	917.4	0.610	5.119
		0.863	4.113	898.44	7.080	969.0	892.9	0.217	7.653 %
	5	0.856	4.814	960.94	0.625	965.3	900.9	0.165	6.826
						954.2	909.1	1.313	5.978
		0.856	4.803	898.44	7.080	969.0	917.4	0.217	5.119

Table 7.A.6: Experimental Results versus SWITCAP Simulation Results (0.8995 V, 966.89 Hz)

7.A.2. Experimental results including swopamp

Biquad	Sample	Amplitude (V)	Deviation (%)	Frequency (fff) (Hz)	Deviation (%)	Frequency (Oscilloscope) (Hz)		Deviation (%)	
BQ1 (BQ1 Fault)	1	0.906	0.264	546.88	0.225	595.2	546.4	0.019	8.182
		1.109	22.04	593.75	8.101	602.4	543.5	1.229	8.669
						588.2	540.5	1.158	9.173
	2	0.906	0.264	546.88	0.225	595.2	546.4	0.019	8.182
		1.109	22.04	593.75	8.101	602.4	543.5	1.229	8.669
						588.2	540.5	1.158	9.173
	3	0.906	0.264	546.88	0.225	595.2	546.4	0.019	8.182
		1.134	24.79	593.75	8.101	602.4	543.5	1.229	8.669
						588.2	540.5	1.158	9.173
	4	0.922	1.46	546.88	0.225	595.2	546.4	0.019	8.182
		1.119	23.14	593.75	8.101	602.4	543.5	1.229	8.669
						588.2	540.5	1.158	9.173
	5	0.906	0.264	546.88	0.225	595.2	546.4	0.019	8.182
		1.122	23.47	593.75	8.101	602.4	543.5	1.229	8.669
						588.2	540.5	1.158	9.173

Table 7.A.7: Experimental Results versus SPECTRE Simulation Results (0.909 V, 595.09 Hz)

Biquad	Sample	Amplitude (V)	Deviation (%)	Frequency (fff) (Hz)	Deviation (%)	Frequency (Oscilloscope) (Hz)	Deviation (%)
BQ1 / BQ1 Fault	1	0.906 / 1.109	0.625 / 20.61	593.75 / 546.88	0.212 / 8.090	595.2 / 546.4	0.031 / 8.170
						602.4 / 543.5	1.241 / 8.658
						588.2 / 540.5	1.145 / 9.162
	2	0.906 / 1.109	0.625 / 20.61	593.75 / 546.88	0.212 / 8.090	595.2 / 546.4	0.031 / 8.170
						602.4 / 543.5	1.241 / 8.658
						588.2 / 540.5	1.145 / 9.162
	3	0.906 / 1.134	0.625 / 24.34	593.75 / 546.88	0.212 / 8.090	595.2 / 546.4	0.031 / 8.170
						602.4 / 543.5	1.241 / 8.658
						588.2 / 540.5	1.145 / 9.162
	4	0.922 / 1.119	1.086 / 22.70	593.75 / 546.88	0.212 / 8.090	595.2 / 546.4	0.031 / 8.170
						602.4 / 543.5	1.241 / 8.658
						588.2 / 540.5	1.145 / 9.162
	5	0.906 / 1.122	0.625 / 23.03	593.75 / 546.88	0.212 / 8.090	595.2 / 546.4	0.031 / 8.170
						602.4 / 543.5	1.241 / 8.658
						588.2 / 540.5	1.145 / 9.162

Table 7.A.8: Experimental Results versus SWITCAP Simulation Results (0.912 V, 595.01 Hz)

Biquad	Sample	Amplitude (V)	Deviation (%)	Frequency (fft) (Hz)	Deviation (%)	Frequency (Oscilloscope) (Hz)	Deviation (%)
BQ2 / BQ2 Fault	1	0.727	2.141	1929.7	0.077	1931, 1923, 1938	0.145, 0.270, 0.508
		1.553	4.593	1531.5	20.57	1515, 1538, 1562	21.43, 20.24, 18.99
	2	0.735	1.057	1929.7	0.077	1931, 1923, 1916	0.145, 0.270, 0.633
		1.538	3.583	1531.5	20.57	1515, 1538, 1562	21.43, 20.24, 18.99
	3	0.735	1.057	1929.7	0.077	1931, 1923, 1938	0.145, 0.270, 0.508
		1.516	2.101	1531.5	20.57	1515, 1538, 1562	21.43, 20.24, 18.99
	4	0.735	1.057	1929.7	0.077	1931, 1923, 1916	0.145, 0.270, 0.633
		1.547	4.189	1531.5	20.57	1515, 1538, 1562	21.43, 20.24, 18.99
	5	0.735	1.057	1929.7	0.077	1931, 1923, 1908	0.145, 0.270, 1.048
		1.540	3.718	1531.5	20.57	1515, 1538, 1562	21.43, 20.24, 18.99

Table 7.A.9: Experimental Results versus SPECTRE Simulation Results (0.742 V, 1928.2 Hz)

Biquad	Sample	Amplitude (V)	Deviation (%)	Frequency (fff) (Hz)	Deviation (%)	Frequency (Oscilloscope) (Hz)		Deviation (%)	
BQ2	1	0.727	1.518	1929.7	0.051	1931	1515	0.119	21.45
BQ2 Fault		1.553	4.997	1531.5	20.59	1923	1538	0.296	20.26
						1938	1562	0.482	19.01
BQ2	2	0.735	0.434	1929.7	0.051	1931	1515	0.119	21.45
BQ2 Fault		1.538	4.243	1531.5	20.59	1923	1538	0.296	20.26
						1916	1562	0.659	19.01
BQ2	3	0.735	0.434	1929.7	0.051	1931	1515	0.119	21.45
BQ2 Fault		1.516	2.752	1531.5	20.59	1923	1538	0.296	20.26
						1938	1562	0.482	19.01
BQ2	4	0.735	0.434	1929.7	0.051	1931	1515	0.119	21.45
BQ2 Fault		1.547	4.853	1531.5	20.59	1923	1538	0.296	20.26
						1916	1562	0.659	19.01
BQ2	5	0.735	0.434	1929.7	0.051	1931	1515	0.119	21.45
BQ2 Fault		1.540	4.378	1531.5	20.59	1923	1538	0.296	20.26
						1908	1562	1.073	19.01

Table 7.A.10: Experimental Results versus SWITCAP Simulation Results (0.738 V, 1928.7 Hz)

Biquad	Sample	Amplitude (V)	Deviation (%)	Frequency (fft) (Hz)	Deviation (%)	Frequency (Oscilloscope) (Hz)	Deviation (%)
BQ3 / BQ3 Fault	1	0.863 / 0.863	0.243 / 0.243	960.94 / 898.44	0.097 / 6.597	965.3 / 900.9	0.353 / 6.342
						961.5 / 917.4	0.042 / 4.626
						972.8 / 892.9	1.133 / 7.173
	2	0.863 / 0.869	0.243 / 0.486	960.94 / 898.44	0.097 / 6.597	965.3 / 900.9	0.353 / 6.342
						972.8 / 909.1	1.133 / 5.489
						957.9 / 892.9	0.416 / 7.173
	3	0.856 / 0.863	0.972 / 0.243	960.94 / 898.44	0.097 / 6.597	965.3 / 900.9	0.353 / 6.342
						961.5 / 917.4	0.042 / 4.626
						972.8 / 909.1	1.133 / 5.489
	4	0.863 / 0.863	0.243 / 0.243	960.94 / 898.44	0.097 / 6.597	965.3 / 900.9	0.353 / 6.342
						961.5 / 877.2	0.042 / 8.805
						980.4 / 892.9	1.923 / 7.173
	5	0.856 / 0.856	0.972 / 0.613	960.94 / 898.44	0.097 / 6.597	965.3 / 900.9	0.353 / 6.342
						961.5 / 909.1	0.042 / 5.489
						972.8 / 917.4	1.133 / 4.626

Table 7.A.11: Experimental Results versus SPECTRE Simulation Results (0.865 V, 961.9 Hz)

Biquad	Sample	Amplitude (V)	Deviation (%)	Frequency (fff) (Hz)	Deviation (%)	Frequency (Oscilloscope) (Hz)	Deviation (%)
BQ3 / BQ3 Fault	1	0.863 / 0.856	4.113 / 4.803	960.94 / 898.44	0.625 / 7.080	965.3 / 900.9; 969.0 / 909.1; 957.9 / 892.9	0.165 / 6.826; 0.217 / 5.978; 0.931 / 7.653
	2	0.869 / 0.869	3.413 / 3.413	960.94 / 898.44	0.625 / 7.080	965.3 / 900.9; 972.8 / 909.1; 957.9 / 892.9	0.165 / 6.826; 0.610 / 5.978; 0.931 / 7.653
	3	0.869 / 0.863	3.413 / 4.113	960.94 / 898.44	0.625 / 7.080	965.3 / 900.9; 976.6 / 917.4; 957.9 / 892.9	0.165 / 6.826; 1.003 / 5.119; 0.931 / 7.653
	4	0.863 / 0.856	4.113 / 4.814	960.94 / 898.44	0.625 / 7.080	965.3 / 900.9; 972.8 / 917.4; 969.0 / 892.9	0.165 / 6.826; 0.610 / 5.119; 0.217 / 7.653 %
	5	0.856 / 0.856	4.814 / 4.803	960.94 / 898.44	0.625 / 7.080	965.3 / 900.9; 954.2 / 909.1; 969.0 / 917.4	0.165 / 6.826; 1.313 / 5.978; 0.217 / 5.119

Table 7.A.12: Experimental Results versus SWITCAP Simulation Results (0.900 V, 966.89 Hz)

7.A.3. Accuracy in predicting the theoretical oscillation parameters

An important factor for evaluating the feasibility of OBT is how accurately we can predict the oscillation parameters (amplitude and frequency) by using the linearized model (DF approach). From the equations achieved by the DF approach, we obtain Table 7.A.13 where the theoretical oscillation results for the three biquads reconfigured as nonlinear closed-loop feedback systems are compiled. From the previous experimental Tables, it can be seen that assuming the DF approach entirely valid in the three cases is not always an "acceptable" alternative. However, these theoretical values can be considered as a good first estimation of the true oscillations.

Type	Frequency (Hz)	Amplitude (A)
BQ1	602.29	0.911
BQ2	2015.70	1.242
BQ3	968.60	0.927

Table 7.A.13: Equation Results (DF Approach)

Table 7.A.13 can be supplemented by a comparison of the simulation values obtained from different simulators. We have used other simulation tools as such SWITCAP [74] or SPECTRE [119]. SWITCAP is a switch-level simulator [74] whereas SPECTRE is a transistor-level or circuit-level simulator [119]. Table 7.A.14 reports these more precise oscillation results. If we compare them for each simulator, we can assert that deviations between them are not critical. We can conclude that there are no significant differences, although the electrical-level simulators give (as expected) better accuracy (see Tables in the previous Sections).

Type	Frequecy (Hz) (SWITCAP)	Frequency (Hz) (SPECTRE)	Deviation (SW-SP)	Amplitude (A) (SWITCAP)	Amplitude (A) (SPECTRE)	Deviation (SW-SP)
BQ1	595.01	595.09	0.01%	0.912	0.909	0.33%
BQ2	1928.70	1928.20	0.03%	1.475	1.484	0.6%
BQ3	966.89	961.87	0.52%	0.899	0.865	3.78%

Table 7.A.14: Simulation Results

Notice, however, from Table 7.A.13 and Table 7.A.14, that there is an error margin between the theoretical data and the simulation data. These differences are especially significant in the oscillation amplitude for BQ2. This filter is a high-pass filter and, therefore, the high-order harmonics may invalidate the premises of the DF linear model.

If we make a mindful analysis of faults and we draw the achieved Monte-carlo window (given by the gray squares) in Fig. 7.A.1, the points (in the frequency and amplitude space) where we have obtained the theoretical results (diamonds in green) in the Figure and the points where the faulty circuits are placed (circles in red), we can see that the theoretical data lie (fall) into the called tolerance window with the exception of the amplitude for the BQ2 case. All these considerations allow us to assert that the DF linear model is not always completely satisfactory but however it is always a good starting point for computing the fault-free oscillation nominal values that, most of the times, have to be complemented by simulation.

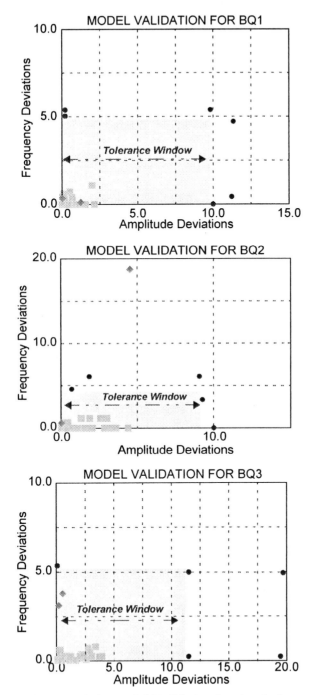

Figure 7.A.1: Model Validation by simulation

7.A.4. Experimental results for the DTMF core

Biquad Group	Biquad	Amplitude (V)	Deviation	Frequency (fff) (Hz)	Deviation	Frequency (Oscilloscope) (Hz)	Deviation
1	n2	0.6600	1.54 %	589.84	0.50 %	588.2	0.22 %
						591.7	0.82 %
						595.2	1.41 %
	l3	0.4560	1.29 %	593.750	0.22 %	595.2	0.03 %
						598.8	0.63 %
						609.8	2.47 %
2	n1	0.4600	2.22 %	691.406	1.11 %	689.7	1.36 %
						694.4	0.68 %
						704.2	0.72 %
	h3	0.7500	2.18 %	1929.69	0.05 %	1923	0.30 %
						1953	1.25 %
						1938	0.48 %
3	l1	0.4560	1.13 %	820.31	0.27 %	819.7	0.35 %
						826.4	0.47 %
						847.5	3.03 %
	h2	0.6700	3.03 %	1179.69	0.18 %	1190	1.05 %
						1163	1.23 %
						1176	0.14 %
4	l2	0.4520	0.29 %	960.94	0.68 %	984.3	1.73 %
						947.0	2.12 %
						965.3	0.23 %
	h1	0.6550	0.71 %	1304.69	0.30 %	1295	0.45 %
						1302	0.09 %
						1289	0.91 %

Table 7.A.15: Experimental results versus SWITCAP simulation results (Sample 1)

Biquad Group	Biquad	Amplitude (V)	Deviation	Frequency (fff) (Hz)	Deviation	Frequency (Oscilloscope) (Hz)	Deviation
1	n2	0.6700	3.08 %	593.75	1.16 %	588.2	0.22 %
						591.7	0.82 %
						595.2	1.41 %
	l3	0.4640	3.06 %	593.750	0.22 %	595.2	0.03 %
						591.7	0.56 %
						602.4	1.23 %
2	n1	0.4600	2.22 %	687.50	1.67 %	689.7	1.36 %
						694.4	0.68 %
						684.9	2.04 %
	h3	0.7500	92.18 %	1929.69	0.05 %	1923	0.30 %
						1916	0.66 %
						1931	0.11 %
3	l1	0.4600	2.02 %	820.31	0.27 %	819.7	0.35 %
						826.4	0.47 %
						833.3	1.30 %
	h2	0.6700	3.08 %	1176.9	0.06 %	1190	1.05 %
						1168	0.82 %
						1142	3.02 %
4	l2	0.4440	1.49 %	960.94	0.68 %	952.4	1.56 %
						970.9	0.35 %
						961.5	0.62 %
	h1	0.6600	1.47 %	1304.9	0.32 %	1299	0.14 %
						1316	1.17 %
						1282	1.45 %

Table 7.A.16: Experimental results versus SWITCAP simulation results (Sample 2)

Biquad Group	Biquad	Amplitude (V)	Deviation	Frequence (fff) (Hz)	Deviation	Frequency (Osciloscope) (Hz)	Deviation
1	n2	0.6650	2.31 %	593.75	1.16 %	588.2	0.22 %
						602.4	2.64 %
						595.2	1.41 %
	I3	0.4560	1.29 %	593.750	0.22 %	595.2	0.03 %
						588.2	1.15 %
						602.4	1.23 %
2	n1	0.4640	3.11 %	687.50	1.67 %	684.9	2.04 %
						694.4	0.68 %
						684.9	2.04 %
	h3	0.7400	0.81 %	1929.69	0.05 %	1923	0.30 %
						1916	0.66 %
						1938	0.48 %
3	I1	0.4560	1.24 %	820.31	0.27 %	819.7	0.35 %
						826.4	0.47 %
						813.0	1.16 %
	h2	0.6650	2.26 %	1171.88	0.49 %	1190	1.05 %
						1163	1.24 %
						1149	2.43 %
4	I2	0.4560	1.18 %	960.94	0.68 %	980.4	1.33 %
						970.9	0.35 %
						961.5	0.62 %
	h1	0.6550	0.70 %	1304.69	0.32 %	1295	0.45 %
						1302	0.09 %
						1282	1.45 %

Table 7.A.17: Experimental results versus SWITCAP simulation results (Sample 3)

Biquad Group	Biquad	Amplitude (V)	Deviation	Frequency (fff) (Hz)	Deviation	Frequency (Oscilloscope) (Hz)	Deviation
1	n2	0.6700	3.08 %	593.75	1.16 %	584.8	0.36 %
						591.7	0.82 %
						595.2	1.41 %
	l3	0.4640	3.07 %	593.750	0.22 %	591.7	0.56 %
						588.2	1.15 %
						602.4	1.23 %
2	n1	0.4640	3.11 %	687.50	1.67 %	689.7	1.36 %
						694.4	0.68 %
						699.3	0.02 %
	h3	0.7500	2.18 %	1929.69	0.05 %	1908	2.08 %
						1969	2.08 %
						1938	0.48 %
3	l1	0.4600	2.02 %	820.31	0.27 %	822.4	0.02 %
						836.1	1.64 %
						809.1	1.64 %
	h2	0.6700	3.08 %	1171.88	0.49 %	1190	1.05 %
						1163	1.24 %
						1185	0.63 %
4	l2	0.4520	0.29 %	960.94	0.68 %	980.4	1.33 %
						936.3	3.22 %
						965.3	0.23 %
	h1	0.6550	0.71 %	1304.69	0.32 %	1295	0.45 %
						1302	0.09 %
						1265	2.75 %

Table 7.A.18: Experimental results versus SWITCAP simulation results (Sample 4)

Biquad Group	Biquad	Amplitude (V)	Deviation	Frequency (fff) (Hz)	Deviation	Frequency (Osciloscope) (Hz)	Deviation
1	n2	0.6650	2.31 %	593.75	1.16 %	595.2	1.41 %
						591.7	0.82 %
						585.2	0.22 %
	l3	0.4600	2.18 %	593.750	0.22 %	595.2	0.03 %
						588.2	1.15 %
						602.4	1.23 %
2	n1	0.4640	3.11 %	687.50	1.67 %	689.7	1.36 %
						694.4	0.68 %
						684.9	2.05 %
	h3	0.7500	2.18 %	1929.69	0.05 %	1923	0.30 %
						1953	1.25 %
						1938	0.48 %
3	l1	0.4600	2.02 %	820.31	0.27 %	819.7	0.35 %
						826.4	0.46 %
						833.3	1.30 %
	h2	0.6700	3.08 %	1179.69	0.18 %	1190	1.05 %
						1163	1.24 %
						1176	0.14 %
4	l2	0.4560	1.18 %	960.94	0.68 %	970.9	0.35 %
						961.5	0.62 %
						952.5	1.55 %
	h1	0.6550	0.70 %	1304.69	0.32 %	1299	0.14 %
						1316	1.17 %
						1282	1.45 %

Table 7.A.19: Experimental results versus SWITCAP simulation results (Sample 5)

References

[1] K. Arabi and B. Kaminska: *"Oscillation-Based Test Strategy (OBTS) for Analog and Mixed-Signal Circuits"*. US Patent Application, 1995.

[2] K. Arabi and B. Kaminska: "Oscillation-Test Strategy for Analog and Mixed-Signal Integrated Circuits". In Proceedings of *the 14th VLSI Test Symposium*, 1996, pp. 476-482.

[3] K. Arabi and B. Kaminska: "Design for Testability of Integrated Operational Amplifiers Using Oscillation-Test Strategy". In Proceedings of *the International Conference on Computer Design, VLSI In Computers and Processors*, 7-9 October 1996, Austin, pp. 40-45.

[4] K. Arabi and B. Kaminska: "A New Technique to Monitor the Electrode and Lead Failures in Implantable Microstimulators and Sensors". In Proceedings of *the 18th Annual International Conference of the IEEE Engineering in Medicine and Biology Society*, 31 October- 3 November 1996, vol. 1, pp. 181-182.

[5] K. Arabi, B. Kaminska and S. Sunter: "Testing integrated operational amplifier based on oscillation-test method". In Proceedings of *the International Mixed-signal Test Workshop*, 15-18 May 1996, Canada, pp. 227-232.

[6] K. Arabi, B. Kaminska and J. Rzeszut: "BIST for D/A and A/D converters". *IEEE Design Test Compu*ters, 1996, vol. 13, no. 4, pp. 40-49.

439

[7] K. Arabi and B. Kaminska: "Testing Analog and Mixed-Signal Integrated Circuits Using Oscillation-Test Method". *IEEE Transactions on Computer-Aided Design of Integrated Circuits and Systems*, July 1997, vol.16, no. 7, pp. 745-753.

[8] K. Arabi and B. Kaminska: "Efficient and Accurate Testing of Analog-to Digital Converters Using Oscillation-Test Method". In Proceedings of *the European Design and Test Conference (ED&TC'97)*, 17-20 March 1997, pp. 348-352.

[9] K. Arabi and B. Kaminska: "Parametric and Catastrophic Fault Coverage of Analog Circuits in Oscillation-Test Methodology". In Proceedings of *the 15th IEEE VLSI Test Symposium*, 27 April-1 May 1997, pp. 166-171.

[10] K. Arabi and B. Kaminska: "Oscillation Built-In Self Test (OBIST) Scheme for Functional and Structural Testing of Analog and Mixed-Signal Integrated Circuits". In Proceedings of *the IEEE International Test Conference*, 1-6 November 1997, pp. 578-586.

[11] I. H. S. Hassan, K. Arabi and B. Kaminska: "Testing Digital to Analog Converters Based on Oscillation-Test Strategy Using Sigma-delta Modulation". In Proceedings of *the International Conference on Computer Design: VLSI in Computers and Processors (ICCD 98)*, 5-7 October 1998, pp. 40-46.

[12] K. Arabi and B. Kaminska: "Design for Testability of Embedded Integrated Operational Amplifiers". *IEEE Journal of Solid-State Circuits*, April 1998, vol. 33, no. 4, pp. 573-581.

[13] K. Arabi, Hassan Ihs, C. Dufaza and B. Kaminska: "Digital Oscillation-Test Method for Delay and Stuck-at Fault Testing of Digital Circuits". In Proceedings of *the International Test Conference*, 18-23 October 1998, pp. 91-100.

[14] K. Arabi and B. Kaminska: "Oscillation-Test Methodology for Low-Cost Testing of Active Analog Filters". *IEEE Transactions on Instrumentation and Measurement*, August 1999, vol. 48, no. 4, pp. 798-806.

[15] K. Arabi and B. Kaminska. *"Oscillation-Based Test Method for testing an at least partially analog circuit"*. US Patent Application, 1999.

[16]K. Arabi, B. Kaminska and J. Rzeszut:"A New Built-in self-Test Approach for D/A and A/D Converters". In Proceedings of *the IEEE International Conference on CAD,* 1994, pp. 491-494.

[17]Banerjee, P. and J.A. Abraham: "Fault Characterization of VLSI MOS Circuits". In Proceedings of *the IEEE International Conference on Circuits and Computers*, 1982, pp. 564-568.

[18]A.H Bratt, R.J. Harvey, A.P. Dorey and A.M.D Richardson: "Design-For-Test Structure to Facilitate Test Vector Application with Low Performance Loss in Non-Test Mode". *Electronics Letters,* August 1993, vol. 29, no. 16, pp. 1438-1440.

[19]Mark Burns and Gordon W. Roberts: *"An Introduction to Mixed-Signal IC Test and Measurement"*. Oxford University Press Inc., New York, 2001.

[20]Devarayanadurg, G. and M. Soma: "Analytical Fault Modeling and Static Test Generation for Analog ICs". *IEEE/ACM International Conference on Computer-Aided Design, ICCAD*, 6-10 November 1994, pp. 44-47.

[21]P. M. Dias, J. E. Franca, N. Paulino: "Oscillation test methodology for a digitally-programmable switched-current biquad". In Proceedings of *theInternational Mixed-signal Test Workshop*, 1996, pp. 221-226.

[22]M. Ehsanian, B. Kaminska, K. Arabi: "A new on-chip digital BIST for Analog-to-Digital Converters". *Microelectronic Reliability*, 1998, vol. 38, no. 3, pp. 409-420.

[23]Fasang, P. P., D. Mulins, and T. Wong: "Design for Testability for Mixed Analog/Digital ASICS". In Proceedings of *the IEEE Custom Integrated Circuit Conference*, 16-19 May, 1998, pp. 16.5/1-16.5/4.

[24]Gielen G., Z. Wang and W. Sansen: "Fault Detection and Input Stimulus Determination for the Testing of Analog Integrated Circuits Based on Power-Supply Current Monitoring", *IEEE/ACM International Conference on Computer-Aided Design, ICCAD*, 6-10 November 1994, pp. 495-498.

[25]G. Huertas, D. Vázquez, A. Rueda and J.L. Huertas: "Oscillation-Based Test Experiments in Filters: a DTMF example". In Proceedings of

the International Mixed-Signal Testing Workshop (IMSTW'99), 15-18 June 1999, British Columbia, Canada, pp. 249-254.

[26]G. Huertas, D. Vázquez, A. Rueda and J.L. Huertas: "Effective oscillation-based test for application to a DTMF filter bank". In Proceedings of *the International Test Conference 1999 (ITC'99)*, 28-30 September 1999, Atlantic City, USA, pp. 549-555.

[27]G. Huertas, D. Vázquez, A. Rueda and J.L. Huertas: "Built-in self-test in mixed-signal ICs: a DTMF example". In Proccedings of *the XIV Design of Circuits and Integrated Systems Conference (DCIS'99)*, November 1999, Palma de Mallorca, pp. 577-582.

[28]G. Huertas, D. Vázquez, A. Rueda and J.L. Huertas: "Built-in self-test in mixed-signal ICs: a DTMF macrocell". In Proceedings of *the International Conference VLSI Design*, January 2000, Calcuta, India, pp. 568-571.

[29]G. Huertas, D. Vázquez, A. Rueda and J.L. Huertas: "A Practical Method for Reading Test Outcomes in Oscillation-Based Test". In Proceedings of *the International Mixed-Signal Testing Workshop (IMSTW'2000)*, 21-23 June 2000, Montpellier (La Grande Motte), France, pp. 135-138.

[30]G. Huertas, D. Vázquez, A. Rueda and J.L. Huertas: "Testing Mixed-Signal Cores". In Proccedings of *the 13th Brazilian Symposium on Integrated Circuit Design (SBCCI2000)*, 18-24 September 2000, Manaus, Brazil, pp. 307-312.

[31]G. Huertas, D. Vázquez, A. Rueda and J.L. Huertas: "Testing Mixed-Signal Cores: Practical Oscillation-Based Test in an Analog Macrocell", In the Proceedings of *the Ninth Asian Test Symposium (ATS 2000)*, Taipei, Taiwan, 4-6 December 2000.

[32]G. Huertas, D. Vázquez, E. Peralías, A. Rueda and J.L. Huertas: "Oscillation-based Test in Oversampling A/D Converters". In Proceedings of *the 7th IEEE International Mixed-Signal Testing Workshop (IMSTW 2001)*, 13-15 June 2001, Atlanta, USA, pp. 35-46.

[33]G. Huertas, D. Vázquez, E. Peralías, A. Rueda and J.L. Huertas: "Practical Oscillation-Based Test in Analog Filters: Experimental Results". In the Proceedings of *the International Workshop on*

Electronic Design, Test, and Applications (DELTA 2002), 29-31 January 2002, New Zealand, pp. 18-24.

[34] G. Huertas, D. Vázquez, E. Peralías, A. Rueda and J.L. Huertas: "Oscillation-based Test in Oversampling A/D Converters". *Microelectronic Journal, Edit. Elsevier Science,* October 2002, vol. 33, no. 10, pp. 799-806.

[35] G. Huertas, D. Vázquez, E. Peralías, A. Rueda and J.L. Huertas: "Oscillation-Based Test in Bandpass Oversampled A/D Converters". In Proceedings of *the International Mixed-Signal Test Workshop*, June 2002, Montreaux (Switzerland), pp. 39-48.

[36] G. Huertas, D. Vázquez, A. Rueda and J.L. Huertas: "Practical Oscillation-Based Test of Integrated Filters". *IEEE Design and Test of Computers,* Nov-Dec 2002, vol. 19, no. 6, pp. 64-72.

[37] G. Huertas, D. Vázquez, E. Peralías, A. Rueda and J.L. Huertas: "Testing Mixed-Signal Cores: Practical Oscillation-Based Test in an Analog Macrocell". *IEEE Design and Test of Computers,* Nov-Dec 2002, vol. 19, n° 6, pp. 73-82.

[38] D. Vázquez, G. Huertas, G. Leger, A. Rueda and J.L. Huertas: "A test interpretation technique for BIST functional testing of analog filters", In Proccedings of *the 4th IEEE Latin American Test Workshop,* Natal, Brazil, February 2003, pp. 223-228.

[39] D. Vázquez, G. Huertas, G. Leger, A. Rueda and J.L. Huertas: "Sinewave signals characterization using first order sigma delta modulators: application to mixed-signal BIST". In Proccedings of *the International Mixed-Signal Testing Workshop,* Sevilla, Spain, June 2003, pp. 171-176.

[40] D. Vázquez, G. Huertas, G. Leger, A. Rueda and J.L. Huertas: "A LP-LV high performance monolitic DTMF receiver with on-chip test facilities". In Proccedings of *the SPIE'S International Symposium on Microtechnologies for the New Millenium,* Maspaloma, Gran Canaria, May 2003, vol. 5117, pp. 298-309.

[41] G. Huertas, D. Vázquez, A. Rueda and J.L. Huertas: "Oscillation-based Test in Bandpass Oversampled A/D Converters". *Microelectronic Journal, Edit. Elsevier Science,* October 2003, vol. 34, no. 10, pp. 927-936.

[42]J. Huang, C.Ong, K. Cheng: "A BIST scheme for on-chip ADC and DAC testing". In Proceedings of *the Design, Automation and Test in Europe Conference (DATE)*, March 2000, Paris, France, pp. 216-220.

[43]J.L. Huertas, A. Rueda and D. Vázquez: "Improving the testability of switched-capacitor filters", *J. Electron. Test., Theory Appl.*, November 1993, vol. 4, no. 4, pp. 299-313.

[44]J.L. Huertas: "Test and design for testability of analog and mixed-signal integrated circuits: Theoretical basis and pragmatical approaches". *Selected Topics in Circuits and Systems, H. Dedieu, Ed. Amsterdam: Elsevier*, 1993.

[45]J.L. Huertas, A. Rueda and D. Vázquez: "Testable switched-capacitor filters". *IEEE Journal of Solid-State Circuits*, July 1993, vol. 28, pp. 719-724.

[46]J.L. Huertas, A. Rueda and D. Vázquez: "A new strategy for testing analog filters". In Proceedings of *the IEEE VLSI Test Symposium*, April 1994, pp. 36-41.

[47]S.H.Lewis, R.Ramachadran and W.M.Snelgrove: "Indirect Testing of Digital-Correction Circuits in A/D Converters with Redundancy", *IEEE Transactions on Circuits and Systems-II*, July 1995, vol. 42, no. 7, pp. 437-445.

[48]C. Mangelsdorff, S.H. Lee, M.Martin, H.Malik, T.Fukuda and H.Matsumoto: "Design for Testability in Digitally-Corrected ADC´s". In Proceedings of *the International Symposium on Solid-State Circuits*, 1993, pp. 70-71.

[49]Marina Santo Zarnik, et al.: "Design of oscillation-based test structures", IJS Tech. Report 7479, 1996.

[50]Marina Santo Zarnik, Franc Novak and Srecko Macek: "Design of oscillation-based test structures for active RC filters". In Proceedings of *the European Design & Test Conference 97,* 17-20 March 1997, Paris, France, pp. 618.

[51]Marina Santo Zarnik, Franc Novak and Srecko Macek: "Oscillation test structures of actives RC filters". In Proceedings of *the IEE Colloquium on Testing mixed signal circuits and systems,* 23 October 1997, London, pp.4/1-4/6.

[52]Marina Santo Zarnik, Franc Novak and Srecko Macek: "Design of oscillation-based test structures for active RC filters", IEE *Proc.-Circuits Devices Syst.,* October 2000, vol. 147, no. 5, pp. 297-302.

[53]M.S. Zarnik and F. Novak: "On oscillation-based test structures of active RC filters". *International Journal of Numerical Modelling, Electronic networks, Devices and Fields,* 2001, vol. 14, no. 3, pp. 283-288.

[54]Marlett, M.J. and J.A. Abraham: "DC IATP-An-Iterative Analog Circuit Test Generation Program for Generating Single Pattern Tests". In Proceedings of *the IEEE Int. Test Conf*erence, 1998, pp. 839-844.

[55]Milor, L. and V. Visanathan: "Detection of Catastrophic Faults in Analog Integrated Circuits". *IEEE Transactions on Computer-Aided Design*, 1989, vol. 8, no. 2, pp. 114-130.

[56]Milor, L. et al: "Optimal Test Set Design for Analog Circuits". *IEEE/ACM International Conference on Computer-Aided Design, ICCAD*, 1990, pp. 294-297.

[57]S. Mir, D. Vázquez, A.Rueda, J.L.Huertas: "Switch-Level Fault Coverage Analysis for Switched-Capacitor Systems". *Instituto de Microelectrónica de Sevilla, Centro Nacional de Microelectrónica, Universidad de Sevilla.*

[58]S. Mir, M. Lubaszewski and B. Courtois: "Fault-based ATPG for linear analog circuits with minimal size multifrequency test sets". *Journal of Electronic Testing: Theory and Applications,* August/October 1996, vol. 9, no. 1/2, pp. 43-57.

[59]S. Mir, A.Rueda, T.Olbrich, E.Peralías and J.L.Huertas: "SWIT-TEST: Automatic Switch-level Fault Simulation and Test Evaluation of Switched-Capacitor Systems". In Proceedings of *the 34th Design Automation Conference*, June 1997, pp. 281-286.

[60]N. Nagi, A. Chatterjee and J.A. Abraham: "Fault simulation of linear analog circuits". *Analog Integrated Circuits and Signal Processing*, 1993, vol. 4, pp. 245-260.

[61]Nigh, P., and W. Maly: "Test Generation for Current Testing". *IEEE Design&Test of Computers*, 1990, vol.7, no. 2, pp. 26-38.

[62] Ohletz M.J.: "Hybrid Built-In-Self-Test (HBIST) structure for Mixed Analog/Digital Integrated Circuits". In Proceedings of *the 2nd European Test Conference*, 1991, pp. 307-316.

[63] T. Olbrich, J. Perez, I.A. Grout, A. Richardson and C. Ferrer: "Defect-Oriented vs schematic-level based fault simulation for mixed-signal ICs". In Proceedings of *the International Test Conference*, 1996, pp. 511-520.

[64] E.J.Peralías, A.Rueda, and J.L.Huertas: "An on-line testing approach for pipelined A/D converters". In Proceedings of *the Int. Mixed Signal Testing Workshop*, June 1995, Grenoble, France, pp. 44-49.

[65] E.J.Peralías, A.Rueda, J.A. Prieto and J.L.Huertas: "DFT & On-line Test of High Performance Data Converters". In Proceedings of *the International Test Conference*, October 1998, pp. 534-540.

[66] M. Renovell, F. Azais and Y. Bertrand: " A DFT technique to fully access embedded modules in analog circuits under test". In Proceedings of *the Int. Mixed Signal Testing Workshop*, June 1995, Grenoble, France, pp. 172-177.

[67] A. Richardson, T. Olbrich, V. Liberali and F. Maloberti: "Design-for-test strategies for analogue and mixed-signal integrated circuits". In Proceedings of *the IEEE 38th Midwest Symposium on Circuits and Systems,* August 1995, Rio de Janeiro, Brazil, pp. 1139-1144.

[68] G. W. Roberts and A. K. Lu: *"Analog Signal Generation for BIST of Mixed-signal Integrated Circuits",* Kluwer Academic Publishers, 1995.

[69] J. Roh and J. Abraham: "A Comprehensive TDM Comparator Scheme for Effective Analysis of Oscillation-based Test". In Proceedings of *the International Mixed-signal Test Workshop*, June 2000.

[70] G. Schafer, H. Sapotta, and W. Denner: "Block-oriented test strategy for analog circuits". In Proceedings of *the European Solid-State Circuits Conference, ESSCIRC*, 1991, pp. 217-220.

[71] M. Soma: "A design-for-test methodology for active analog filters". In Proceedings of *the IEEE International Test Conference*, 1990, pp. 83-192.

[72]M. Soma and V. Kolarik: "A design-for-test technique for switched-capacitor filters". In Proceedings of *the IEEE VLSI Test Symposium*, April 1994, pp. 42-47.

[73]S. K. Sunter and N. Nagi: "A simplified polynomial-fitting algorithm for DAC and ADC BIST". In Proceedings of *the International Test Conference*, 1997, pp.389-95.

[74]*"Users' Manual for SWITCAP2 Version 1.1"*. Department of Electrical Engineering and Center for Telecommunications Research, Columbia University, February 1992.

[75]Adam Osseiran: *"A Guide To The IEEE-1149.4 Test Standard"*. Kluwer Academic Publishers, 1999.

[76]M.F. Toner and G. W. Roberts: "A BIST Scheme for an SNR Test of a Sigma-Delta ADC". In Proceedings of *the International Test Conference,* 1993, pp. 805-814.

[77]D.Vázquez , A. Rueda and J.L Huertas: "On-line error detection for continuous-time MOS-FET-C filters". In Proceedings of *the European Solid-State Circuits Conference, ESSCIRC*, September 1993, Sevilla, Spain, pp. 206-209.

[78]D.Vázquez , A. Rueda and J.L Huertas: "A new Strategy for Testing Analog Filters". In Proceedings of *the IEEE VLSI Test Symposium*, April 1994, pp. 36-41.

[79]D.Vázquez , A. Rueda, J.L Huertas and A.M.D Richardson: "Practical Dft strategy for fault diagnosis in active analogue filter". *Electronics Letters.*, July 1995, vol. 31, no. 15, pp. 1221-1222.

[80]D.Vázquez , A. Rueda and J.L Huertas: "A solution for the on-line test of analog ladder filters". In Proceedings of *the VLSI Test Symposium*, 1995, pp. 48-53.

[81]D.Vázquez , A. Rueda and J.L Huertas: "A DFT Methodology for Fault Diagnosis in Active Analog Filters". In Proceedings of *the International Mixed-Signal Testing Workshop*, June 1995, Grenoble, France, pp. 124-129.

[82]D.Vázquez, A.Rueda and J.L. Huertas: "Reducing the impact of DfT on the performance of analog integrated circuits: Improved

sw-opamp design". In Proceedings of *the VLSI Test Symposium*, 1996, pp. 42-47.

[83]D.Vázquez, A.Rueda and J.L. Huertas: "Fully-differential SW-opamp for testing analog embeded modules". In Proceedings of *the International Mixed-Signal Testing Workshop*, 1996, Quebec, Canada, pp. 204-209.

[84]D.Vázquez, A.Rueda, J.L Huertas and E. Peralías: "A High-Q Bandpass Fully Differential SC Filter with Enhanced Testability". *IEEE Journal of Solid-State Circuits*, July 1998, vol. 33, no. 7, pp. 976-986.

[85]Diego Vázquez, Gloria Huertas, Gildas Leger, Adoracion Rueda and Jose L. Huertas: "On-Chip Evaluation of Oscillation-Based-Test Output Signals". In Proccedings of *the XIV Design of Circuits and Integrated Systems Conference (DCIS'01)*, Porto, Portugal, November 2001, pp. 179-184.

[86]D. Vázquez, G. Huertas, A.Rueda and J.L. Huertas: "A simple and Secure Start-up Circuitry for Oscillation-Based-Test Application". *Analog Integrated Circuits and Signal Processing, Kluwer Academic Publishers*, 2002, vol. 32, no. 2, pp. 187-190.

[87]D. Vázquez, G. Huertas, G. Leger, A. Rueda and J.L. Huertas: "Low-Cost On-Chip Measurements for Oscillation-Based-Test in Analog Integrated Filters". In Proceedings of *the 3rd IEEE Latin American Test Workshop (LATW02)*, February 2002, Montevideo, Uruguay, pp. 89-93.

[88]D. Vázquez, G. Huertas, G. Leger, A. Rueda and J.L. Huertas: "Practical Solutions for the Apliccation of the Oscillation-Based-Test: Start-Up and On_Chip Evaluation". In Proceedings of *the VTS 2002 Symposium*, April - May 2002, Monterey, California, pp. 433-438.

[89]D. Vázquez, G. Huertas, G. Leger, A. Rueda and J.L. Huertas: "Practical Solutions for the Application of the Oscillation-Based-Test in Analog Integrated Circuits", In Proceedings of *the IEEE International Symposium on Circuits and Systems (ISCAS 2002)*, May 2002, Scottsdale, Arizona, pp. 589-592.

[90]D. Vázquez, G. Huertas, G. Leger, A. Rueda and J.L. Huertas: "On-chip Evaluation of Oscillation-Based-Test Output Signals for Switched-Capacitor Circuits". *International Journal of Analog Integrated Circuits and Signal Processing, Edit. Kluwer Academic Publishers*, November 2002, vol. 33, no. 2, pp. 201-211.

[91]R. de Vries, T. Zwemstra, E.M. Bruls and P.P.L. Regtien: "Built-in Self Test Methodology for A/D Converters". In Proceedings of *the European Design and Test Conference*, March 1997, pp.353-358.

[92]Wagner, K.D., and T.W. Williams: "Design for Testability for Mixed Signal Integrated Circuits". In Proceedings of *the IEEE International Test Conf*erence, 1998, pp. 823-829.

[93]C. L. Wey: "Built-In Self-Test Structure for Analog Circuit Fault Diagnosis". *IEEE Transactions on Instrumentation and Measurement,* 1990, vol. 39, no. 3, pp. 517-521.

[94]B.R. Wilkins, et al: "Towards a mixed-signal testability bus standard". *ETC.*, April 1993.

[95]Q.F. Wilson and D.B. Day: "Practical Automatic Test Program Generation Constraint". In Proceedings of *the Automatic Test Conference and Workshop*, 1987.

[96]Mike W.T. Wong: "On the Issues of Oscillation Test Methodology". In Proceedings of *the Instrumentation and Measurement Technology Conferance (IMTC'99)*, 1999, vol. 3, pp. 1409-1414.

[97]Mike W.T. Wong: "On the Issues of Oscillation Test Methodology". *IEEE Transaction on Instrumentation and Measurement*, April 2000, vol. 49, no. 2, pp. 240-245.

[98]K. Y. Ko, N. S. Gorla, Mike Wong and Y. S. Lee: "Improving fault diagnostic resolution of an oscillation-based test methodology scheme for the threshold detector circuit". *International Journal Electronics,* 2001, vol. 88, no. 2, pp. 175-187.

[99]Arthur R. Bergen, Leon O. Chua, Alistair I. Mees and Ellen Szeto: "Error bounds for general describing functions problems". *IEEE Transactions on Circuits and Systems, CAS*, June 1982, vol. 29, no. 6, pp. 345-354.

[100]Aram Budak: *"Passive and Active Network Analysis and Synthesis"*. Houghton Mifflin Company.

[101]P. E. Fleischer, K. R. Laker: "A Family of Active Switched Capacitor Biquad Building Blocks", *Bell Syst. Tech. J*, December 1979, vol 58, pp. 2235-2269.

[102]Paul E. Fleischer, A. Ganesan, and Kenneth R. Laker: "A switched Capacitor Oscillator with Precision Amplitude Control and Guaranteed Start-up". *IEEE Journal of Solid -State Circuits*, April 1985, vol. SC-20, no. 2, pp. 641-647.

[103]*" Selected papers on Integrated Analog Filters"*. Edited by Gabor C. Temes. A series published for The IEEE Circuits and Systems Society. IEEE Press.

[104]John E. Gibson: *"Nonlinear Automatic Control"*. International Student Edition. McGraw-Hill Book Company.

[105]S. J. G. Gift: "Multiphase Sinusoidal Oscillator Using Inverting-Mode Operational Amplifiers". *IEEE Transactions on Instrumentation and Measurement*, August 1998, vol. 47, no. 4., pp. 986-991.

[106]S. A. Jantzi, W. M. Snelgrove and P. F. Ferguson: "A Fourth-Order Bandpass Sigma-Delta Modulator". *IEEE Journal of Solid-State Circuits*, March 1993, vol. 28, no. 3, pp. 282-291.

[107]Stephen. A. Jantzi, Kenneth W. Martin and Adel S. Sedra: "Quadrature Bandpass $\Sigma\Delta$ Modulation for Digital Radio". *IEEE Journal of Solid-State Circuits*, December 1997, vol. 32, no. 12, pp. 1935-1950.

[108]B. Linares-Barranco, A. Rodriguez-Vázquez, E. Sánchez-Sinencio and J.L Huertas. "CMOS OTA-C high -frequency sinusoidal oscillators". *IEEE Journal of Solid-State Circuits*, February 1991, vol. 26, no. 2, pp. 160-165.

[109]Alistair I. Mees: "Limit cycle stability". *J. Inst. Maths Applics*, 1973, vol. 11, pp. 281-295.

[110]Alistair I. Mees and Arthur R. Bergen: "Describing functions revisited". *IEEE Transactions on Automatic Control*, August 1975, vol. AC-20, no. 4, pp. 473-478.

[111]Alistair I. Mees: *"Dynamics of Feedback Systems"*. John Wiley, Chichester, 1981.

[112]Alistair I. Mees: "Describing functions: Ten years on". *IMA Journal of Applied Mathematics*, 1984, vol. 32, pp. 221-233.

[113]MITEL Semiconductor: *"DTMF Receiver Test Cassette"*. Issue 3. September 1989.

[114]Katsuhiko Ogata: *"Ingenieria de Control Moderna"*. Prentice/ Hall International.

[115]Jose Manuel de la Rosa: "Modelado y Diseño de Moduladores ΣΔ Paso de Banda para Comunicaciones Digitales usando Circuitos de Corrientes en Conmutación". *P.H. Thesis*, December 2002, Sevilla, Spain.

[116]Shanthi Pavan and Yannis P. Tsividis: "An Analytical Solution for a Class of Oscillators, and Its Application to Filter Tunning". *IEEE Transactions on Circuits and Systems-I: Fundamental Theory and Applications,* May 1998, vol. 45, no. 5, pp. 547-556.

[117]Shanthi Pavan and Yannis P. Tsividis: "An Analytical Solution for a Class of Oscillators, and Its Application to Filter Tunning". In Proceedings of *the IEEE International Symposium on Circuits and Systems,* 31 May- 3 June 1998, vol. 2, pp. 339-342.

[118]Steven R. Norsworthy, Richard Schreier and Gabor C. Temes: " *Delta- Sigma Data Converters. Theory, Design and Simulation"*. IEEE PRESS, New York 1997.

[119]*"Spectre User Guide"*. Product Version 4.4.1, February 1997. Cadence Design Systems, Inc.

[120]D. Vázquez, G. Huertas and J.L. Huertas: *"Circuitry for providing safe initial conditions in systems based on operational amplifiers"*. Patent application P200100996.

[121]Diego Vázquez, María J. Avedillo, Gloria Huertas, José M. Quintana, Manfred Pauritsch, Adoración Rueda and José L. Huertas: "A Low-Voltage Low-Power High Performance Fully Integrated DTMF Receiver". In Proceedings of *the 27th European Solid-State Circuits Conference (ESSCIRC 2001)*, Villach, Austria, 18-20 Septiembre 2001, pp. 368-371.

[122]D.Vázquez, G. Huertas, A.Rueda and J.L. Huertas: *"DTMF Silicon Demonstrator Report"*. Internal Report of the European Sprit Project ASTERIS.

[123]M. Vidyasagar: *"Nonlinear Systems Analysis"*, Prentice-Hall, New Jersey 1978.

[124] *"Using SIMULINK: Dynamic System Simulation for MATLAB"*. The MathWorks Inc, December 1996.

[125]J. Bayard and M. Ayachi: "OTA- or CFOA-based LC sinusoidal oscillators - analysis of the magnitude stabilization phenomenon". *IEEE Transactions on Circuits and Systems I: Fundamental Theory and Applications,* August 2002, vol. 49, no. 8, pp. 1231-1236.

[126] *"MATLAB: The Language of Technical Computing"*. The MathWorks Inc, December 1996.

[127]F. Medeiro, B. Perez-Verdu, A. Rodriguez-Vázquez: *"Top-down design of high-performance sigma-delta modulators"*. Kluwer Academic Publishers, 1999.

[128]Simon Haykin: *"Communication Systems"*. John Wiley & Sons, Inc.

[129]E. Romero, G. Peretti and C. Marqués: "Oscillation Test Strategy: a Case Study". In Proceedings of *the Latin American Test Workshop,* Uruguay, February 2002, pp. 94-98.

[130]E. Peralías, A. Rueda and J.L. Huertas: "New BIST Schemes for Structural Testing of Pipelined Analog to Digital Converters". *Journal of Electronic Testing: Theory and Applications*, 2001, no. 17, pp. 373-383.

Printed in the United Kingdom
by Lightning Source UK Ltd.
116434UKS00006B/72